Sustainable Innovations and Digital Circular Economy

Rubee Singh · Vikas Kumar
Editors

Sustainable Innovations and Digital Circular Economy

Springer

Editors
Rubee Singh
Institute of Business Management
GLA University
Mathura, Uttar Pradesh, India

Vikas Kumar
University of Portsmouth
Portsmouth, UK

ISBN 978-981-96-1063-1   ISBN 978-981-96-1064-8 (eBook)
https://doi.org/10.1007/978-981-96-1064-8

This Springer imprint is published by the registered company Springer Nature Singapore Pte Ltd.
The registered company address is: 152 Beach Road, #21-01/04 Gateway East, Singapore 189721, Singapore

Preface

In today's rapidly evolving world, the drive toward sustainability is no longer an option but a necessity. As our planet faces pressing challenges such as climate change, resource depletion, and environmental degradation, the need for innovative solutions to balance economic growth with environmental stewardship has become paramount. One such solution lies in the transition from traditional linear economic models to a more sustainable and regenerative circular economy (CE). This shift offers a blueprint for industries, governments, and societies to rethink how they design, produce, and consume goods, with a focus on minimizing waste, maximizing resource efficiency, and fostering regeneration. In parallel, the digital revolution has transformed nearly every aspect of modern life, from how businesses operate to how individuals interact with technology. With advancements such as artificial intelligence, blockchain, the Internet of Things (IoT), and data analytics, the digital era presents unprecedented opportunities to enhance efficiency, transparency, and innovation. When paired with circular economy principles, these technologies offer powerful solutions for addressing sustainability challenges and unlocking new avenues for economic and environmental resilience.

This book, *Sustainable Innovations and Digital Circular Economy*, seeks to explore the integration of these two powerful forces—sustainability and digital transformation. It brings together diverse perspectives from industry experts, researchers, and thought leaders who collectively delve into how digital tools and innovations can drive the adoption and success of circular economic practices across various sectors and geographies. The book begins by examining the foundational aspects of the digital circular economy, focusing on how decentralized recycling systems, waste management innovations, and digital tools can reshape the way resources are managed. By leveraging digital platforms and networks, waste recovery and recycling processes can become more efficient, localized, and scalable, contributing to the reduction of environmental impacts. Building on this, the book also describes the role of organizational efficiency and sustainable development in driving the digital circular economy forward. Digital technologies enable organizations to embed sustainability into their core operations, facilitating not only resource conservation but also innovation and competitive advantage. This integration of digital tools into business strategies is vital for companies that aim to meet the growing demand for environmentally responsible practices while maintaining economic viability.

Throughout the book, there is a recurring emphasis on the importance of collaboration—whether it's between businesses, governments, communities, or international bodies. Sustainable innovation in the digital circular economy is not the sole responsibility of any single entity but requires a collective effort from all stakeholders. By fostering collaboration and creating synergies among these players, sustainable practices can be amplified and scaled across industries and regions. The book also explores how global policy frameworks, such as the Sustainable Development Goals (SDGs), serve as powerful catalysts for driving digital circular economy initiatives. Strategic alignment with these global goals encourages businesses to adopt responsible, forward-thinking approaches that contribute to a more sustainable and equitable future. The barriers to adopting circular business models, particularly in emerging markets and developing economies, are also explored. Transitioning to sustainable, circular practices is not without challenges. Economic, social, and political factors often play significant roles in shaping the success or limitations of these models. This book provides insights into how businesses in different regions navigate these challenges and how digitalization can offer solutions to overcome them.

Recognizing that the shift toward a digital circular economy has profound implications for future generations, the book dedicates a significant portion of its focus to human capital and organizational strategies. Sustainable workplaces, empowered employees, and human resource management (HRM) practices are essential for fostering a culture of sustainability. As organizations increasingly adopt circular practices, it is critical that employees are equipped with the skills, knowledge, and motivation to contribute to these changes. The role of HRM strategies in building sustainable work environments becomes central to this transformation, supporting both organizational goals and employee engagement. Furthermore, the book examines sector-specific approaches to the digital circular economy, addressing industries such as manufacturing, healthcare, and small and medium-sized enterprises (SMEs). Each sector faces unique challenges and opportunities when transitioning to circular practices, and digital innovations provide tailored solutions to enhance product circularity, resource efficiency, and sustainability outcomes. By presenting case studies and real-world examples, the book illustrates how organizations across various sectors are adopting digital technologies to drive sustainable innovation.

Ultimately, this book serves as a comprehensive resource for academics, professionals, and policymakers seeking to understand the critical role of digital technologies in shaping the future of the circular economy. It offers a deep dive into the strategies, challenges, and innovations that will define the next generation of sustainable development. By exploring the intersection of digitalization and circularity, this book provides a roadmap for a future where economic growth is decoupled from resource depletion, and where sustainability is not just a goal but a guiding principle for businesses and societies alike.

Mathura, India Rubee Singh
Birmingham, UK Vikas Kumar

Acknowledgments

We extend our heartfelt gratitude to all those who have contributed to this book and our utmost appreciation to the SDGs professionals, researchers, and scholars whose vigorous dedication to advancing personal care serves as a sustaining book source. Your vital insights and firsthand experiences have enhanced the material of this book, guaranteeing its importance and usefulness in attending to real-world difficulties. To the scientists at the center of the expert system, whose groundbreaking duty remains to push the borders of opportunity in healthcare and improve technical improvements, create the keystone of this book, providing a glance into the transformative possibility of Sustainable Innovations and Digital Circular Economy.

Moreover, we further stretch our innermost appreciation to the data contributors who were used during data collection and the simulation and whose interest, passion, and commitment to driving favorable change in the digital circular economy serve as the supreme inspiration behind this book. It is our honest hope that the understandings contained within the book will influence purposeful discussion, foster ingenious options, and catalyze concrete progression in the direction of developing a more sustainable and equitable circular economy and innovative field for all. We are immensely thankful to the entire team at Springer Nature for bringing out this publication in a very professional and timely manner.

Rubee Singh
Vikas Kumar

About This Book

This book, *Sustainable Innovations and Digital Circular Economy*, explores the transformative power of digital technologies in fostering sustainability and driving circular economic practices. With contributions from leading experts, it examines the role of decentralized recycling, waste management, and innovation in achieving organizational efficiency and sustainable development. The book also explores how global policy frameworks, such as the Sustainable Development Goals, serve as powerful catalysts for driving digital circular economy initiatives. The chapters delve into the constraints organizations face in promoting technological innovations and highlight the synergy between stakeholders in creating sustainable business models. Divided into three parts, the book first discusses the foundational aspects of the digital circular economy, including stakeholder collaboration and strategic policies aligned with the global Sustainable Development Goals (SDGs). Part II focuses on sustainability for future generations, exploring the role of HR strategies, barriers to circular business models, and the impact of digitalization on organizational performance. Part III examines how technology can drive circular practices across various sectors, particularly focusing on small and medium enterprises (SMEs), ethical business practices, and sustainability. With real-world case studies and actionable insights, this book is an essential resource for academics, industry professionals, and policymakers interested in advancing sustainability through the digital circular economy.

Contents

About the Editors

Rubee Singh is Assistant Professor of Human Resources at the Institute of Business Management, GLA University India. She is Post-doctoral Researcher (D. Lit.) at the Department of Management Studies, Kumaun University, Nainital, India. She holds a Ph.D. in Management from Noida International University India, and an MBA in HR from Dr. A. P. J. Abdul Kalam Technical University India. She has more than 10 years of experience in research and teaching. Apart from teaching and research, she has handled many administrative positions in premier institutes affiliated with A.K.T.U. She has conducted a number of FDPs, Conferences, Workshops, and HR Conclaves in the areas of Sustainability, CSR, AI, HR, Human Factors, Women Empowerment and Human rights. She is United Nations Volunteer and a contributing author for Voices of Youth (UNICEF). She is an honorary life member of Women Economic Forum and National Secretary of IATM (Youth Wing) International Anti-Terrorism Movement India. Dr. Singh is a Managing Editor of IJARSH India. She is serving on the Editorial Board of more than 30 international journals. She serves as Advisory Board Member of the *Heliyon Journal*, Elsevier, Research in Globalization, and reviewer of many reputed journals like the *European Journal of Innovation Management* (EJIM), *British Food Journal*, *Sustainability* etc. She is a life member of International Association of Engineers, Thailand, IMRTC USA, ISRD UK etc. She is a bestselling author with significant experience as excellent Editor, eminent Researcher, great Academician, Global Ambassador (Children's) and Social Activist. She has presented many papers at national and international conferences, published 20 academic books, and more than 75+ research papers in SCI, ABDC-A, Scopus and UGC index journals. Her areas of research are Sustainability, Digital HRM, Sustainable HRM, Circular Economy, Digital CE, Net Zero, Sustainable Innovation, Digital Innovation, Eco-Innovation, Industry 4.0, Artificial Intelligence, Blockchain and Corporate Social Responsibility. Being a thought leader, her book titled *Training & Development* was added to the MBA syllabus of "The Institute of Aeronautical Engineering" (Jawaharlal Nehru Technological University, Hyderabad). Dr. Singh has been honored with many prestigious awards, which include Best Researchers Award of the Year 2022 by GLA University, Child Ambassador Award of The Year 2022, Iconic Faculty Award 2019, International Young Researchers Award 2021, Young Scientist Award 2020,

Nari Ratan Award 2023, India Book of Records-2020 and Asia Book of Records Holder for 2021. The list of awards and appreciation is long, but Dr. Singh has always prioritized her work over recognition, making her unique and commendable in her field.

Vikas Kumar is Deputy Vice-Chancellor Research and Innovation at University of Portsmouth, UK. He is also Professor of Operations and Supply Chain Management. Professor Kumar has over a decade of teaching and research experience. Prior to joining BCU, he was Director of Research and Scholarship at Bristol Business School, University of the West of England. Professor Kumar has also worked in India, Ireland and Hong Kong. He earned his Ph.D. in Management from the University of Exeter, UK. He did his graduation (B.Tech.) from the National Institute of Advanced Manufacturing Technology Ranchi (Formerly NIFFT), India. He has published over 300 journal articles and conference papers including A*/A as per ABDC and 4*/3* as per ABS-indexed journals. He has an excellent research track record with 19000+ citations on Google Scholar and H-index of 74. He serves on the editorial board of several international journals and has guest-edited several special issues in high impact journals. He is also a reviewer of more than 15 international journals including ABS 4/4* and ABS 3* journals. Professor Kumar works very closely with industries and has generated research funding in the excess of £1 million from various research agencies such as Innovate UK, EPSRC, British Council, British Academy, Newton Fund, and Science Foundation of Ireland. He has worked on several international collaborative projects with researchers from several UK universities and researchers from Brazil, Vietnam, Thailand, Indonesia, India, Mexico, Taiwan, Columbia, Peru, Costa Rica and Turkey. His research interests are Net Zero Emissions, Circular Economy, Sustainability, Industry 4.0, Digital Supply Chains, Sustainable Supply Chains, and Operational Excellence.

Part I
Digital Technologies Integrated Sustainable Innovations

Chapter 1
Decentralised Recycling and Waste Management for Digital Circular Economy

Narayanage Jayantha Dewasiri, Rubee Singh, and Vikas Kumar

Abstract This chapter focuses on decentralised recycling and waste management in a digital circular economy. It provides insight into the potentially valuable applications of blockchain, the Internet of Things (IoT), artificial intelligence (AI), and big data as solutions to the problem. This study assesses how decentralised systems mitigate the issues characteristic of linear waste management approaches, minimise environmental impact, and increase community participation. It also assesses the limitations many players have in facilitating technological advancement in waste management. The study is underpinned by two approaches: Preferred Reporting Items for Systematic Reviews and Meta-Analyses (PRISMA) and the Population, Interest, Context, Outcome (PICO) framework. This study conducted a systematic literature review based on the research questions and interests in the roles of digital technologies to compile the research findings concerning decentralised waste management using 32 sample studies. The PICO approach was used to identify the pragmatic reality and potential of the organisations, especially in South Asia's waste management sector and other emerging economies. The investigation reveals that decentralised recycling results in economic, environmental, and social benefits, including reduced transportation costs, increased community employment opportunities, decreased landfills, and increased community participation. However, some barriers include regulatory and policy issues, technologies, funding constraints, and cultural issues. Lastly, the study also shows that digital technologies can enhance waste management,

N. J. Dewasiri (✉)
Department of Accountancy and Finance, Faculty of Management Studies, Sabaragamuwa University of Sri Lanka, Belihuloya, Sri Lanka
e-mail: jayantha@mgt.sab.ac.lk

Adjunct Faculty, University Center for Research and Development (UCRD), Chandigarh University, Punjab, India

R. Singh
Institute of Business Management, GLA University, Mathura, India

Department of Management Studies, Kumaun University, Nainital, Uttarakhand, India

V. Kumar
University of Portsmouth, Portsmouth, UK
e-mail: Vikas.Kumar@port.ac.uk

R. Singh and V. Kumar (eds.), *Sustainable Innovations and Digital Circular Economy*,
https://doi.org/10.1007/978-981-96-1064-8_1

tracking, monitoring, and sorting wildly, but this needs infrastructure and support to be effective.

Keywords Decentralised recycling · Waste management · Digital circular economy · Blockchain technology · Sustainable innovation

1.1 Introduction

Waste management is one of the biggest challenges the world is facing today, fuelled by the rise of linear economic models (Das et al., 2019). These linear operation models see resources as being procured, used, and then disposed of, resulting in uncontrolled waste generation and drawdown on stocks of natural resources. This linear model of manufacture-use-dispose has been one of the leading causes of the deterioration of the environment, pollution of natural resources, destruction of biodiversity, and overloading of already exhausted dumping sites. Especially in larger cities, the increasing pile of waste and shortcomings of centralised systems for waste removal have established the need for new technologies that can effectively replace misconceptions of current practices. A shift towards a circular economy is a crucial solution to these problems. Building a circular economy is a new development model that detaches business processes from the depletion of valuable resources and focuses on restoring and reusing resources in the market. This is a model where waste is not dumped over a landfill in the traditional sense of the word; instead, it is recycled and put back to utile use within that particular production process. Recycling, reuse, and remanufacturing are fundamentals of the circular economy; these concepts are critical to efforts to minimise the harm done to the environment (Khan et al., 2022). With the increasing shift to green economies, it is essential to take stock of waste management measures regarding environmental sustenance, efficiency in using natural resources, and financial viability.

This chapter explores the intersection of two significant trends shaping modern waste management: decentralisation and a circular digital economy as the main topics. Decentralised recycling means the change from large-scale processing plants to smaller-scale localised systems that allow communities or enterprises to handle waste effectively. This approach eliminates the need for many transportation systems and enables localised decisions to help improve the most sustainable form of waste management. Subsequently, cities and regions can decline the centralised facilities, decrease the greenhouse effect, cut out the logistical costs, and amplify the engagement of those living in that area towards waste reduction.

The chapter will also look at the digital circular economy, which is the application of digital technologies like blockchain, IoT, and AI to the circular economy. The main feature of these technologies is the enhanced ability to track, sort, and handle recycled waste, making the process more efficient and transparent. The potentiality of digitalisation of waste management physically brings new opportunities to work with the economic circulation of resources and close the loop, invoking the concept of the recipient in terms of materials.

The following sections in this chapter will help uncover how decentralised recycling and digital technologies can drive innovation within the waste management sphere and make a shift towards the progress of a circular economy more effective. The chapter will also identify the latest trends in technology and present examples of best practices and successful implementation of decentralised recycling to illustrate the possibility of a radical breakthrough in waste management in developed and developing nations.

1.2 Methodology

This chapter utilises the PRISMA checklist and PICO framework to provide a structured methodological approach for reviewing the decentralised recycling and waste management in the digital circular economy. Applying these two frameworks means the research systematically and explicitly covers the area of interest to reveal the current literature's state and main stakeholders' experiences in implementing decentralised systems.

1.2.1 PRISMA Framework

The PRISMA framework adheres to carry out an SLR on decentralised recycling, waste management, and the digital circular economy. PRISMA seeks to establish a precise, definitive method for systematic review that accurately reflects recent trends and discoveries as highlighted by Karunarathna et al. (2024), Dewasiri et al. (2024), Rathnasiri et al. (2024), Jayawardena and Dewasiri (2023), Dewasiri et al. (2023), Mishra et al. (2024), and Dewasiri et al. (2022). The first step in the PRISMA process was to consider what literature might be relevant. Sources indexed in Scopus, Web of Science, and Google Scholar were retrieved using decentralised recycling, waste management, digital circular economy, blockchain waste tracking, AI in recycling, and IoT waste management. The search was limited to literature written in the last ten years (from 2014 to 2024) to capture current technological innovation and emerging regulations. Only those publications in English were considered, such as articles from scholarly journals, conference proceedings, and technical reports.

In the first stage, nonrelevant studies were excluded, focusing on the titles and abstracts of studies that were identified as discussing decentralised waste management or the implementation of digital technologies. We concentrate on articles that offer quantitative evidence, qualitative theory, or real-world examples of decentralised waste systems in any geographic region. Many studies were dropped at this stage based on their focus, specifically on geographical areas of interest, such as South Asia and emerging markets where these technologies are either piloted or adopted.

In extracting and synthesising data, information considered relevant for subsequent analysis was collected from each article. This comprised the study aims and scope, study area, critical findings on decentralised recycling systems, the potential use of technology in enhancing waste management, and factors or issues of concern. The synthesis also produced common patterns like economic aspects of decentralisation, environmental prospects, and social effects of communal waste management. They included policy areas, technological support or deficiency, and cultural resistance issues. This final sample consisted of thirty-two (32) seminal studies.

1.2.2 PICO Framework

In line with PRISMA, the PICO model, which included population, interest, context, and outcome, was used to identify the implementation challenges and opportunities, as highlighted by Frandsen et al. (2020).

Population (P): This study targets government and private entities involved in decentralised recycling and waste management, start-ups and other actors of decentralised recycling, and local communities covering the population.

Interest (I): The paper's core interest is to enhance the adoption of technological developments in decentralised waste management systems. This also involves discovering how blockchains, IoT, AI, and big data can improve waste management efficiency, raise recycling rates, and increase transparency.

Context (Co): South Asia and other emerging markets form the study's central area of interest. As these areas are experiencing rapid urbanisation, waste disposal remains a significant challenge due to inadequate appropriate infrastructure.

Outcome (O): The expected result of this study is to define the challenges and possibilities for developing decentralised waste management through digital platforms. The survey examines several advantages of such systems, including the short-term and long-term benefits related to economic impacts, environmental changes, and social enfranchisement.

1.3 Findings and Discussion

1.3.1 Decentralised Recycling and Waste Management: Concepts and Principles

Decentralised recycling and waste management is a system of the social organisation of the processing and recycling processes different from the extensive facilities that occupy large spaces as waste-processing centres for large regions, and there is no central point for coordination and execution. A decentralised waste collection, sorting, and recycling system occurs locally, for example, at blocks, small clusters of houses, street level, business premises level, etc. (Tripathi, 2024). This proximity minimises the long haul, fuel consumption, and greenhouse gas emissions accompanying transportation. It also provides an opportunity for developing more adaptable solutions to specific environments necessary for some areas of the country or certain population groups.

Some general characteristics of decentralised waste management systems are increased population participation, increased population autonomy, and better fit with local environmental and socio-economic conditions (Massoud et al., 2019). In contrast, distributed waste management has shortcomings because it can be centralised systems, which may not be very responsive to the equilibrium of waste materials within districts or the total quantity of waste produced in a given region. Local institutions can arrange separate collection and disposal for their residents according to their community circumstances; many residents can compost organic waste, recyclable wastes can be processed in their facilities, and regional materials can be reused. Third, decentralised systems promote invention at the community level since individuals can adapt approaches best for their context.

One of the most significant differences between decentralised and centralised systems is the scale of an enterprise. Some types of systems are centralised, in which a vast and remote facility handles the waste of several areas; such a system implies complex transfers, increased transport expenses, and more extended periods for the waste to be treated. Decentralised systems function on scales much smaller than centralised systems, which positively impacts the environment and costs of operations. Decentralised waste management is also practical because members of the public and local authorities are directly involved in their waste streams to enhance recycling and minimise landfill use.

1.3.1.1 Role in Circular Economy

Organised recycling and decentralised waste management platforms are most compatible with the circular economy principles that focus on reducing resource use and recycling and reusing assets (Singh et al., 2023). These processes can be better managed through decentralisation so that wastes are returned to the loop and recycled as valuable resources. The sort collected in this context is not a waste in the traditional sense of the term. Still, the opportunities for its recovery and reuse are acknowledged, and effective decentralised systems are in a position to support this circular pattern.

It can be noted that three philosophical concepts of circular economy, namely reduce, reuse, and recycle, all converge with the ideas of decentralisation (Kouhizadeh et al., 2020). This is because, as local systems, they enjoy the privilege of proximity to waste generation points. Reduction can be achieved optimally since systems can reduce their waste output. Reuse is promoted as localised facilities may be better positioned to look for the material they can recycle, especially when dealing directly with establishments within their immediate vicinity. Accordingly, recycling is optimised in decentralised systems since waste cannot be transported long distances before arriving at a recycling centre. This helps reduce the emission of carbons and, simultaneously, enhances the chances of the material being processed more efficiently and without being contaminated to enhance the recycling rate.

Experience in decentralised digital and physical system implementations helps to understand their applicability and efficiency in the framework of the digital circular economy. For instance, waste management policies at the community level in some of Europe's cities have led to a drastic decrease in landfill usage as people sort and process waste at the community level (Malinauskaite et al., 2017). Likewise, in decentralised systems, experiments in India have effectively managed composting through alternative centres in both urban and rural capacities and have complemented the municipal waste system by providing organic fertilisers. When properly executed, these examples demonstrate that decentralisation can potentially change the current narrative on waste and recycling and the economy.

Hence, decentralised systems are connected with the circular economy to show that communities and businesses can close the loop. This, in turn, helps save resources and positively impacts the local economy because it generates new employment in recycling, repairing, and recovery firms. Thus, decentralised waste management systems drive out today's waste issues and help to achieve future circular economy objectives.

1.3.2 Digital Technologies Enabling Decentralised Waste Management Towards Digital Circular Economy

As known, decentralised waste management systems are critical to comprising emerging technologies that determine the success of the digital transformation era. Technologies like blockchain, the IoT, AI, and computing in the cloud provide unique ways to address waste streams locally and improve resource recovery efficiency in the digital circular economy (Cheah et al., 2022). They uplift operational features and make recycling more accountable and transparent processes. This section focuses on how each technology contributes towards decentralised waste management in a digital circular economy.

1.3.2.1 Blockchain for Transparent Waste Tracking

Though it has been developed to implement secure and transparent digital transactions, blockchain technology can play an indispensable role in waste management, including transparency and traceability of waste flows (Gopalakrishnan et al., 2021). Blockchain can establish tamper-proof accounts of waste collection, sorting, and processing in decentralised recyclable systems. Waste production, transportation, or recycling can all be captured on this blockchain, and it is observable and auditable by all concerned parties.

Blockchain means that waste's origin, type, and destination details are renditions that cannot be manipulated, while the recyclable supply chain is transparent. In this regard, disclosure may assist in confronting fraud issues, where waste is usually misreported or practised inappropriately. For instance, manufacturers, recyclers, and local government can determine if a given type of material has been recycled to meet the set environmental standards and trace the material from collection, sorting, and, ultimately, recycling stages. At the same time, blockchain might facilitate the issuance of unique tokens of digital waste, which would stimulate participants' effective sorting and recycling of waste (Alnuaimi et al., 2023).

Decentralised waste management acknowledges that multiple participants, from the waste generator to the recycler, can engage in blockchain-based operations, thereby leveraging blockchain's decentralised architecture. Smart contracts in blockchain can also automate payment and other activities within the waste management ecosystem to reduce the general workload.

1.3.2.2 Internet of Things (IoT) for Waste Monitoring

Another transformative technology in the decentralisation of waste management currently underway is the introduction of IoT (Jiang et al., 2023). Other connected items, such as smart sensors and connected waste bins, will be able to provide information on the volumes, types of waste, and frequency of collections. These sensors are essential in collecting information that, in turn, enables one to collect waste efficiently, avoid occasions where containers are overfilled, and ultimately enhance the efficiency of particular recycling processes.

For instance, intelligent bins containing full sensors can detect that the bins are full and send a notification to the nearby waste collection teams (John et al., 2022). This reduces undue collections while avoiding situations where the bins are full of waste. This is because there is no more waste to collect, thereby cutting the costs of littering the environment with waste. When waste services may be few and far between or provided by the communities in a decentralised system, this level of definition is essential in cost efficiency and performance.

Collected via IoT sensors, data can provide vivid information on waste production, allowing communities and municipal districts to create proper waste reduction initiatives and find out the receding rates of recycling to develop the correct programmes for usage. This assertiveness is paramount since decentralised systems depend on resident and business involvement in data collection. As such, the IoT technologies exhibit real-time information in managing waste operations, helping local waste managers eliminate or minimise inefficiencies in waste management services, thus developing a viable framework of waste management systems.

1.3.2.3 Artificial Intelligence (AI) and Machine Learning for Waste Sorting

AI, specifically machine learning, is also seen in the sorting step, which is usually manual and contains many errors. AI techniques consist of complicated analytical and robotic methods for correctly sorting and separating waste materials. They can sort different materials, including plastics, metals, and papers, and skip the process of having to be operated on manually to be sorted out to the various recycling streams (Wilts et al., 2021). This is perhaps one of the biggest strengths of using the decentralised recycling centres where local communities or individual businesses can recycle by managing small plants; using AI sorting systems will enable one to sort immense piles of waste quickly and with less chance of contamination. One of the great benefits of using machine learning is that the algorithms can be trained to perform with increasing accuracy as more data on previous sorting is obtained, enhancing the equipment's ability to identify complex or mixed waste types for sorting.

For example, waste sorting robots are used in intelligent recycling centres to separate waste that is reusable from that which is not by applying artificial intelligence, hence increasing the efficiency and cleanness of recycled products (Kshirsagar et al., 2022). This eliminates cases of rejected material in the later stages of the recycling business, a typical problem with centralised systems. When using AI in decentralised systems, communities are more capable of returning valuable materials to the circular economy at a higher rate than relying on stakeholders.

1.3.2.4 Cloud Computing and Big Data for Efficient Waste Management

Flexible and reliable information technologies, such as cloud computing and big data analytics, serve as the real-time data management and processing platform in decentralised waste disposal systems (Mishra & Kumar, 2020). Built on cloud platforms, IoT devices, AI systems, and blockchain networks can store, analyse, and even trade freely collected data so diverse stakeholders can make informed decisions based on refined real-time data.

Cloud hubs enable the increase of decoupled waste management systems, with easy data storage and elaborate computation services at relatively minimal costs since the need to invest in physical servers is minimal. For example, within the waste management sector, cloud computing allows the tracking of waste levels, collection routes, and recycling rates across decentralised networks without the need to be there.

Cloud is an on-demand access to real-time information that enables organisations to respond more quickly to new events, such as an increase in waste or breakdown of equipment.

However, when implemented with cloud infrastructure, big data analytics goes a notch higher in enhancing optimality in waste management. Using big data from IoT and other sources, big data analytics can help specify tendencies in waste production, recognise improvement opportunities in collection or sorting, and predict future tendencies. Such a level of predictive knowledge is beneficial for data-decentralised systems where, locally, any community or company may be unable to afford a waste audit. Waste management operators can use customer benefits to increase efficiency, increase recycling results, and reduce costs in the long term due to constant data analysis and usage.

Therefore, tools like blockchain, IoT, AI, and cloud computing are lynchpins of decentralised waste management solutions. These technologies provide transparency, real-time monitoring, automation, and data, increasing waste management's sustainability efficiency, which will pave the way towards a digital circular economy. As these innovations develop, they help more and more in the decade-long revolution for decentralised recycling systems for the circular digital economy.

1.3.3 Benefits of Decentralised Recycling for the Digital Circular Economy

Decentralised recycling systems present several meaningful advantages that make them the best option for circling sustainability within the digital circular economy. These can be aligned with various functional areas, including economic profit, promoting environmental stewardship, improved waste disposal, and stakeholders' interests in community advancement and economic growth. Since waste management is gradually adopting new digital technologies, decentralised systems present a promising way to mitigate the negative externalities characteristic of centralised, traditional solutions.

1.3.3.1 Economic Benefits

The economic aspect is an instant benefit because reduced logistics and transportation are one of the primary goals for decentralised recycling (de-Oliveira-Neto et al., 2017). As indicated in centralised collection systems, wastes are conveyed long distances to extensive centralisation facilities, meaning high fuel, personnel, and vehicle maintenance costs. Thus, decentralised systems reduce the transportation distance drastically, thus cutting down the operational cost of recycling programmes to the extent that even the recycling companies or municipalities may not find the services affordable. These lowered transportation costs also mean a more financially feasible recycling system, as money can be reinvested to enhance local amenities and equipment.

Furthermore, decentralised recycling systems promote business development in their regions, which are often small- and medium-sized recycling and waste-processing companies. These regional initiatives may rely on nearby waste flows to provide recycling services to meet the consumers' needs. The decentralised systems help generate local markets for waste material, resulting in new business and economic activity and employment opportunities in waste collection, sorting, and recovery. This local employment generation is even more important for low- and middle-income countries with limited access to 'proper' employment.

Moreover, the decentralised plan promotes a circular economy since it optimises resource usage efficiently on a small-scale and community level. Material continuation within specific local economies enhances product circulation, minimising the need for raw material importation and thus enhancing resource consumption sustainability.

1.3.3.2 Environmental Impact

Decentralised recycling has a significant positive impact in that it helps reduce the carbon emissions associated with waste management (de Souza & Drumond, 2022). Because decentralised systems decentralise waste processing and disposal closer to the processing and disposal point, they reduce the emissions resulting from transportation, which pollutes the air and causes global warming in systems of centralised waste management. Since decentralised systems slash transport distances over which fossil fuels are utilised, they directly support the fight against global emissions of greenhouse gases towards net-zero economies.

In addition, decentralised recycling systems can avoid vast quantities of waste disposed of in landfills, which is one environmental concern. This is more so because landfills are the largest producer of methane gas, which is known to be a greenhouse gas that affects soils and waters. As a result of handling waste at the source, a decentralised system enhances the possibility of recycling materials, decreasing the proportion of waste that goes to landfill sites and the possibility of recovering valuable resources that would otherwise be wasted. This saves on utilising natural resources and reduces the harm of environmentally unfriendly landfills, including toxicity leachate and loss of habitat.

1.3.3.3 Social and Community Impact

Apart from the technical advantages, decentralised systems can also have profound social implications by including people in tackling waste issues (Singh et al., 2022). In centralised systems, waste collection is usually a distant, concealed affair, which results in no answerability and public participation. However, a decentralised system brings waste management closer to the people, making them more aware of the waste problems and involving them in recycling. This community participation leads to accountability, and other people will be willing to embrace programmes meant to control waste and recycle.

In decentralised waste management systems, communities also better understand integrated waste management, which is waste literacy. The more people know why and how waste is being sorted, collected, recycled, and utilised within the vicinity, the more likely they will engage in proper waste management or the 3Rs: reducing, reusing, and recycling. Local strategies supporting this awareness formed part of global educational campaigns improving behavioural change towards sustainable consumption.

Besides, it allows social innovation because decentralised systems permit individuals and local communities to design the most suitable waste disposal strategies. For instance, enterprises in recycling cooperatives, homemade composting programmes, or waste-to-energy come from bottom-up developments considering specific local communities, groups, and residents' needs and abilities. These efforts enhance environmental performance and advance social capital, facilitating closer cooperation among residents, local governments, and businesses.

Lastly, decentralised systems can be more accountable and transparent because waste management remains within the community. When waste disposal is done at the community level, one can check whether or not the recycling process is done correctly. These activities foster transparency between councils and the people, strengthening their backing of sound waste management policies.

1.3.4 Key Challenges in Implementing Decentralised Recycling and Waste Management

Although decentralised systems have many benefits for advancing the digital circular economy, a few key challenges hinder the implementation of a decentralised recycling system. These include regulatory and policy issues, technological lag, economic barriers, and socio-cultural issues. Overcoming obstacles to universal adoption demands a concerted effort by governments, industries, and local communities combined with investment designed to improve infrastructure and take advantage of technological breakthroughs. This section presents the main drawbacks of decentralised recycling and waste management systems.

1.3.4.1 Regulatory and Policy Barriers

A key challenge affecting the rollout of decentralised recycling systems is the absence of a well-developed regulatory framework and standards to address their distinct requirements (Rathnasiri & Gunathilaka, 2024). Centralised models are often seen by traditional waste management policy, as extensive waste collection, sorting, and disposal happens at large-scale facilities. In many cases, these regulations may not be flexible enough to enable the evolution of smaller, community-based recycling initiatives, which can lead to decentralised systems without specific legal guidance or financial support (Rathnasiri et al., 2024).

Policies encouraging waste reduction, recycling, and decentralisation are either absent or insufficiently enforced in most regions, especially developing countries. The absence of regulatory support can further result in fragmentation—in other words, various local authorities or municipalities are working in silos without any coordination. However, this leads to difficulties for such decentralised systems to garner traction and operate in a legally ambiguous setting, thereby dissuading private sector investment and hindering the ability of local communities to scale their recycling efforts.

Additionally, there is currently no standardisation in handling waste, making the interoperability of decentralised systems impossible. The absence of clear regulations that outline how waste should be collected, sorted, or processed can give rise to inconsistencies in practices, which, in turn, will affect the efficiency of the recycling process as well. The government needs to have policies supporting decentralised systems and having them meet environmental standards and follow best practices in different regions.

1.3.4.2 Technological Gaps and Infrastructure Needs

Successful decentralisation of recycling is contingent upon sufficient digital technologies and supporting infrastructure. The only problem is a vast digital divide in many developing countries; hence, accessing advanced technologies like blockchain, IoT, and AI is brutal. The digital divide represents a significant barrier to the widespread proliferation of decentralised waste management systems, which require access to these technologies to achieve efficiency, transparency, and scale.

For instance, IoT devices and AI technologies require internet connectivity and access to data analytics platforms that might be unavailable in underserved areas (Dhar-Dwivedi et al., 2024). Many areas also do not have enough infrastructure to support decentralised waste processing (e.g., scale recycling facilities, collection points, and composter units). This infrastructure deficit hampers communities' ability to set up and support effective recycling systems, as often when competing with well-established centralised facilities.

In addition, implementing these technologies and constructing requisite infrastructure can be too expensive for local governments or community-based organisations. Many communities may be unable to afford the upfront investment needed to establish decentralised systems without external funding or public–private partnerships. Consequently, decentralised recycling might persist only in more developed regions, aggravating global inequalities in waste management.

1.3.4.3 Economic and Financial Constraints

The economic viability of decentralised recycling systems is a significant challenge for their execution. Decentralised systems can reduce the long-term cost of transporting goods to landfills or other waste sites; however, startup operations will require sizeable capital to establish local recycling operations and implement digital technology. Also, these initiatives may reduce the ability of communities or municipalities to financially provide for them, especially where there is economic incapacity in such locations.

Moreover, the financial hurdle of ensuring the long-term sustainability of decentralised systems can arise alongside initial investment costs (Dewasiri, 2024). While centralised centres will benefit from the economies of scale, smaller facilities may not be able to and are thus likely to incur higher per-unit processing costs. Due to the unstable nature of financial support, decentralised systems can have difficulty running operations, especially during economic crises or varying market demand for recyclable materials.

However, as the economic incentive to participate in a decentralised recycling system is either low or absent, community participation can be challenging to sustain. Funding that may not always be available locally is required to give incentives such as recyclable rebates that offer incentives for individuals and businesses to recycle. To remain financially viable in the long term, decentralised systems must look to innovative financing mechanisms, including green bonds and public–private partnerships, that governments and industry stakeholders must explore.

1.3.4.4 Socio-cultural and Behavioural Issues

Decentralised recycling systems are sensitive to socio-cultural factors and public attitudes towards waste management. In many communities, there may be resistance to decentralised systems because people are unfamiliar with them or do not trust them. The success of decentralised systems largely hinges on behavioural change because these systems often rely on active community participation in waste sorting, recycling, and composting. However, ingrained ways to waste are defined as using their recyclables with regular waste or depending on expert services.

Additionally, communities new to digital technologies may be reticent to incorporate digital tools into waste management, such as IoT sensors or blockchain tracking systems. Specifically in low literacy zones, privacy or data security concerns and the perceived complexity of these technologies can paralyse their acceptance. Addressing these challenges demands funding public education and training programmes in decentralised systems' environmental benefits and how digital tools can help improve their effectiveness.

Furthermore, a regional-specific cultural mindset towards waste management can differ significantly from one to another. It will influence people's and the community's willingness to participate in decentralised projects. Some waste can be seen in some areas as a low-priority issue, and social stigma is commonly associated with waste collection and recycling activities. To defeat these cultural barriers, tailored outreach aimed at local values and customs and the input of community leaders who can lead the charge against decentralised waste management are needed.

1.3.5 Case Studies and Best Practices

The practical application of decentralised waste management systems is illustrated using examples from different geographical regions. The case studies in this chapter show the potential and problems of decentralisation, especially when linked to digital technologies. By analysing successful initiatives and assessing their implementation, best practices can be extracted that can be leveraged by future initiatives aiming to scale decentralised recycling systems for a digital circular economy.

1.3.5.1 Successful Decentralised Waste Management Initiatives

There are many other examples of the success of decentralised waste management from localised recycling and reduction in many cities worldwide. The case studies show how communities can leverage decentralised systems to design more sustainable waste management solutions.

There is one such case in point in effect, the Alappuzha Municipality in Kerala, India, which is also referred to as the 'Cleanest Town in India.' How Alappuzha implemented a decentralised waste management system where individual households are responsible for organic waste through composting (Gopalan et al., 2021). Waste was heavily reduced in landfills when citizens were encouraged to segregate their waste at their source by the municipality, which supplied compost bins. Having empowered local communities to manage their organic waste, Alappuzha has minimised waste collection costs, reduced landfill dependency, and improved environmental results. Through this, the community became aware of and engaged with sustainability's cultural development.

Waste Wise Cities Campaign (led by UN-Habitat) encourages decentralised recycling in low-income urban areas of developing countries, which is another example (Citaristi, 2022). Small-scale enterprises in cities such as Nairobi, Kenya, are responsible for the collection and recycling of waste materials on a local scale in community-led recycling activities. Recovered valuable waste stream materials and provided income opportunities to informal waste workers. Localising recycling efforts and using mobile technology to organise the collection, these communities have enhanced resource recovery rates and spurred resource recovery in places lacking infrastructure.

The successful performance of these examples highlights the necessity for (i) community involvement, (ii) government support, and (iii) localised solutions in obtaining decentralised waste management successes. These initiatives show how waste at the source can be addressed and that community ownership can enable widespread and sustainable waste management practices consistent with the principles of the circular economy.

1.3.5.2 Lessons Learned from Digital Integration

Integrating digital technologies into decentralised waste management systems provides beneficial insights into how technology can improve efficiency, transparency, and scale. Those key takeaways from these implementations create the opportunities and challenges of digitising waste management.

One of the most important lessons learned is blockchain technology's critical role in ensuring transparency and accountability across the recycling process. In the Plastic Bank effort in countries such as Haiti and the Philippines, blockchain provides proof that waste is entered, taken care of, presumed until it is sure, and then ultimately recycled. The plastic waste is gathered through this system, and recyclers receive digital tokens. In addition to encouraging ethical recycling practices, blockchain is utilised to reward local communities for engaging in waste management efforts. However, it has a scalability issue: it relies on access to digital infrastructure, which is largely unavailable in some areas.

An important note to learn is how the Internet of Things (IoT) technologies can optimise waste collection and recycling operations. IoT-enabled bins in cities like Barcelona, Spain, record real-time waste levels in WhatsApp and Facebook Messenger so water companies can prioritise waste collection routes and have fewer unnecessary trips. The benefits of this operation include reducing carbon emissions and operational costs based on waste collection. The lesson here is to spend money on infrastructure and data management capabilities to integrate IoT to increase the waste management efficiency of decentralised systems.

It has also proved effective for automating recycling processes, especially in processes of sorting facilities. For example, AI-controlled robots in Norwegian recycling plants can sort waste materials at high levels, as they pass more for recyclable material throughput and reduce material contamination in the recycling stream. It reduces manual labour and reliance. This improves the precision of waste sorting, essential for decentralised systems without access to large-scale processing facilities. However, the high initial cost of putting AI technologies in place continues to serve as a barrier to implementation, particularly in regions with lower incomes where resources are scarce.

1.4 Conclusions

This chapter highlights the critical role of decentralised recycling and waste management in promoting the digital circular economy. While traditional waste management systems may serve diverse challenges and opportunities, decentralisation enabled by digital innovations is a path to a more sustainable, inclusive, and environmentally friendly future (Dewasiri, 2024).

Decentralised recycling and waste management are among the primary benefits of its ability to localise the waste handling process, meaning it does not rely on an extensive centralised system. Localisation minimises logistical and transportation costs, lowers carbon emissions, and empowers communities to reclaim their waste streams. Decentralisation also offers new economic opportunities (particularly for small-scale recycling businesses) that create jobs and generate income for local economies. Technology in decentralised sanitation is also influenced in terms of its environmental impact as decentralised systems reduce direct resource consumption, minimise waste diversion to landfills, and endorse recycling practices consistent with a circular economy.

Still, the path to full decentralised system implementation is not paved with roses. Technology outpaces regulatory and policy frameworks, and without clear guidelines, there is little incentive to promote decentralised models. Another hurdle is the technological gap between technology and natural infrastructure in places not endowed with it. A remaining concern is ensuring equitable access to the digital tools of decentralised systems like IoT devices, blockchain for waste tracking, and AI-driven sorting technologies. Additionally, due to economic constraints, initial investment costs may prove a constraint to communities and businesses wishing to transition to decentralised systems. Socio-cultural factors like lack of willingness to change or waste literacy can accelerate progress.

However, integrating digital technologies like blockchain, IoT, AI, and cloud computing has already shown its ability to revolutionise waste management. These innovations will grow as they mature and drive increased transparency, efficiency, and scalability of decentralised recycling systems, making them more commonplace and accessible.

1.5 Implications and Future Directions

Emerging technologies promise to advance the evolution of decentralised recycling and waste management with new and efficient ways to collect, process, and recycle waste. Through innovative digital tools, governments, businesses, and communities can work together to improve waste management's workflow, sustainability, resiliency, and efficiency. Consequently, this paper looks at possible future developments in this space, technological innovations, policy recommendations, and strategic actions to develop resilient systems that fit the circular economy model.

1.5.1 Policy Recommendations and Strategic Actions

Strong policy support and strategic actions are needed to fully realise the potential for decentralised waste management systems with the digital circular economy. Governments, businesses, and communities drive these changes.

Governments must implement comprehensive policies promoting decentralised waste management to achieve the above. It includes having clear rules that support recycling initiatives on a small scale, giving communities and businesses subsidies to enter into decentralised systems or even regulations to make designers their designs, and ensuring that systems conform to environmental regulations. Additionally, policymakers must significantly prioritise the development of digital infrastructure, especially in less served areas, so that these advanced technologies (IoT, AI, blockchain) are accessible to all communities. Public–private partnerships could also be encouraged by governments which provide environments for investment in emerging waste management technologies to allow innovation while improving the long-term longevity of decentralised systems.

Innovative technologies and their integration into decentralised systems depend heavily on the actions of businesses. These populate the market with advanced waste sort technology, such as AI-driven robots or IoT-equipped monitoring tools, to increase waste sorting and recycling efficiency and scalability of decentralised systems. Circular economy principles can be adopted by businesses as well by designing products that are easier to recycle and extending producer responsibility (EPR) programmes, where manufacturers take on turning from when a product is taken from their hands to guarantee products for the entire lifecycle of the product including from where the product will become waste. Businesses can also collaborate with local communities to advance the use of decentralised systems and discover new forms of economic action.

Decentralised waste management systems rely upon communities for local implementation and daily operation. However, for decentralised recycling picking to be effective, community engagement, education, and awareness programmes must help encourage residents to participate in waste sorting and recycling. Local leaders serve as champions for waste literacy campaigns, run recycling drives, and promote pro-environmental behavioural change in a culture of sustainability. Communities can tap into that work to co-create waste management solutions appropriate to specific community needs and priorities in partnership with industry and government.

To facilitate decentralised waste management, a total strategy involving support for technological innovation through underlying policies, community involvement, and business investment is required. Stakeholders that claim a share in the circular economy can create an enabling environment for decentralised recycling systems that can support the digital circular economy in its broader aspects.

1.5.2 Building Resilient Systems for a Sustainable Future

The long-term success of the digital circular economy will require resilient waste management systems to flourish in the context of future environmental, economic, and social challenges. The possibility of decentralising waste management through systems which foster reliance on large-scale, centralised waste management facilities and allow local communities to self-administer their waste streams provides opportunities to promote resilience.

Designing decentralised systems that are adaptable, flexible, and capable of responding to changing conditions can make for a more resilient waste management with a form of resilience as part of their core. It covers the development of local recycling infrastructure for different kinds of waste, such as organic and electronic waste. It supports these systems with robust digital technologies for real-time monitoring and data-driven decision-making. For example, suppose a recycling system employs blockchain technology in a decentralised system. In that case, it is possible to have transparency and traceability in the recycling process, allowing the recycling system to respond to regulatory change, market fluctuations, or supply chain disruptions.

Building social resilience on top of technological adaptability is critical to the success of decentralised systems. In other words, this is about building strong community engagement and ownership of waste management initiatives to ensure communities can withstand external pressures of climate change, economic downturn/recession, or political instability. A decentralised approach empowers the people to take control of their waste management systems, creating a sense of community responsibility and solidarity among the people, making the whole system much more resilient.

Finally, the design of waste management systems that are environmentally sustainable and resource-conscious should be improved significantly to enhance environmental resilience. A set of decentralised systems of waste reduction, waste recycling, and waste recovery promotes the sustainable use of materials by reducing ecological impacts and stress on natural resources. In addition, these systems can play an essential role in more extensive efforts to meet climate goals by eliminating greenhouse gas emissions resulting from waste transportation and landfill use.

References

Alnuaimi, E., Alsafi, M., Alshehhi, M., Debe, M., Salah, K., Yaqoob, I., Zemerly, M. J., & Jayaraman, R. (2023). Blockchain-based system for tracking and rewarding recyclable plastic waste. *Peer-to-Peer Networking and Applications, 16*(1), 328–346. https://doi.org/10.1007/s12083-022-01413-5

Cheah, C. G., Chia, W. Y., Lai, S. F., Chew, K. W., Chia, S. R., & Show, P. L. (2022). Innovation designs of industry 4.0 based solid waste management: Machinery and digital circular economy. *Environmental Research, 213*, 113619. https://doi.org/10.1016/j.envres.2022.113619

Citaristi, I. (2022). United Nations human settlements programme—UN-habitat. In *The Europa directory of international organizations 2022* (pp. 240–243). Routledge. https://doi.org/10.4324/9781003292548

Das, S., Lee, S. H., Kumar, P., Kim, K. H., Lee, S. S., & Bhattacharya, S. S. (2019). Solid waste management: Scope and the challenge of sustainability. *Journal of Cleaner Production*, 228, 658–678. https://doi.org/10.1016/j.jclepro.2019.04.323

de Oliveira Neto, G. C., Correia, A. D. J. C., & Schroeder, A. M. (2017). Economic and environmental assessment of recycling and reuse of electronic waste: Multiple case studies in Brazil and Switzerland. *Resources, Conservation and Recycling, 127*, 42–55. https://doi.org/10.1016/j.resconrec.2017.08.011

de Souza, L. C. G., & Drumond, M. A. (2022). Decentralised composting as a waste management tool connects with the new global trends: A systematic review. *International Journal of Environmental Science and Technology, 19*(12), 12679–12700. https://doi.org/10.1007/s13762-022-04504-1

Dewasiri, N. J. (2024). Sustainable strategies for Net Zero: An Asia pacific perspective. In R. Singh, & D. Crowther (Eds.), *Transition towards a sustainable future. approaches to global sustainability, markets, and governance*. Springer. https://doi.org/10.1007/978-981-97-5756-5_1

Dewasiri, N. J., Baker, H. K., Banda, Y. K. W., & Rathnasiri, M. S. H. (2022). The dividend decision model: A possible solution for the dividend puzzle. In S. Rana, Sakshi, & J. Singh (Eds.), *Exploring the latest trends in management literature* (Review of Management Literature, Vol. 1, pp. 249–267), Emerald Publishing Limited, Leeds. https://doi.org/10.1108/S2754-586520220000001013

Dewasiri, N. J., Karunarathne, K. S. S. N., Menon, S., Jayarathne, P. G. S. A., & Rathnasiri, M. S. H. (2023). Fusion of artificial intelligence and blockchain in the banking industry: Current application, adoption, and future challenges. In A. Saini, & V. Garg (Eds.), *Transformation for sustainable business and management practices: Exploring the spectrum of industry 5.0* (pp. 293–307). Emerald Publishing Limited, Bingley. https://doi.org/10.1108/978-1-80262-277-520231021

Dewasiri, N. J., Dharmarathna, D. G., & Choudhary, M. (2024). Leveraging artificial intelligence for enhanced risk management in banking: A systematic literature review. In R. Singh, S. Khan, A. Kumar, & V. Kumar (Eds.), *Artificial Intelligence enabled management: An emerging economy perspective* (pp. 197–213), De Gruyter, Berlin. https://doi.org/10.1515/9783111172408

Dhar Dwivedi, A., Singh, R., Kaushik, K., Rao Mukkamala, R., & Alnumay, W. S. (2024). Blockchain and artificial intelligence for 5G-enabled Internet of Things: Challenges, opportunities, and solutions. *Transactions on Emerging Telecommunications Technologies, 35*(4), e4329. https://doi.org/10.1002/ett.4329

Frandsen, T. F., Nielsen, M. F. B., Lindhardt, C. L., & Eriksen, M. B. (2020). Using the full PICO model as a search tool for systematic reviews resulted in lower recall for some PICO elements. *Journal of Clinical Epidemiology, 127*, 69–75. https://doi.org/10.1016/j.jclinepi.2020.07.005

Gopalakrishnan, P. K., Hall, J., & Behdad, S. (2021). Cost analysis and optimisation of Blockchain-based solid waste management traceability system. *Waste Management, 120*, 594–607. https://doi.org/10.1016/j.wasman.2020.10.027

Gopalan, R. B., Babu, B. V., Sugunan, A. P., Murali, A., Ma, M. S., Balasubramanian, R., & Philip, S. (2021). Community engagement to control dengue and other vector-borne diseases in Alappuzha municipality, Kerala, India. *Pathogens and Global Health, 115*(4), 258–266. https://doi.org/10.1080/20477724.2021.1890886

Jayawardena, N. S., & Dewasiri, N. J. (2023). Food Acquisition and consumption issues of South Asian countries: A systematic literature review and future research agenda. *FIIB Business Review*. https://doi.org/10.1177/23197145231194113

Jiang, P., Zhang, L., You, S., Van Fan, Y., Tan, R. R., Klemeš, J. J., & You, F. (2023). Blockchain technology applications in waste management: Overview, challenges and opportunities. *Journal of Cleaner Production, 421*, 138466. https://doi.org/10.1016/j.jclepro.2023.138466

John, J., Varkey, M. S., Podder, R. S., Sensarma, N., Selvi, M., Santhosh Kumar, S. V. N., & Kannan, A. (2022). Intelligent prediction and monitoring of waste disposal systems using IoT and cloud for IoT-based intelligent cities. *Wireless Personal Communications, 122*(1), 243–275. https://doi.org/10.1007/s11277-021-08897-z

Karunarathna, K. S. S. N., Dewasiri, N. J., Singh, R., & Rathnasiri, M. S. H. (2024). What does artificial intelligence–powered ChatGPT bring to academia? A review. In R. Singh, S. Khan, A. Kumar, & V. Kumar (Eds.), *Artificial intelligence enabled management: An emerging economy perspective* (pp. 3–13). De Gruyter, Berlin. https://doi.org/10.1515/9783111172408

Khan, S., Ali, S. S., & Singh, R. (2022). Determinants of remanufacturing adoption for circular economy: A causal relationship evaluation framework. *Applied System Innovation, 5*(4), e62. https://doi.org/10.3390/asi5040062

Kouhizadeh, M., Zhu, Q., & Sarkis, J. (2020). Blockchain and the circular economy: Potential tensions and critical reflections from practice. *Production Planning & Control, 31*(11–12), 950–966. https://doi.org/10.1080/09537287.2019.1695925

Kshirsagar, P. R., Kumar, N., Almulihi, A. H., Alassery, F., Khan, A. I., Islam, S., Rothe, J. P., Jagannadham, D. B. V., & Dekeba, K. (2022). Artificial intelligence-based robotic technique for reusable waste materials. *Computational Intelligence and Neuroscience, 2022*(1), 2073482. https://doi.org/10.1155/2022/2073482

Malinauskaite, J., Jouhara, H., Czajczyńska, D., Stanchev, P., Katsou, E., Rostkowski, P., Thorne, R. J., Colon, J., Ponsá, S., Al-Mansour, F., & Anguilano, L. (2017). Municipal solid waste management and waste-to-energy in the context of a circular economy and energy recycling in Europe. *Energy, 141*, 2013–2044. https://doi.org/10.1016/j.energy.2017.11.128

Massoud, M. A., Mokbel, M., Alawieh, S., & Yassin, N. (2019). Towards improved governance for sustainable solid waste management in Lebanon: Centralised versus decentralised approaches. *Waste Management & Research, 37*(7), 686–697. https://doi.org/10.1177/0734242X19836705

Mishra, A., & Kumar Ray, A. (2020). IoT cloud-based cyber-physical system for efficient solid waste management in smart cities: A novel cost function based route optimisation technique for waste collection vehicles using dustbin sensors and real-time road traffic informatics. *IET Cyber-Physical Systems: Theory & Applications, 5*(4), 330–341. https://doi.org/10.1049/iet-cps.2019.0110

Mishra, U., Rathnasiri, M. S. H., & Dewasiri, N. J. (2024). Net-zero economy and energy sustainability: A Nepalese perspective. In R. Singh, S. Khan, A. Kumar, S. Luthra, & H. Chokshi (Eds.), *Net zero economy, corporate social responsibility and sustainable value creation*. CSR, Sustainability, Ethics & Governance (pp. 145–161). Springer, Cham. https://doi.org/10.1007/978-3-031-55779-8_9

Rathnasiri, M. S. H., Dewasiri, N. J., & Kumar, A. A. (2024). Policy framework and implementation strategies for Sri Lanka's transition to a net-zero economy. In R. Singh, S. Khan, A. Kumar, S. Luthra, & H. Chokshi (Eds.), *Net zero economy, corporate social responsibility and sustainable value creation*. CSR, Sustainability, Ethics & Governance (pp. 43–60). Springer, Cham. https://doi.org/10.1007/978-3-031-55779-8_3

Rathnasiri, M. S. H., & Gunathilaka, D. H. P. M. (2024). Net zero policies for sustainable future: Issues and challenges in the Asia pacific region. In R. Singh, & D. Crowther (Eds.), *Transition towards a sustainable future*. Approaches to global sustainability, markets, and governance. Springer. https://doi.org/10.1007/978-981-97-5756-5_12

Singh, R., Khan, S., & Centobelli, P. (2022). Investigating the Interplay between social performance and organisational factors supporting circular economy practices. *Sustainability, 14*(24), e16781. https://doi.org/10.3390/su142416781

Singh, R., Khan, S., & Dsilva, J. (2023). A framework for assessment of critical factor for circular economy practice implementation. *Journal of Modelling in Management, 18*(5), 1476–1497. https://doi.org/10.1108/JM2-06-2021-0145

Tripathi, N. G. (2024). Plastic waste crisis: How smart cities can lead the change. In M. Goel, & N. G. Tripathi (eds.), *Plastic pollution*. Springer. https://doi.org/10.1007/978-981-97-5528-8_10

Wilts, H., Garcia, B. R., Garlito, R. G., Gómez, L. S., & Prieto, E. G. (2021). Artificial intelligence in sorting municipal waste is an enabler of the circular economy. *Resources, 10*(4), e28. https://doi.org/10.3390/resources10040028

Chapter 2
Innovation for Sustainability: The Role of Digital Circular Economy in Driving Organizational Efficiency and Sustainable Development

Gönül Gül Ekşi and Cumali Kılıç

Abstract This paper explores the critical relationship between innovation and sustainability, emphasizing the role of the digital circular economy (DCE) in advancing sustainability goals. Sustainability, defined by its environmental, social, and economic dimensions, requires an integrated, interdisciplinary approach to address global challenges. Innovation serves as a key driver in this process, enabling businesses to adapt their models, processes, and products to meet sustainability objectives. By integrating sustainability into organizational strategies and governance frameworks, firms can enhance operational efficiency and long-term competitiveness while minimizing environmental impact. The paper highlights the importance of sustainable innovation governance, which aligns organizational goals with broader sustainability objectives through effective policies and practices. Additionally, it explores how digital technologies within the circular economy (CE) facilitate resource efficiency, reduce waste, and support closed-loop systems that contribute to sustainable development. This study addresses sustainability and innovation in terms of organizational processes, strategies, and business models, drawing on literature to assess their impact. The theoretical framework evaluates both past studies and the implications of adopting digital circular practices, offering insights and recommendations for future business sustainability initiatives.

Keywords Sustainability practices · Innovation strategies · Digital transformation · Digital circular economy

---

G. G. Ekşi
Department of Management and Organization, Vocational School, Karatekin University, Çankırı, Turkey
e-mail: gonulguleksi@karatekin.edu.tr

C. Kılıç (✉)
Department of Business Administration, Faculty of Economics and Administrative Sciences, Artuklu University, Mardin, Turkey
e-mail: cumalikilic@artuklu.edu.tr

R. Singh and V. Kumar (eds.), *Sustainable Innovations and Digital Circular Economy*,
https://doi.org/10.1007/978-981-96-1064-8_2

2.1 Introduction

Sustainability is a multifaceted concept that integrates environmental, social, and economic dimensions striving to meet present needs without compromising the ability of future generations to fulfill their own. This definition aligns with the core principles of sustainable development, emphasizing the necessity of balancing these three pillars to ensure the long-term viability of ecosystems, societies, and economies (Yarime et al., 2012; Schoolman et al., 2011). Sustainability science has consequently emerged as a vital interdisciplinary field, addressing the complex interactions between human systems and natural ecosystems by drawing on diverse academic knowledge to tackle sustainability challenges (Shrivastava et al., 2020; Gatti & Seele, 2013). Also, sustainability extends beyond traditional paradigms by leveraging digital technologies to enhance resource efficiency, reduce waste, and promote the regenerative use of materials, contributing to the overarching goals of sustainability in a technologically integrated world (Tekin & Aslan, 2022; Dale & Newman, 2005).

The interdisciplinary nature of sustainability science is crucial, as it facilitates a comprehensive understanding of the socio-environmental dynamics that shape sustainable development (Lang et al., 2012; Kajikawa, 2008). Sustainability science not only addresses ecological and environmental issues but also integrates social and economic dimensions, promoting a more cohesive approach to problem-solving (Konys, 2019). When integrated into corporate strategies, sustainability fosters the adoption of practices that are environmentally responsible, socially equitable, and economically viable. This holistic approach is essential for driving innovation and ensuring long-term profitability (Dyllick & Hockerts, 2002), enhancing not only operational efficiency but also the creation of sustainable business models capable of adapting to evolving market demands (Olusola-Christwealth, 2023). Within the framework of the digital circular economy, this integration becomes even more impactful and plays a pivotal role in promoting sustainable practices by leveraging digital tools to advocate for environmentally friendly solutions (Zambika, 2022).

The integration of sustainability into business practices is increasingly recognized as vital for long-term viability and competitiveness. This perspective aligns with scholarly contributions that highlight how innovations foster sustainability (Havemo, 2023). The relationship between innovation and sustainability is now a critical focus of research and practice, particularly in light of contemporary business and environmental challenges (Adams et al., 2016). Innovation, broadly defined as the introduction of new ideas, products, or processes, is instrumental in advancing sustainability goals across multiple sectors (Kahn, 2018). This relationship is multifaceted, encompassing technological advancements, business model transformations, and strategic management practices, all of which contribute to sustainable development (Lehoux et al., 2018). Innovation takes on an even greater significance within the digital circular economy and digital technologies not only accelerate the development of sustainable solutions but also enable the efficient reuse and recycling of resources, creating closed-loop systems that reduce environmental impact while maintaining economic growth.

A key insight into the relationship between innovation and sustainability is the mediating role of innovation performance in enhancing sustainability outcomes (Zhou et al., 2023). This highlights that fostering a culture of innovation within organizations can lead to improved sustainability metrics, reinforcing the notion that innovation is not only a tool for gaining competitive advantage but also a crucial element for achieving sustainability objectives (Xu et al., 2020). Moreover, the strategic orientation of an organization significantly influences its innovation capabilities, which in turn drives sustainable business growth (Khan & Naeem, 2018). This connection underscores the importance of integrating quality management practices with innovation strategies to achieve long-term sustainability (Rocha et al., 2007). Firms that prioritize both quality and innovation are better equipped to adapt to evolving market demands and regulatory pressures, thus enhancing their sustainability performance.

In addition to organizational strategies, the nature of innovation itself—whether in products, services, or processes—has a direct impact on sustainability (Faisal et al., 2022). This aligns with the Triple Bottom Line theory, which emphasizes balancing economic, environmental, and social outcomes in business practices. By implementing innovative services and processes, companies can not only improve operational efficiency but also contribute to broader sustainability objectives (Miemczyk & Luzzini, 2019). Additionally, effective governance frameworks that promote sustainable innovation can help align organizational goals with sustainability objectives (Varadarajan, 2017). Governance frameworks are further strengthened by digital technologies that facilitate transparency, resource optimization, and circular business models, driving sustainability through more efficient and sustainable innovation practices.

Consequently, sustainability is a complex and dynamic concept that requires an integrated approach to address the interconnected challenges of environmental degradation, social inequality, and economic instability. The ongoing development of sustainability science highlights the importance of interdisciplinary collaboration and community engagement in achieving sustainable outcomes. As businesses increasingly recognize the significance of sustainability, fostering innovation becomes essential for reaching long-term sustainability goals. By incorporating innovative practices into their operations, organizations can not only enhance their competitive advantage but also contribute to a more sustainable future. Thus, this study explores the relationship between sustainability and innovation, focusing on organizational processes, strategies, and business models in the perspective of digital circular economy. These relationships are examined with a focus on how digital technologies and circular practices can further enhance sustainability efforts and theoretical insights and practical recommendations are also provided as key outcomes of this research.

2.2 A Theoretical Approach

2.2.1 Sustainable Innovations and Digital Circular Economy

Sustainable innovations refer to the development and implementation of new ideas, products, processes, or business models that aim to address environmental, social, and economic challenges while promoting sustainability (Boons et al., 2013). This concept is increasingly recognized as essential for organizations seeking to enhance their competitive advantage while contributing positively to society and the environment (McWilliams & Siegel, 2011). Sustainable innovations encompass a wide range of practices, including eco-innovation, green technologies, and sustainable business models, all of which are designed to minimize negative impacts on the planet and improve the quality of life for current and future generations (Singh et al., 2022).

One of the critical aspects of sustainable innovations is their ability to integrate environmental considerations into the innovation process (Singh et al., 2023). In particular, eco-innovation focuses specifically on creating products and processes that reduce environmental impacts, enhance resource efficiency, and promote sustainability (Oduro, 2024). As well, sustainable innovations are often driven by the need for organizations to adapt to changing market demands and regulatory pressures (Demirel & Kesidou, 2019). As consumers become more environmentally conscious, businesses are compelled to innovate in ways that align with sustainability goals (Evans et al., 2017). Examples like the adoption of electronic human resource management systems can enhance operational efficiency and contribute to sustainable competitive advantages by optimizing resource utilization and minimizing waste (Alqarni, 2023) and this highlights the role of technology in facilitating sustainable innovations and improving organizational performance.

Organizations that foster a culture of sustainability are more likely to engage in innovative behaviors that align with sustainability objectives (Galpin et al., 2015). Cultivating an intrapreneurial mindset among employees can lead to sustainable innovative behavior, as individuals are encouraged to creatively solve problems and transform constraints into opportunities (Duradoni & Fabio, 2019). Furthermore, the concept of sustainability-oriented innovation emphasizes the need for businesses to reconcile economic, environmental, and social goals (Cillo et al., 2019). This involves not only creating new products and services but also rethinking existing processes and business models to align with sustainability objectives. In addition to product and process innovations, sustainable innovations can also manifest in the form of business model innovations (Zott & Amit, 2015). These innovations focus on integrating sustainability into the core of business operations, thereby creating value for both the organization and its stakeholders (Evans et al., 2017). Companies may adopt circular economy principles, which emphasize resource efficiency and waste reduction, as part of their business models. This shift not only enhances sustainability but also opens new avenues for growth and profitability.

Collaborative efforts among various stakeholders, including businesses, governments, and non-governmental organizations, can facilitate the development of innovative solutions to complex sustainability challenges (Loureiro et al., 2020; Moons et al., 2021). Also, the impact of regulatory frameworks on sustainable innovations is significant. Governments play a crucial role in shaping the innovation landscape by establishing policies and regulations that encourage sustainable practices (Ashford & Hall, 2011). Environmental regulations can incentivize firms to adopt sustainable innovations by promoting the development of cleaner technologies and practices (Ramanathan et al., 2017). This regulatory support is essential for fostering an environment conducive to sustainable innovation.

Sustainable innovations are increasingly recognized as essential for addressing global environmental challenges and fostering economic resilience (Faivre et al., 2017). The concept of the circular economy plays a pivotal role in this context, promoting resource efficiency, waste reduction, and the regeneration of natural systems (Singh et al., 2022). By shifting from a linear model of production and consumption—characterized by a "take-make-dispose" approach—to a circular model, businesses can enhance sustainability and create economic value while minimizing environmental impact (Aithal & Aithal, 2023; Castro et al., 2022). The integration of sustainable design thinking and social innovation is crucial in overcoming barriers to the circular economy, as it encourages a holistic approach that considers environmental, economic, and social dimensions (Deniz, 2021).

Digital technologies are instrumental in facilitating the transition to a circular economy. The interplay between digitalization and circular business models is profound, as digital tools optimize processes, enhance transparency, and foster innovation (Turcan et al., 2023). Technologies such as the Internet of Things (IoT), blockchain, and big data analytics enable organizations to track resources throughout their lifecycle, thereby improving resource management and reducing waste (Çetin et al., 2021; Magrini et al., 2021). This digital transformation not only supports the circular economy but also aligns with broader sustainability goals, as it encourages companies to rethink their resource management strategies and adopt more sustainable practices (Jakubelskas, 2023). Also, as awareness of sustainability issues grows, consumers are increasingly making purchasing decisions that prioritize durable, repairable, and recyclable products (Bucur, 2023). This shift in consumer behavior is essential for driving demand for circular products and services, thereby incentivizing businesses to adopt circular practices (Raj & Aithal, 2020).

In conclusion, sustainable innovations encompass a broad spectrum of practices aimed at addressing environmental, social, and economic challenges. By integrating sustainability into the innovation process, organizations can enhance their competitive advantage while contributing positively to society and the environment. The interplay between eco-innovation, organizational culture, collaboration, and regulatory frameworks highlights the complexity of sustainable innovations and their critical role in achieving sustainable development goals. Thus, sustainable innovations, particularly within the framework of the circular economy, are crucial for achieving long-term sustainability and economic resilience. The integration of digital technologies enhances the effectiveness of circular practices, while consumer engagement plays a significant role in driving this transition. As organizations continue to explore and implement circular business models, the synergy between sustainability and digitalization will be essential for fostering a more sustainable future.

2.2.2 Sustainable Businesses and Digital Circular Economy

One significant avenue through which innovation fosters sustainability is the development of sustainable business models (Havemo, 2023). Sustainable businesses are organizations that operate with a commitment to environmental stewardship, social responsibility, and economic viability (Bahadur & Waqqas, 2013). The concept of sustainable business encompasses a broad spectrum of practices and strategies aimed at minimizing negative impacts on the environment while maximizing positive contributions to society and the economy (Bocken et al., 2014).

At the core of sustainable businesses is the integration of sustainability into their business models (Khan et al., 2022). Sustainable business models are designed to create, deliver, and capture value in ways that are environmentally sound, socially equitable, and economically viable (Schaltegger et al., 2016). This involves rethinking traditional business practices to align with sustainability goals, such as reducing resource consumption, minimizing waste, and promoting ethical labor practices. Businesses that adopt sustainable models not only enhance their reputation but also improve their operational efficiency and profitability (Behringer & Szegedi, 2016).

Corporate Social Responsibility (CSR) plays a significant role in the framework of sustainable businesses. CSR initiatives are voluntary actions taken by companies to address social and environmental issues, and they are increasingly seen as integral to achieving sustainable development (Behringer & Szegedi, 2016). By embedding CSR into their core strategies, businesses can foster trust and loyalty among stakeholders, including customers, employees, and investors. This alignment between CSR and sustainable business practices is essential for creating shared value and ensuring long-term success (Marczewska & Kostrzewski, 2020).

Sustainable businesses often leverage innovation as a key driver of their sustainability efforts (Singh et al., 2023). Business model innovation, particularly in the context of sustainability, involves developing new ways of delivering value that prioritize environmental and social considerations (Nosratabadi et al., 2019). This can include adopting circular economy principles, which emphasize resource efficiency and waste reduction, as well as exploring new technologies that enhance sustainability (Bocken, 2015). The ability to innovate sustainably is increasingly recognized as a competitive advantage, enabling businesses to adapt to changing market demands and regulatory pressures (Belz & Binder, 2015).

Moreover, the concept of sustainable entrepreneurship is gaining traction as a means of fostering sustainable business practices (Muntean, 2018). This entrepreneurial approach emphasizes the importance of aligning business objectives with sustainability goals, thereby contributing to broader societal and environmental outcomes (Li & Kang, 2018). Also, effective leadership is crucial for fostering a culture of sustainability within organizations, as leaders set the tone for how sustainability is prioritized and integrated into decision-making processes (Bocken et al., 2019). Leaders who champion sustainability can significantly influence their organizations' performance and commitment to sustainable practices (Rudnicka, 2017).

Collaborative approaches, such as co-creation and knowledge sharing, enable businesses to leverage diverse perspectives and resources, ultimately leading to more innovative and impactful sustainability solutions (Rezaee et al., 2017). Engaging with various stakeholders, including suppliers, customers, and local communities, can enhance the effectiveness of sustainability initiatives (Lamolinara, 2023). Furthermore, businesses that prioritize sustainability in their strategic planning are better positioned to navigate risks and seize opportunities in a rapidly changing environment (Bateh et al., 2014). This strategic alignment not only enhances operational resilience but also fosters innovation and competitiveness in the marketplace (Joyce, 2016).

Sustainable businesses are increasingly recognized as vital players in the transition toward a more sustainable economy, particularly through the lens of the circular economy (Pla-Julian & Guevara, 2019). The CE paradigm emphasizes the importance of resource efficiency, waste reduction, and the continuous cycling of materials, which aligns closely with the principles of sustainable business practices (Awan & Sroufe, 2022). Sustainable businesses are characterized by their commitment to balancing economic growth with social equity and environmental protection, often referred to as the triple bottom line (TBL) approach (Belz & Binder, 2015; Sargani et al., 2020; Singh et al., 2021). This holistic view encourages businesses to innovate in ways that not only enhance profitability but also contribute positively to society and the environment (Hizarci-Payne, 2020; Shepherd & Patzelt, 2011).

Digital technologies are transforming the landscape of sustainable businesses by enabling the implementation of circular economy models (Demirel & Danışman, 2019; Turcan et al., 2023). For instance, digital platforms can streamline supply chain operations, allowing businesses to optimize resource use and minimize waste (Angelis et al., 2018). This digital transformation is crucial for small and medium-sized enterprises (SMEs), which often face unique challenges in adopting sustainable practices due to limited resources (Demirel & Danışman, 2019). By leveraging digital solutions, SMEs can enhance their competitiveness while contributing to sustainability goals.

Additionally, sustainable entrepreneurship involves the pursuit of business opportunities that not only generate economic returns but also address social and environmental challenges (Hizarci-Payne, 2020; Hudek & Hojnik, 2020). This form of entrepreneurship is increasingly seen as a catalyst for innovation, driving the development of new products and services that align with sustainable development goals (Barna et al., 2023; Sargani et al., 2020). The motivations behind sustainable entrepreneurship can vary, with both opportunity-driven and necessity-driven entrepreneurs contributing to sustainability initiatives (Rojas-Cruz, 2024). Educational institutions also play a pivotal role in promoting sustainable entrepreneurship by integrating sustainability concepts into their curricula, thereby nurturing a new generation of entrepreneurs committed to sustainable practices (Fanea-Ivanovici & Baber, 2022; Parra, 2013).

Consequently, sustainable businesses, particularly those embracing the principles of the circular economy, are essential for achieving long-term economic, social, and environmental sustainability. The integration of digital technologies enhances the effectiveness of these businesses by optimizing resource management and fostering innovation. Furthermore, sustainable entrepreneurship serves as a critical driver of this transformation, highlighting the importance of education and motivation in cultivating a sustainable business ecosystem. By adopting sustainable business models, leveraging innovation, and fostering a culture of sustainability, organizations can create value for themselves and society while addressing the pressing challenges of our time. The interplay between leadership, collaboration, and strategic alignment further underscores the complexity of sustainable business practices and their critical role in shaping a more sustainable future.

2.2.3 Sustainable Organizations and Digital Circular Economy

Sustainable organizations are entities that actively integrate sustainability principles into their operations, strategies, and cultures, aiming to balance economic success with social equity and environmental stewardship (Camilleri, 2017). This multifaceted approach to sustainability encompasses various dimensions, including environmental management, social responsibility, and economic viability, thereby ensuring that the organization can thrive while contributing positively to society and the planet (Espinosa & Walker, 2017; Sheehy & Farneti, 2021).

At the core of sustainable organizations is the commitment to environmental sustainability, which involves minimizing negative impacts on the environment through responsible resource management, waste reduction, and the adoption of eco-friendly practices (Ahmad, 2015). Organizations that prioritize environmental sustainability not only enhance their operational efficiency but also improve their reputation among stakeholders, including customers, employees, and investors (Musgrave, 2011). This alignment between environmental practices and organizational performance is crucial for long-term success in a competitive marketplace.

Sustainable organizations recognize the importance of social sustainability, which encompasses the well-being of employees, communities, and other stakeholders. This dimension of sustainability involves promoting fair labor practices, diversity, and community engagement (Singh et al., 2019). With a positive organizational culture that values social responsibility, sustainable organizations can enhance employee satisfaction and retention, ultimately contributing to improved organizational performance (Jensen, 2017). The integration of social sustainability into business practices is increasingly seen as essential for building trust and loyalty among stakeholders.

The concept of sustainable enterprise excellence further emphasizes the need for organizations to develop strategies that support sustainability while achieving operational excellence (Sony, 2019). This involves creating a framework that aligns sustainability goals with business objectives, ensuring that sustainability is embedded in the organization's core functions and decision-making processes (Edgeman et al., 2013). Organizations that adopt such frameworks are better positioned to navigate the complexities of the modern business landscape, characterized by rapid changes in consumer preferences, regulatory requirements, and environmental challenges (Iqbal & Ahmad, 2020).

Collaboration and stakeholder engagement are also vital components of sustainable organizations (Hoque et al., 2018). Collaborative approaches, such as co-creation and knowledge sharing, enable organizations to leverage diverse perspectives and resources, ultimately leading to more innovative and impactful sustainability solutions (Abbate et al., 2019; Craig & Allen, 2013). This approach enables organizations to make informed decisions and continuously refine their sustainability strategies, ensuring that they remain responsive to evolving stakeholder expectations and environmental challenges.

Furthermore, the role of stakeholders in shaping sustainable organizations cannot be overstated. Customers, employees, and community members increasingly demand sustainable practices, compelling organizations to adapt and innovate (Singh et al., 2019). Engaging stakeholders in sustainability initiatives not only enhances organizational reputation but also drives innovation and improves operational efficiency (Schwaninger, 2015). For example, organizations that actively involve their employees in sustainability efforts often experience increased morale and productivity, as individuals feel a sense of ownership and purpose in their work (Manuti et al., 2022).

Sustainable organizations adopt practices that align with the principles of sustainability, which encompass not only ecological considerations but also social equity and economic growth (Hariram et al., 2023). The digital circular economy provides a transformative framework for sustainable organizations, enabling them to optimize resource use and minimize waste through innovative digital solutions (Singh et al., 2023). Digital technologies facilitate the implementation of circular economy principles by enhancing transparency and efficiency across supply chains (Cagno et al., 2021; Magrini et al., 2021). Organizations can leverage digital platforms to track materials throughout their lifecycle, enabling better resource management and reducing environmental impact (Turcan, 2023). This integration of digital tools not only supports the transition to a circular economy but also enhances the resilience and sustainability of business models by fostering collaboration among stakeholders (Khong, 2023).

Ultimately, sustainable organizations are pivotal in advancing the principles of sustainability through effective leadership, stakeholder engagement, and the adoption of digital circular economy practices. By integrating sustainability into their core operations and leveraging digital technologies, these organizations can optimize resource use, enhance their competitive advantage, and contribute to a more sustainable future. As the global economy continues to evolve, the synergy between sustainability and digitalization will be essential for organizations seeking to thrive in an increasingly complex and interconnected world.

2.3 Discussion

The intersection of innovation, sustainability, and the digital circular economy represents a transformative shift in how businesses and organizations approach long-term value creation. Through the adoption of sustainable business models and organizational practices, companies are recognizing that sustainability is not just a corporate responsibility but a source of competitive advantage. In particular, the circular economy, reinforced by digital technologies, offers a pathway for businesses and organizations to align economic, social, and environmental goals.

Sustainable businesses have emerged as key drivers of this transformation. By integrating sustainability into their core models, these businesses are creating frameworks that reduce resource consumption, promote ethical labor practices, and minimize environmental impacts, all while maintaining profitability and efficiency. Innovation plays a central role in this transition, enabling businesses to explore new technologies and strategies that prioritize sustainability. The incorporation of digital tools like the internet of things, big data analytics, and blockchain enhances transparency and efficiency, particularly in resource management. These advancements help businesses reduce waste, optimize operations, and respond to growing regulatory and market pressures for sustainability.

Furthermore, CSR initiatives are being increasingly integrated into sustainable business strategies, fostering trust among stakeholders and reinforcing the alignment between economic performance and social responsibility. Sustainable entrepreneurship also emerges as a powerful force, with entrepreneurs driving innovation that tackles environmental and social challenges while achieving economic success. This blend of entrepreneurship and sustainability positions businesses to lead the transition toward a more circular, resource-efficient economy.

The digital circular economy enhances these efforts by offering organizations innovative ways to manage resources and reduce environmental impact. Digital platforms facilitate the tracking of materials throughout their lifecycle, enabling more efficient use of resources and reducing waste. This technological integration not only improves organizational efficiency but also fosters greater resilience in the face of economic and environmental challenges.

In this evolving landscape, businesses and organizations that embrace the synergy between sustainability, innovation, and digitalization will be better positioned to address the global challenges of resource scarcity, climate change, and social inequality. The shift toward a sustainable future demands a new approach to value creation—one that balances profitability with responsibility toward the planet and society. The role of leadership, strategic alignment, and the adoption of digital tools will be critical for businesses and organizations to remain competitive and resilient in this rapidly changing environment.

In conclusion, the integration of sustainability into business and organizational practices, supported by the digital circular economy, represents a profound shift toward long-term, holistic value creation. Sustainable businesses and organizations that prioritize innovation, collaboration, and leadership will lead the way in addressing the pressing environmental and social challenges of our time, while simultaneously enhancing their operational success.

2.4 Concluding Remarks

The integration of sustainability and digital circular economy principles into business and organizational practices represents a critical shift in the way economic, social, and environmental goals are pursued. As demonstrated in this paper, the intersection of innovation, digitalization, and sustainability has the potential to transform how businesses create value, engage stakeholders, and manage resources. Sustainable business models and organizational strategies that prioritize ethical, resource-efficient, and environmentally sound practices are not just necessary responses to global challenges but also avenues for gaining competitive advantage.

The success of this transition hinges on the role of leadership, collaboration, and strategic foresight. Organizations that foster a culture of sustainability, engage in meaningful partnerships, and leverage digital tools will be better positioned to adapt to the evolving demands of a global economy. At the same time, the importance of stakeholder engagement, particularly in fostering a shared commitment to sustainability, cannot be overstated.

Thus, this paper contributes to the growing body of literature on sustainability, innovation, and the digital circular economy by offering a comprehensive exploration of how these domains intersect within the context of businesses and organizations. Theoretically, it expands the understanding of sustainable business models, the role of leadership in fostering sustainability, and the importance of stakeholder engagement in achieving long-term success. By integrating concepts such as CSR, sustainable entrepreneurship, and the digital circular economy, this research provides a holistic framework that links sustainability with technological advancements and innovation.

A key theoretical contribution is the extension of the resource-based view (RBV) by demonstrating how digital tools can optimize resource management and enhance organizational performance. The integration of these digital technologies into sustainability models introduces new ways to conceptualize resource efficiency, transparency, and innovation, thus advancing the circular economy literature. Also, this research highlights the interplay between sustainability and entrepreneurship, showing how sustainable entrepreneurship serves as a catalyst for innovation and organizational change. By aligning sustainability with competitive advantage, this paper also contributes to theories of strategic management, emphasizing that sustainability is not merely a compliance issue but a core driver of long-term success.

From a practical standpoint, this paper provides valuable insights for business leaders, policymakers, and entrepreneurs seeking to navigate the complexities of sustainability in an increasingly digital and interconnected economy. The discussion of sustainable business models offers a blueprint for organizations looking to embed sustainability into their operations while maintaining profitability and competitiveness. Practical implications include strategies for implementing circular economy principles through digital transformation, such as leveraging IoT for resource tracking, big data for decision-making, and blockchain for supply chain transparency. These technologies enable businesses to minimize waste, optimize resources, and enhance stakeholder engagement, ultimately driving both economic and environmental gains.

For practitioners, this paper emphasizes the importance of leadership in fostering a culture of sustainability and promoting the alignment of sustainability goals with business objectives. Leaders who prioritize sustainability can cultivate stronger relationships with stakeholders, improve employee engagement, and build resilient organizations that are adaptable to environmental, social, and regulatory pressures. Moreover, the findings highlight the importance of collaboration across industries, communities, and supply chains, suggesting that businesses that engage in partnerships and co-creation with diverse stakeholders will be better positioned to innovate and thrive in a rapidly changing market.

For policymakers, the paper underscores the need for policies that incentivize sustainable practices, promote digital infrastructure, and support the transition to a circular economy. By encouraging the adoption of digital technologies, policymakers can accelerate the shift toward more sustainable and resource-efficient business ecosystems, fostering economic growth that is aligned with sustainability goals.

To sum up, the theoretical and practical implications of this research demonstrate the centrality of sustainability and digital innovation in shaping the future of businesses and organizations. By integrating these principles into their operations, organizations can not only achieve sustainability goals but also secure a competitive edge in a dynamic global marketplace.

Also, while this paper offers valuable insights into the intersection of sustainability, innovation, and the digital circular economy, several avenues for future research could deepen and expand the findings presented here. First, more empirical studies are needed to examine the long-term impact of digital technologies on the sustainability of businesses and organizations. Research that investigates the specific effects of tools like IoT, big data, and blockchain on resource efficiency and waste reduction, across different industries and regions, would provide a richer understanding of the digital circular economy's potential.

Additionally, future studies could explore the role of SMEs in adopting sustainable practices through digital transformation. While this paper touches on the challenges and opportunities faced by SMEs, more research is needed to understand how digital tools can help overcome resource constraints and scale sustainability efforts in smaller organizations. A comparative analysis of large corporations and SMEs in their approach to integrating sustainability and digital technologies would offer valuable insights into the varying strategies for achieving circular economy goals.

Likewise, there is a need to explore the societal and policy implications of the digital circular economy. Research could investigate how governments and regulatory bodies can create policies that better facilitate the integration of digital technologies into sustainability frameworks, particularly in developing economies. This line of inquiry could also examine the socio-economic impact of widespread adoption of digital circular economy models, assessing how these changes affect job creation, skill development, and economic growth on a broader scale.

Lastly, the evolving concept of sustainable entrepreneurship provides fertile ground for future research. While this paper outlines the role of sustainable entrepreneurship in fostering innovation, future studies could investigate how education systems can better prepare entrepreneurs for sustainability challenges. The effectiveness of sustainability-focused entrepreneurial education programs, as well as the motivations of entrepreneurs to pursue sustainability goals, could offer practical insights for shaping future business ecosystems.

As a closing thought, the ongoing development of digital technologies, coupled with increasing pressure to address environmental and social challenges, presents numerous opportunities for future research. By focusing on these areas, future studies can further illuminate the pathways through which sustainability, innovation, and the digital circular economy will shape the business landscape in the coming decades.

References

Abbate, T., Codini, A. P., & Aquilani, B. (2019). Knowledge co-creation in open innovation digital platforms: Processes, tools and services. *Journal of Business & Industrial Marketing, 34*(7), 1434–1447. https://doi.org/10.1108/JBIM-09-2018-0276

Adams, R., Jeanrenaud, S., Bessant, J., Denyer, D., & Overy, P. (2016). Sustainability-oriented innovation: A systematic review. *International Journal of Management Reviews, 18*(2), 180–205. https://doi.org/10.1111/ijmr.12068

Ahmad, S. (2015). Green human resource management: Policies and practices. *Cogent Business and Management, 2*(1), 1030817. https://doi.org/10.1080/23311975.2015.1030817

Aithal, S., & Aithal, P. (2023). Importance of circular economy for resource optimization in various industry sectors—a review-based opportunity analysis. *International Journal of Applied Engineering and Management Letters (IJAEML), 7*(2), 191–215. https://doi.org/10.47992/ijaeml.2581.7000.0182

Alqarni, K. (2023). The effect of electronic human resource management systems on sustainable competitive advantages: The roles of sustainable innovation and organizational agility. *Sustainability, 15*(23), 16382. https://doi.org/10.3390/su152316382

Angelis, R., Howard, M., & Miemczyk, J. (2018). Supply chain management and the circular economy: Towards the circular supply chain. *Production Planning and Control, 29*(6), 425–437. https://doi.org/10.1080/09537287.2018.1449244

Ashford, N. A., & Hall, R. P. (2011). The importance of regulation-induced innovation for sustainable development. *Sustainability, 3*(1), 270–292. https://doi.org/10.3390/su3010270

Awan, U., & Sroufe, R. (2022). Sustainability in the circular economy: Insights and dynamics of designing circular business models. *Applied Sciences, 12*(3), 1521. https://doi.org/10.3390/app12031521

Bahadur, W., & Waqqas, O. (2013). Corporate social responsibility for a sustainable business. *Journal of Sustainable Society, 2*(4), 92–97.

Barna, C., Zbuchea, A., & Stanescu, S. (2023). Social economy enterprises contributing to the circular economy and the green transition in Romania. *Ciriec-España Revista De Economía Pública Social y Cooperativa, 107*, 47–69. https://doi.org/10.7203/ciriec-e.107.21738

Bateh, J., Horner, D., Broadbent, A., & Fish, D. (2014). Towards a theoretical integration of sustainability: A literature review and suggested way forward. *Journal of Sustainability Management (JSM), 2*(1), 35–42. https://doi.org/10.19030/jsm.v2i1.8756

Behringer, K., & Szegedi, K. (2016). The role of CSR in achieving sustainable development—theoretical approach. *European Scientific Journal (ESJ), 12*(22), 10. https://doi.org/10.19044/esj.2016.v12n22p10

Belz, F., & Binder, J. (2015). Sustainable entrepreneurship: A convergent process model. *Business Strategy and the Environment, 26*(1), 1–17. https://doi.org/10.1002/bse.1887

Bocken, N. M., Short, S. W., Rana, P., & Evans, S. (2014). A literature and practice review to develop sustainable business model archetypes. *Journal of Cleaner Production, 65*, 42–56. https://doi.org/10.1016/j.jclepro.2013.11.039

Bocken, N. (2015). Sustainable venture capital—catalyst for sustainable start-up success? *Journal of Cleaner Production, 108*, 647–658. https://doi.org/10.1016/j.jclepro.2015.05.079

Bocken, N., Boons, F., & Baldassarre, B. (2019). Sustainable business model experimentation by understanding ecologies of business models. *Journal of Cleaner Production, 208*, 1498–1512. https://doi.org/10.1016/j.jclepro.2018.10.159

Boons, F., Montalvo, C., Quist, J., & Wagner, M. (2013). Sustainable innovation, business models and economic performance: An overview. *Journal of Cleaner Production, 45*, 1–8. https://doi.org/10.1016/j.jclepro.2012.08.013

Bucur, L. (2023). Exploring the role of consumers in promoting a circular economy: Increasing awareness and engagement. In *Proceedings of the international conference on business excellence*, (Vol. 17(1), pp. 38–47). https://doi.org/10.2478/picbe-2023-0006

Cagno, E., Neri, A., Negri, M., Bassani, C., & Lampertico, T. (2021). The role of digital technologies in operationalizing the circular economy transition: A systematic literature review. *Applied Sciences, 11*(8), 3328. https://doi.org/10.3390/app11083328

Camilleri, M. A. (2017). Corporate sustainability and responsibility: Creating value for business, society and the environment. *Asian Journal of Sustainability and Social Responsibility, 2*(1), 59–74. https://doi.org/10.1186/s41180-017-0016-5

Castro, A., Castro, J., Burgos, R., & Suqilanda, E. (2022). Circular economy and its impact on environmental sustainability. *Centro Sur, 6*(4). https://doi.org/10.37955/cs.v6i4.286

Cillo, V., Petruzzelli, A., Ardito, L., & Giudice, M. (2019). Understanding sustainable innovation: A systematic literature review. *Corporate Social Responsibility and Environmental Management, 26*(5), 1012–1025. https://doi.org/10.1002/csr.1783

Craig, C., & Allen, M. (2013). Sustainability information sources: Employee knowledge, perceptions, and learning. *Journal of Communication Management, 17*(4), 292–307. https://doi.org/10.1108/jcom-05-2012-0035

Çetin, S., Wolf, C., & Bocken, N. (2021). Circular digital built environment: An emerging framework. *Sustainability, 13*(11), 6348. https://doi.org/10.3390/su13116348

Dale, A., & Newman, L. (2005). Sustainable development, education and literacy. *International Journal of Sustainability in Higher Education, 6*(4), 351–362. https://doi.org/10.1108/14676370510623847

Demirel, P., & Danışman, G. (2019). Eco-innovation and firm growth in the circular economy: Evidence from European small and medium sized enterprises. *Business Strategy and the Environment, 28*(8), 1608–1618. https://doi.org/10.1002/bse.2336

Demirel, P., & Kesidou, E. (2019). Sustainability-oriented capabilities for eco-innovation: Meeting the regulatory, technology, and market demands. *Business Strategy and the Environment, 28*(5), 847–857. https://doi.org/10.1002/bse.2286

Deniz, D. (2021). Sustainable design thinking and social innovation for beating barriers to circular economy. *WIT Transactions on Ecology and the Environment, 253*. https://doi.org/10.2495/sc210191

Duradoni, M., & Fabio, A. (2019). Intrapreneurial self-capital and sustainable innovative behavior within organizations. *Sustainability, 11*(2), 322. https://doi.org/10.3390/su11020322

Dyllick, T., & Hockerts, K. (2002). Beyond the business case for corporate sustainability. *Business Strategy and the Environment, 11*(2), 130–141. https://doi.org/10.1002/bse.323

Edgeman, R., Bollingtoft, A., Eskildsen, J., Kallehave, P., & Kjærgaard, T. (2013). Sustainable enterprise excellence and the continuously relevant and responsible organization. *International Journal of Social Ecology and Sustainable Development, 4*(4), 65–76. https://doi.org/10.4018/ijsesd.2013100105

Espinosa, A., & Walker, J. (2017). *Complexity approach to sustainability, A: Theory and application* (Vol. 5). World Scientific. https://doi.org/10.1142/q0060

Evans, S., Vladimirova, D., Holgado, M., Fossen, K., Yang, M., Silva, E., & Barlow, C. (2017). Business model innovation for sustainability: Towards a unified perspective for creation of sustainable business models. *Business Strategy and the Environment, 26*(5), 597–608. https://doi.org/10.1002/bse.1939

Faisal, F., Muhammad, K., & Ghani, E. (2022). The effect of innovation in business model, services and technology on firm sustainability: An examination using triple bottom line theory. *The Economics and Finance Letters, 9*(2), 170–179. https://doi.org/10.18488/29.v9i2.3095

Faivre, N., Fritz, M., Freitas, T., De Boissezon, B., & Vandewoestijne, S. (2017). Nature-Based solutions in the EU: Innovating with nature to address social, economic and environmental challenges. *Environmental Research, 159*, 509–518. https://doi.org/10.1016/j.envres.2017.08.032

Fanea-Ivanovici, M., & Baber, H. (2022). Sustainability at universities as a determinant of entrepreneurship for sustainability. *Sustainability, 14*(1), 454. https://doi.org/10.3390/su14010454

Galpin, T., Whitttington, J. L., & Bell, G. (2015). Is your sustainability strategy sustainable? Creating a culture of sustainability. *Corporate Governance, 15*(1), 1–17. https://doi.org/10.1108/CG-01-2013-0004

Gatti, L., & Seele, P. (2013). Evidence for the prevalence of the sustainability concept in european corporate responsibility reporting. *Sustainability Science,* 9(1), 89–102. https://doi.org/10.1007/s11625-013-0233-5

Hariram, N. P., Mekha, K. B., Suganthan, V., & Sudhakar, K. (2023). Sustainalism: An integrated socio-economic-environmental model to address sustainable development and sustainability. *Sustainability, 15*(13), 10682. https://doi.org/10.3390/su151310682

Havemo, E. (2023). Sustainable trajectories for business model innovation. *Journal of Business Models, 11*(1), 58–67. https://doi.org/10.54337/jbm.v11i1.7179

Hizarci-Payne, A. K. (2020). Sustainable entrepreneurship. In S. Idowu, R. Schmidpeter, N. Capaldi, L. Zu, M. Del Baldo, & R. Abreu (Eds.), *Encyclopedia of sustainable management*. Springer. https://doi.org/10.1007/978-3-030-02006-4_23-1

Hoque, N., Rahman, A., Molla, R., Noman, A., & Bhuiyan, M. (2018). Is corporate social responsibility pursuing pristine business goals for sustainable development? *Corporate Social Responsibility and Environmental Management, 25*(6), 1130–1142. https://doi.org/10.1002/csr.1527

Hudek, I., & Hojnik, B. B. (2020). Impact of entrepreneurship activity sustainable development. *Problemy Ekorozwoju, 15*(2), 175–183. https://doi.org/10.35784/pe.2020.2.17

Iqbal, Q., & Ahmad, N. (2020). Sustainable development: The colors of sustainable leadership in learning organization. *Sustainable Development, 29*(1), 108–119. https://doi.org/10.1002/sd.2135

Jakubelskas, U. (2023). Circular economy practices as a tool for sustainable development in the context of renewable energy: What are the opportunities for the EU? *Oeconomia Copernicana, 14*(3), 833–859. https://doi.org/10.24136/oc.2023.025
Jensen, P. (2017). People can't believe we exist!: Social sustainability and alternative nonprofit organizing. *Critical Sociology, 44*(2), 375–388. https://doi.org/10.1177/0896920517691106
Joyce, A. (2016). The triple layered business model canvas: A tool to design more sustainable business models. *Journal of Cleaner Production, 135*, 1474–1486. https://doi.org/10.1016/j.jclepro.2016.06.067
Kahn, K. B. (2018). Understanding innovation. *Business Horizons, 61*(3), 453–460. https://doi.org/10.1016/j.bushor.2018.01.011
Kajikawa, Y. (2008). Research core and framework of sustainability science. *Sustainability Science, 3*(2), 215–239. https://doi.org/10.1007/s11625-008-0053-1
Khan, B., & Naeem, H. (2018). The impact of strategic quality orientation on innovation capabilities and sustainable business growth. *International Journal of Quality and Reliability Management, 35*(8), 1568–1598. https://doi.org/10.1108/ijqrm-07-2017-0126
Khan, S., Ali, S. S., & Singh, R. (2022). Determinants of remanufacturing adoption for circular economy: A causal relationship evaluation framework. *Applied System Innovation, 5*(4), 62. https://doi.org/10.3390/asi5040062
Khong, I. (2023). The circular economy's performance and the impact of digitalization. *International Transactions on Education Technology (ITEE), 2*(1), 18–23. https://doi.org/10.33050/itee.v2i1.389
Konys, A. (2019). Towards sustainable entrepreneurship holistic construct. *Sustainability, 11*(23), 6749. https://doi.org/10.3390/su11236749
Lamolinara, B. (2023). Sustainable versus circular business models in agribusiness: A comparative bibliometric analysis. *Revista de Economia e Sociologia Rural, 61*(spe). https://doi.org/10.1590/1806-9479.2023.275416
Lang, D., Wiek, A., Bergmann, M., Stauffacher, M., Martens, P., Moll, P., … & Thomas, C. (2012). Transdisciplinary research in sustainability science: Practice, principles, and challenges. *Sustainability Science, 7*(S1), 25–43. https://doi.org/10.1007/s11625-011-0149-x
Lehoux, P., Silva, H. P., Sabio, R. P., & Roncarolo, F. (2018). The unexplored contribution of responsible innovation in health to sustainable development goals. *Sustainability, 10*(11), 4015. https://doi.org/10.3390/su10114015
Li, W., & Kang, R. (2018). Transforming sustainability challenges into competitive advantages for enterprises: A perspective of sustainable business models. https://doi.org/10.36689/uhk/hed/2018-01-053
Loureiro, S. M. C., Romero, J., & Bilro, R. G. (2020). Stakeholder engagement in co-creation processes for innovation: A systematic literature review and case study. *Journal of Business Research, 119*, 388–409. https://doi.org/10.1016/j.jbusres.2019.09.038
Magrini, C., Nicolas, J., Berg, H., Bellini, A., Vincenti, N., Campadello, L., Bonoli, A., et al. (2021). Using internet of things and distributed ledger technology for digital circular economy enablement: The case of electronic equipment. *Sustainability, 13*(9), 4982. https://doi.org/10.3390/su13094982
Manuti, A., Giancaspro, M., & Callea, A. (2022). Sustainable careers and flourishing organizations. *Sustainability, 14*(19), 11898. https://doi.org/10.3390/su141911898
Marczewska, M., & Kostrzewski, M. (2020). Sustainable business models: A bibliometric performance analysis. *Energies, 13*(22), 6062. https://doi.org/10.3390/en13226062
McWilliams, A., & Siegel, D. S. (2011). Creating and capturing value: Strategic corporate social responsibility, resource-based theory, and sustainable competitive advantage. *Journal of Management, 37*(5), 1480–1495. https://doi.org/10.1177/0149206310385696
Miemczyk, J., & Luzzini, D. (2019). Achieving triple bottom line sustainability in supply chains: The role of environmental, social and risk assessment practices. *International Journal of Operations and Production Management, 39*(2), 238–259. https://doi.org/10.1108/IJOPM-06-2017-0334

Moons, I., Daems, K., & Velde, L. (2021). Co-creation as the solution to sustainability challenges in the greenhouse horticultural industry: The importance of a structured innovation management process. *Sustainability, 13*(13), 7149. https://doi.org/10.3390/su13137149
Muntean, M. (2018). Business intelligence issues for sustainability projects. *Sustainability, 10*(2), 335. https://doi.org/10.3390/su10020335
Musgrave, J. (2011). Moving towards responsible events management. *Worldwide Hospitality and Tourism Themes, 3*(3), 258–274. https://doi.org/10.1108/17554211111142211
Nosratabadi, S., Mosavi, A., Shamshirband, S., Zavadskas, E., Rakotonirainy, A., & Chau, K. (2019). Sustainable business models: A review. *Sustainability, 11*(6), 1663. https://doi.org/10.3390/su11061663
Oduro, S. (2024). Eco-innovation and SMEs' sustainable performance: A meta-analysis. *European Journal of Innovation Management, 27*(9), 248–279. https://doi.org/10.1108/ejim-11-2023-0961
Olusola-Christwealth, D. (2023). The development of sustainable business models: Concept, elements and importance. *Vilnius University Proceedings, 37*, 75–81. https://doi.org/10.15388/vgisc.2023.11
Parra, S. (2013). Exploring the incorporation of values for sustainable entrepreneurship teaching/learning. *Journal of Technology Management and Innovation, 8*(1), 11–20. https://doi.org/10.4067/s0718-27242013000100002
Pla-Julian, I., & Guevara, S. (2019). Is circular economy the key to transitioning towards sustainable development? Challenges from the perspective of care ethics. *Futures, 105*, 67–77. https://doi.org/10.1016/j.futures.2018.09.001
Raj, K., & Aithal, P. (2020). A new concept of circular model of management for achieving sustainable success and growth. *International Journal of Management Technology and Social Sciences (IJMTS), 5*(1), 1–17. https://doi.org/10.47992/ijmts.2581.6012.0079
Ramanathan, R., He, Q., Black, A., Ghobadian, A., & Gallear, D. (2017). Environmental regulations, innovation and firm performance: A revisit of the porter hypothesis. *Journal of Cleaner Production, 155*, 79–92. https://doi.org/10.1016/j.jclepro.2016.08.116
Rezaee, Z., Zhen, K., & Ha, M. (2017). Progress toward business sustainability in Asia in the aftermath of 2015 Hong Kong stock exchange requirements. *International Journal of Sustainability Management and Information Technologies, 3*(4), 40. https://doi.org/10.11648/j.ijsmit.20170304.12
Rocha, M., Searcy, C., & Karapetrovic, S. (2007). Integrating sustainable development into existing management systems. *Total Quality Management and Business Excellence, 18*(1–2), 83–92. https://doi.org/10.1080/14783360601051594
Rojas-Cruz, L. (2024). Understanding the link: The competencies and motivations of nascent entrepreneurs to engage in sustainable entrepreneurship. *Management Research the Journal of the Iberoamerican Academy of Management, 22*(2), 134–158. https://doi.org/10.1108/mrjiam-10-2023-1468
Rudnicka, A. (2017). Understanding sustainable business models. *Journal of Positive Management, 7*(4), 52. https://doi.org/10.12775/jpm.2016.022
Sargani, G., Zhou, D., Raza, M., & Wei, Y. (2020). Sustainable entrepreneurship in the agriculture sector: The nexus of the triple bottom line measurement approach. *Sustainability, 12*(8), 3275. https://doi.org/10.3390/su12083275
Schaltegger, S., Lüdeke-Freund, F., & Hansen, E. (2016). Business models for sustainability. *Organization and Environment, 29*(3), 264–289. https://doi.org/10.1177/1086026616633272
Schoolman, E., Guest, J., Bush, K., & Bell, A. (2011). How interdisciplinary is sustainability research? Analyzing the structure of an emerging scientific field. *Sustainability Science, 7*(1), 67–80. https://doi.org/10.1007/s11625-011-0139-z
Schwaninger, M. (2015). Organizing for sustainability: A cybernetic concept for sustainable renewal. *Kybernetes, 44*(6/7), 935–954. https://doi.org/10.1108/k-01-2015-0008

Sheehy, B., & Farneti, F. (2021). Corporate social responsibility, sustainability, sustainable development and corporate sustainability: What is the difference, and does it matter? *Sustainability, 13*(11), 5965. https://doi.org/10.3390/su13115965
Shepherd, D., & Patzelt, H. (2011). The new field of sustainable entrepreneurship: Studying entrepreneurial action linking "what is to be sustained" with "what is to be developed." *Entrepreneurship Theory and Practice, 35*(1), 137–163. https://doi.org/10.1111/j.1540-6520.2010.00426.x
Shrivastava, P., Smith, M., O'Brien, K., & Zsolnai, L. (2020). Transforming sustainability science to generate positive social and environmental change globally. *One Earth, 2*(4), 329–340. https://doi.org/10.1016/j.oneear.2020.04.010
Singh, A., Sushil, S., Kar, S., & Pamucar, D. (2019). Stakeholder role for developing a conceptual framework of sustainability in organization. *Sustainability, 11*(1), 208. https://doi.org/10.3390/su11010208
Singh, R., Dsilva, J., Centobelli, P., & Tripathi, V. (2021). A review on sustainable HRM: Evaluating sustainability for organizational development. *Empirical Economics Letters, 20*(2), 62–69.
Singh, R., Khan, S., & Centobelli, P. (2022). Investigating the interplay between social performance and organisational factors supporting circular economy practices. *Sustainability, 14*(24), 16781. https://doi.org/10.3390/su142416781
Singh, R., Khan, S., & Dsilva, J. (2023). A framework for assessment of critical factor for circular economy practice implementation. *Journal of Modelling in Management, 18*(5), 1476–1497. https://doi.org/10.1108/JM2-06-2021-0145
Sony, M. (2019). Implementing sustainable operational excellence in organizations: An integrative viewpoint. *Production & Manufacturing Research, 7*(1), 67–87. https://doi.org/10.1080/21693277.2019.1581674
Tekin, N., & Aslan, O. (2022). Analysis of pre-service science teachers' biodiversity images according to sustainable environmental awareness. *Present Environment and Sustainable Development,* 63–78. https://doi.org/10.47743/pesd2022161005
Turcan, I. (2023). Digitalization and its role in the development of circular economy business models. *The 5th Economic International Conference "Competitiveness and sustainable development",* 103–109. https://doi.org/10.52326/csd2023.16
Turcan, I., Turcan, R., & Stratila, A. (2023). Digitalization and its role in the development of circular economy business models. *Competitiveness and Sustainable Development*, 103–109. https://doi.org/10.52326/csd2023.16
Varadarajan, R. (2017). Innovating for sustainability: A framework for sustainable innovations and a model of sustainable innovations orientation. *Journal of the Academy of Marketing Science, 45*, 14–36. https://doi.org/10.1007/s11747-015-0461-6
Xu, K., Loh, L., & Chen, Q. (2020). Sustainable innovation governance: An analysis of regional innovation with a super efficiency slack-based measure model. *Sustainability, 12*(7), 3008. https://doi.org/10.3390/su12073008
Yarime, M., Trencher, G., Mino, T., Scholz, R., Olsson, L., Ness, B., … & Rotmans, J. (2012). Establishing sustainability science in higher education institutions: Towards an integration of academic development, institutionalization, and stakeholder collaborations. *Sustainability Science, 7*(S1), 101–113. https://doi.org/10.1007/s11625-012-0157-5
Zambika, H. (2022). The importance of sustainable procurement in public institutions. *Journal of Economics Management and Trade,* 11–21. https://doi.org/10.9734/jemt/2022/v28i830426
Zhou, S., Rashid, M., Zobair, S., Sobhani, F., & Siddik, A. (2023). Does ESG impact firms' sustainability performance? The mediating effect of innovation performance. *Sustainability, 15*(6), 5586. https://doi.org/10.3390/su15065586
Zott, C., & Amit, R. (2015). Business model innovation: Toward a process perspective. In *The Oxford handbook of creativity, innovation, and entrepreneurship* (pp. 395–406).

Chapter 3
Examining the Constraints Faced by Organisations in Promoting Technological Innovations for a Sustainable Future in South Asia

Mananage Shanika Hansini Rathnasiri and Pasan Wathsala Perera

Abstract This chapter investigates constraints faced by organisations in promoting technological innovations for sustainable development in South Asia. The study, conducted using a systematic literature review with Preferred Reporting Items for Systematic Reviews and Meta-Analyses (PRISMA) and Population, Interest, Context, Outcomes (PICO) approaches, focuses on various obstacles detrimental to organisational initiatives in sustainability using 35 sample studies. The findings show that regulations, funding, technology, and culture impact innovation as it is implemented in South Asia's energy, agriculture, and manufacturing sectors. The research acknowledges the areas of improvement, especially highlighting the need for improving government and private partnerships, favourable legal conditions, and improvement of institutional capacity. Against this background, organisations have to undertake collaborative strategies consistent with the international sustainable development framework to address the challenges faced and encourage the promotion of newer technologies. Alleviating such constraints would indeed enhance the status and further the missions of key organisations in a far-reaching manner in South Asia and the larger region for a more sustainable course and future.

Keywords Sustainable development · Technological innovations · Organisational constraints · South Asia · Systematic literature review

M. S. H. Rathnasiri (✉)
Department of Marketing Management, Faculty of Management Studies, Sabaragamuwa University of Sri Lanka, Belihuloya, Sri Lanka
e-mail: shanika@mgt.sab.ac.lk

P. W. Perera
Faculty of Business Management, SLTC Research University, Padukka, Sri Lanka
e-mail: Pasanwathsala2@gmail.com

R. Singh and V. Kumar (eds.), *Sustainable Innovations and Digital Circular Economy*,
https://doi.org/10.1007/978-981-96-1064-8_3

3.1 Introduction

3.1.1 Overview of Technological Innovation for Sustainability

Technology is central to the search for sustainable solutions to the environmental, social, and economic problems pervading the world today (Gamage et al., 2024). With the global awareness of climate change, depletion of resources, and ecological degradation, the call for developing new technologies to foster sustainable development has been realised. Sustainable technologies include a variety of technology solutions that address issues such as renewable resources, waste minimisation, innovative uses of resources, and the generation of eco-friendly products and services. These innovations aim to make negative impacts on the environment while at the same time improving the sustainable economic returns of businesses. Through such green aspects introduced within specific technology, organisations ensure they are part of developing systems that satisfy present and future human needs.

Given the current dynamic industrialisation and urbanisation, there is a high requirement for technological advancement for sustainability activities, especially in achieving environmental and resource endowment in South Asia (Wang et al., 2024). These ideas present opportunities for change in varying fields, such as energy efficiency or carbon emissions. However, the diffusion of such solutions can be challenging, and several barriers may hinder its attainment. It is, therefore, essential to apprehend such constraints to advance technological innovation towards building a sustainable future.

3.1.2 Importance of Promoting Sustainability Through Innovation in South Asia

South Asia, the region hosting more than one-fifth of the world's population and is among the fastest-growing areas, has specific prospects and demands for successful sustainability trends (Srivastava, 2020). Agriculture, industry, and services are identified as vital for the region's economy, which is tightly connected with natural resources and ecosystems. In addition, other factors that contribute to environmental pressures in the area include urbanisation, poverty, inequality, and climate vulnerability. Therefore, South Asian countries are now between a rock and a hard place regarding environmental deterioration, resource depletion, and sustainable economic development, and there is a need to come up with technological advancements.

Generally, the South Asian region could act as a model for all the other developing areas in the world; thus, promoting sustainable technological advancement should be emphasised. Technological advancement in power sources, water harvesting, and soil conservation can act as sustainable solutions to the region's environmental challenges. Further, innovation can act as a springboard for creating new jobs. It thus can boost the economy, strengthen its competitiveness, and contribute to the quality of life of millions of people. Nevertheless, the region has several challenges, including inadequate or sometimes non-existent regulatory and policy environments, limited financial capacity, and cultures that do not readily support promoting sustainable technologies (Rathnasiri et al., 2024).

3.1.3 Scope and Objectives of the Chapter

This chapter discusses critical organisations' challenges in advocating for technological advancement for sustainability in South Asia, namely, the institutional, financial, legal, and cultural factors hindering the generation and implementation of sustainable technologies. The chapter also focuses on identifying the contribution of the various stakeholders, including the governments, non-government organisations (NGOs), international firms, and domestic firms, in promoting innovation. This chapter seeks to provide a multi-faceted understanding of the elements constraining the region's technological evolution by discussing these organisations' limitations and barriers.

Hence, this chapter's primary purpose is to guide organisations in mitigating these constraints to support introducing sustainable technologies. Furthermore, it would present a guide on what policymakers, heads of industries and organisations, and development agencies should embrace to promote innovation for sustainable development. The chapter advances the literature on sustainability and technology in developing areas, particularly South Asia.

3.2 Methodology

3.2.1 Systematic Review Framework (PRISMA)

This chapter adopts PRISMA to achieve a comprehensive systematic analysis of the available literature. PRISMA can be aimed at guaranteeing high methodological quality, reporting standards, and reproducibility of systematic reviews. Specifically, PRISMA guides the researchers in finding relevant literature, evaluating their quality, and generalising the findings non-biasedly as highlighted by Karunarathna et al. (2024), Dewasiri et al. (2024), Rathnasiri et al. (2024), Jayawardena and Dewasiri (2023), Rathnasiri and De Silva (2023), Dewasiri et al. (2023), Mishra et al. (2024), and Dewasiri et al. (2022).

The systematic review used three academic databases: Scopus, Web of Science, and Google Scholar. These databases were selected because they comprehensively cover academic articles and the scientific output of peer-reviewed journals, conference proceedings, and other relevant technological innovation and sustainability publications. Keywords used for the search included the following: technological innovation, sustainability, South Asia, organisational constraint, and barriers to innovation. Boolean operators were applied to get only the best and most appropriate studies for the search. The inclusion criteria for the systematic review were as follows.

First, we only considered articles published within the last 15 years to be recently published and most likely to reflect current issues; second, we only considered studies related to South Asia; and third, we only considered articles that focused on the different organisations' roles, such as governments, non-government organisations (NGOs), and private organisations, in promoting innovations. Exclusion criteria were also used to exclude specific papers that did not suitably factor sustainability or were not aligned with organisational limitations. Also, work published in languages other than English was not considered because it was convenient for analysis in one language.

3.2.2 PICO Framework

In addition to PRISMA, the PICO was used to enrich the parameters of the systematic review. Because of its flexibility and utility in framing the identification of the PICO framework in qualitative research, the essential features of a study's focus are defined as the population (key organisations), the interest (technological innovations), the context (South Asia, sustainability), and the outcomes in question.

PRISMA and PICO together offered a solid methodological background for this chapter. PRISMA provides a structured and guided approach to possible bias in the systematic review, whereas PICO facilitates scrutiny of population, interest, context, and outcomes peculiar to South Asia. This combination helps execute this study since it allows a dual examination of organisations' limits in championing technological innovation for sustainability. Seventy (70) relevant studies were initially identified through database searches, and after applying the inclusion and exclusion criteria, thirty-five (35) studies were deemed appropriate for detailed analysis.

3.3 Results and Findings

3.3.1 Contextual Background

South Asia is one of the most highly populated territories in the world, economically highly diverse, and unites India, Pakistan, Bangladesh, Sri Lanka, Nepal, Bhutan, and Maldives (Rathnasiri et al., 2024). These characteristics and contextual factors make this region vulnerable to numerous sustainable development challenges intensified by population, urban, and industrial growth. These are problems such as pollution of the environment, air, and water, cutting down trees, reduced/elimination of bio-diversity, and expulsion of resources. Climate change compounds these issues since many

South Asian countries are susceptible to climatic shocks such as floods, droughts, and cyclones, incidents that upset livelihoods and economies (Rhaman et al., 2022).

One of South Asia's significant barometers of sustainability issues is balancing economic development and environmental degradation. Today, the area remains highly dependent on fossil fuels for energy production due to coal and oil, which significantly contribute to greenhouse gas emissions. Food security and employment sources such as agriculture suffer sustainability challenges such as water, soil quality, and excessive chemical fertiliser and pesticide use. Solving these reasons for sustainability imbalances calls for transitioning to more sustainable technologies and closing sustainability gaps throughout the economy.

3.3.1.1 Key Sectors Involved in Technological Innovations

Technological solutions are essential in overcoming sustainability issues in South Asia, and the energy, agriculture, and manufacturing industries are leading these changes.

Energy: Given the current banking on fossil energy forms and the increasing electricity consumption frenzy, the energy sector's role may be critical to South Asia. Renewable energy technologies help achieve carbon emission reduction and energy transitions, as technological development in renewable energy resources like solar, wind, and hydropower is essential (Dewasiri, 2024). India and Bangladesh have begun laying down large-scale projects to boost renewable energy sources.

Agriculture: Agriculture is another sector where, based on the analyses, sustainability improvements require technology. Here, new methods of farming, water and crop management, and organic agriculture are appropriate for enhancing crop production while exhibiting environmental sensitivity (Dewasiri, 2024). For example, that is using drip irrigation technologies in water-defunct areas and climate-smart agriculture technologies to counter climate change impacts on farming.

Manufacturing: The growth of the South Asian manufacturing industry is undoubtedly one of the primary sources of pollution and natural resource exploitation (Huang et al., 2023). Green technologies in manufacturing, including efficient energy use and waste control procedures, are vital for reducing the impact industrial processes bring about. Campaigns to increase the understanding of the circular economy model, which seeks to reduce waste, are also growing in the region's manufacturing industries (Khan et al., 2022).

3.3.1.2 The Role of Key Organisations in Promoting Sustainability

Several national, intergovernmental, non-governmental, private, and multinational stakeholders are responsible for advocating technology for sustainability in South Asia. These stakeholders are involved in developing the enabling conditions that facilitate innovation and the removal of barriers to the uptake of sustainable technologies.

Governments: National governments in South Asia are the central agents of change. They must create supporting policies, regulation(s), and incentive structures that will enable the use of green technologies (Ul-haq et al., 2020). International governments may control technology through direct financial contributions to research and development (R&D), providing subsidies for renewable energy technologies, or setting stringent environmental standards that force cleaner technologies.

NGOs: Pressure groups are essential to the societal urge for social acceptance and promote sustainability and awareness of innovative technologies. Non-governmental organisations are usually involved in transferring the concept of implementing sustainable development at the community level. For example, power is produced mainly through renewable resources, adequate power created for rural areas, renewable energy options, and organic farming in South Asian NGOs.

Private Sector: The energy manufacturing and agriculture industries, specifically with the help of private companies, continue to invest in technological innovation for sustainability. This has led to corporate responsibility practices and corresponding investments to enhance cleaner production in organisations because of the emerging market preferences for eco-friendly goods and services. Nonetheless, the ability of the private sector to fund large-scale sustainable innovations remains restricted by financial and regulatory factors.

Multilateral Institutions: The World Bank, other multilateral development banks, and international organisations such as the Asian Development Bank and the United Nations support sustainable development in South Asian countries. These institutions are vital in financing infrastructural development, enhancing capacity development measures, and managing knowledge of sustainable technologies. They are critical stakeholders that must be incorporated into the effort to solve some of the cross-border sustainability issues witnessed in the area.

3.3.2 Findings from Systematic Literature Review (PRISMA)

The review showed many studies on different aspects of technology for sustainable development in South Asia. Findings from across the environment and utilitarianism, environmental science, engineering, economics, and management analysis were used to establish the existing constraints in organisations when practising sustainable innovations. Detailed discussions supported the four principal areas: renewable energy, sustainable agriculture, green manufacturing, and waste management industries' strategies and dilemmas in achieving sustainability.

The sample studies covered countries in the South Asian region, allowing a regional analysis. The findings from this literature review are categorised into critical themes: regulatory, financial, technological, social, and institutional natures that define the diffusion and adoption of sustainability innovations.

3.3.2.1 Classification of Studies Based on Key Themes

The findings were categorised into five broad themes, showing that organisations face multiple challenges when promoting sustainable technological advancements.

- **Regulatory Constraints**: One analysed pattern is the absence of favourable regulations that encourage people and companies to invest in sustainable technologies. Most of the studies emphasised that current policies are either outmoded or not effectively implemented, especially in issues related to green energy, waste management, and green production (Rathnasiri et al., 2024). India and Bangladesh have enacted policies to encourage clean energy, but the problems persist due to unpredictable policies and bureaucratic procedures. However, low awareness of universal environmental standards and a lack of robust enforcement systems act as barriers that discourage organisational investment in sustainable innovations.
- **Financial Constraints**: The findings showed that financial issues were the most frequently mentioned challenges affecting technological sustainability advancements (Ahmed & Elfaki, 2024). Findings pointed out that high initial investment is a barrier to green technologies because organisations and tiny and medium-sized enterprises (SMEs) cannot afford them. Lack of means of attracting affordable funding, including green loans and investment capital for sustainability projects, was also common. The literature, therefore, postulates that while South Asian lenders are gradually incorporating environmental sustainability into their portfolios, there is still a significant gap in applying it towards environmental financing, thereby limiting organisations' access to funds for green innovations.
- **Technological Constraints**: The technological factors, and more specifically, the accessibility of sustainable technology, were cited more often than others. Most South Asian organisations have significant problems with technical experience, infrastructural conditions, and low levels of R&D, which are essential for developing innovative technologies or integrating commercial ones (Rathnasiri & Gunathilaka, 2024). For example, modern renewable energy sources like solar energy have improved. However, the high costs of new technology and low technological development in many South Asian countries pose significant barriers. Further, technical training programmes are deficient, which restrains the prospect of organisations employing and deploying these technologies effectively.
- **Social Constraints**: Other peculiarities related to the social and cultural aspects were identified as critical social barriers to technological advancement towards sustainable development. The findings pointed out that there are usually features resulting in a lack of awareness of attitude towards change in sustainability culture. This is seen in food production as farmers are more likely to refrain

from embracing sustainable practices due to that traditional way of farming. Also, there is low awareness, or even ignorance, of the public in sustainable innovations, resulting in minimal support and uptake (Singh et al., 2022).

- **Institutional Constraints**: Institutional constraints are ordinarily the organisational culture, management, and administration of the significant institutions that advocate for sustainability. In many organisations, sustainability management is less advanced due to fragmented governance, lack of strategic focus, or insufficient internal and external communications. The literature found that a general lack of practical cooperation between government agencies, private enterprises, and NGOs hinders the knowledge and resources needed for innovation.

3.3.3 Constraints Facing Organisations

Regulatory Barriers

South Asian organisations have often lamented the major challenge of disparity in policies and regulations between the various countries (Rathansiri et al., 2024). While the respective countries' leaders have paid lip service to sustainability, there is still a significant gap between policy formulation and practice. Lack of policy consistency, changes in policy focus, and breakdown of policies of different government bodies put barriers in how organisations want to invest in and champion technological advancement. For instance, ambiguity in tariffs and unstable government policies have challenged the development of renewable energy in Pakistan (Khan et al., 2020). Also, preservation of the environment, which should help to make the environment sustainable, is usually poorly complied with, thus leaving polluters to go scot-free.

The bureaucratic delay accompanying the approval of new innovative projects adds to these regulatory barriers. This is due to long waiting times resulting from cumbersome approval mechanisms governing permits and licences for sustainability projects that such organisations might wish to undertake. It stifles innovation and efficiency within specific renewable energy sectors, where business and development could be lost within the needed bureaucratic process. The absence of simplified mechanisms for enabling fast-track and comparatively easy innovations with a positive sustainability focus makes the operating climate for green pioneers and financiers unwelcoming.

Financial Challenges

Lack of financial resources is one of the main challenges South Asian organisations experience during sustainability innovation (Song et al., 2023). Credit is slightly constrained, especially for SMEs with inadequate funds to finance sustainable investments. Due to the risks mentioned above associated with sustainability projects, the banks and other financial institutions in the region still have a poor inclination towards providing loans in sustainability projects to embrace green investment opportunities despite understanding the long-term benefits of the provision of loans. This constrained capital hinders organisational innovation by restraining the implementation of advanced technology that supports sustainability affairs.

However, high research and development costs exacerbate the research challenges beset by sustainable technologies. Since most of the breakthrough solutions belong to such sectors as funding for renewable energy production, waste disposal, and efficient agricultural production, lots of money has to be spent to pursue meaningful research, which is scientifically and financially beyond the reach of many organisations. Another hurdle in sustainability is that government funding for R&D in sustainability is also low in most South Asian countries, and even the total cost has to be borne by the private sector (Dewasiri, 2024). Many organisations face the challenge of not being able to take new technologies to the market or increase the size of existing solutions.

Technological Infrastructure

The lack of infrastructure technology is another substantial challenge organisations face when encouraging sustainability initiatives. Many South Asian countries lack sufficient physical infrastructure to support adopting more sophisticated technological solutions (Bashir et al., 2021). For instance, investments in renewable power depend on upgraded transmission, distribution, and storage infrastructure, most of which is either nascent or completely missing across the region. An insufficient enabling environment limits the full mainstreaming of clean technologies, and the use of fossil energy sources is persistent.

There are apparent infrastructure issues and voids regarding digital and workforce skills development. Most South Asian organisations cannot provide the technical knowledge needed when implementing and supporting new technologies, particularly in rural areas (Rathnasiri et al., 2024). This skills deficiency hinders the ability of organisations to implement new, better solutions that may improve sustainability. In addition, it is now urgent to stress and promote training programmes and capacity strengthening that enhance the needed knowledge and competencies for efficient and effective management and operation of sustainable technologies among the workforce. In essence, these human capital deficits will not be closed if the region's technological advances are to be fully leveraged.

Cultural and Social Barriers

Cultural and social factors influence technology adoption to promote sustainability (Al-Emran & Griffy-Brown, 2023). Lack of implemental change within organisational cultures and communities is often viewed as a problem. Literature evidence suggests that culture change is difficult in large contemporary organisations across different industries, especially those that may stem from traditional sectors like agriculture and manufacturing; decision-makers are often unmotivated to change from an old habit to more sustainable ways. Such reliance is usually based on faster returns on investment rather than the long-term viability of a venture, combined with ignorance of the potential advantages of green technologies.

Organisational culture and primarily cultural and social factors define the use of technology in rural or relatively developed countries. For example, the farmer may resist adopting new sustainable farming practices because of an ingrained traditional way of farming, even if new technologies could be efficient for productivity and environmental conservation. Similarly, social and organisational culture can slow down the embracement of innovative ideas and solutions when the top leadership has not fully committed to supporting sustainability. Addressing these cultural and social barriers implies that sustainable development has to be championed and marketed to the people so that they take up sustainable practices by default.

Institutional and Organisational Constraints

This chapter affirms that limited inter-organisation cooperation has emerged as one of the most pressing obstacles to sustainable organisations (Adomako & Nguyen, 2024). In particular, the efficient interaction between different levels of governing, business, non-governmental organisations, and overseas organisations is very significant in creating practical innovations. Still, these entities work independently most of the time. Such fragmentation leads to potential synergies in knowledge exchange, resource sharing, and cooperation to advance technology agendas that are hardly recognised and exploited. In some nations, such as Sri Lanka, the lack of vibrant public–private partnerships denies organisations the potential to pool from both sectors to achieve big sustainable development projects (Weeraratne, 2019).

The next hurdle concerns the relatively weak institutional environment for innovation and sustainability. Most organisations, especially the public ones, do not have the strategic leadership, adequate governance structures, and technical know-how to support and drive sustainability initiatives correctly. Applying sustainability in organisations is a thorny issue because entities are poorly connected vertical structures, and bureaucratic structures hinder communication between departments. Also, few institutions have the proper mechanisms for measuring the success of sustainability programmes, which increases the chances of implementing mediocre programmes, hence denying society remarkable changes.

Leadership challenges are another factor associated with the scale of success of sustainable innovation (Domínguez-Escrig et al., 2019). Indeed, if top managers are only partially committed to the sustainability objectives, they may not have the vision or motivation to guide their organisations in innovative and sustainable directions. In many South Asian organisations, sustainability is seen as an add-on activity, not a strategic business initiative. This lack of leadership focus on sustainability angles means innovation gets downplayed, and short-term financial solutions triumph over long-term sustainability solutions. The solution to this challenge requires robust leadership, on the one hand, and a clear, futuristic orientation, on the other, which causes organisations to push towards sustainable innovation.

3.4 Case Studies of Key Organisations

3.4.1 Government Initiatives and Challenges

The government remains an influential actor in shaping policies or providing a framework within which technologies for sustainable development are encouraged or discouraged. Another analysis of the wind energy potential, particularly in India, is also helpful in understanding the progress accomplished through the National Solar Mission initiated to spur the use of solar power-related technology (Bansal, 2021). Despite this, the mission has recorded some success. However, problems such as disparities in regulating fee-charging across the state and national levels compromise the effort through bureaucratic inertia. For instance, several years could be spent securing approval for renewable energy projects—this hinders domestic and foreign investment. Also, many State-owned and operated electricity distribution companies are not financially buoyant enough to underwrite such expensive innovations, hence a significant challenge in implementing government-led sustainability programmes.

Likewise, challenges are institutional; despite having a dedicated ministry for sustainable agriculture and renewable energy, Sri Lanka's government struggles to scale up projects, many of which fail to coordinate between departments or secure adequate funding. To date, no concerted effort has been made to integrate sustainable policies across all tiers of government and, more significantly, to ensure that they are consistent with the private sector players' goals. Furthermore, due to limited technical know-how and inadequate human resources, the recommendations made in this report do not contemplate acquiring new technologies in public institutions.

3.4.2 Multinational Corporations (MNCs)

South Asian companies like Unilever and Siemens have adopted green technologies because of their international environmental conservation policies. However, these companies are faced with restraints that are peculiar to the region across the value chain. For example, Unilever, which has committed to decreasing the use of single-use plastics and instead engaging in circular economy systems within South Asia, has found that the weak structures of waste management and the volatility of policies between the South Asian countries affect its operations. More recently, some Unilever initiatives like recycling with local NGOs have been effective in Indonesia; however, overall regulatory and infrastructural problems like the disorganisation of waste processing facilities hindered the expansion of these measures (de Arruda Pollice & Scarcelli, 2024).

The manufacturing giant Siemens, which has already pioneered energy-efficient systems and technologies for the production line, is financially hampered by prohibitively expensive technologies in the South Asian region. Very few industries, particularly SMEs, can afford to purchase these technologies without significant state support in terms of subsidies or grants. Also, Siemens faces issues connected with the digital competencies of the population and their skills to work in the corresponding factory, as local employees frequently do not have enough skills to maintain and use modern equipment due to their insufficient training. These barriers highlight the challenges even the most prominent organisations encounter while advocating for regional sustainability innovations.

3.4.3 Startups and Innovation Hubs

Specifically, new startups are increasingly part of South Asia innovation, especially renewables and sustainable farming. Bangladeshi startup SimpliSolar is one exciting example that provides people with the needed energy through cheap solar home systems and responds to energy and environmental problems (Govindarajan et al., 2023). Nevertheless, the startup has issues with its finances—it lacks venture capital and has no governmental subsidies. Still, the widespread use of their solar systems has excellent social effects since the high initial costs for installation and the shortage of available financing options for users hinder the expansion of the company's activities. Further, the regulations are complex for a startup and entail burdensome complexities that affect its market advancement schemes, especially in import regulations regarding solar components and the general tariff instability status.

3.4.4 Successes and Limitations in Promoting Technological Innovations for Sustainability

Notable successes have highlighted sustainability via technology adoption; however, the challenges that governments, multinational enterprises (MNEs), startups, and NGOs elicit the region's dynamism. While governments have been able to conceptualise good policies, such as the targets for renewable energy in India, the implementation process slows things down due to inconsistent regulations. As we see from Siemens and Unilever, despite these MNCs being committed to green investment and technological advances, they are hampered by the infrastructure and ability to finance the change. Still, if one turns to current startups and innovational NGO projects, they highlight the financing problems, deficiency and lack of correct regulation, and excellent resistance of culture.

In general, the analysed case represents a call for a better correspondence of the currently stimulated technological advancements by looking at the management of environmental responsibilities as an integrated system. As the abovementioned barriers suggest, cooperation between governments, private companies, and civil society organisations is necessary for addressing. Moreover, continued efforts on the recipient end towards strengthening organisational capacities, refining rules and policies, and developing better and more robust funding models are critical for the continuity and up-scaling of proven interventions in the South Asian region.

3.5 Mitigating Strategies and Policy Recommendations

3.5.1 Insights from the Literature on Overcoming Identified Barriers

The systematic literature review yields several findings on how organisations can overcome the limitations of the challenges in supporting technological innovation for sustainability. This study has several strategic implications, one of which is that successful organisations use multi-stakeholder strategies with governments, the private sector, civil society, and international institutions. It makes it easier for organisations to draw together and collect their resources, knowledge, and objectives in dealing with regulatory, financial, and institutional problems. For instance, the two emerging economies of India and Bangladesh have immense experience partnering between the public sector and private players to mitigate issues related to regulation or finance for their renewable energy sectors.

The other understanding derived from the literature is that strategic planning requires vision and commitment from organisational leadership over the long term. Firms treating sustainability as a strategic management issue are likely to overcome organisational culture resistance to change and consequently deliver technological solutions more efficiently than their counterparts who do not. Furthermore, these organisations engage the concept of sustainability in their research and development (R&D) procedures, which provides a sustainable stream of innovations towards better technologies in the protection of the environment and sustainable use of resources.

3.5.2 Policy Recommendations for Creating a Supportive Regulatory Environment

The governments of South Asian nations have a chance to become the driving forces of change by creating a more unified and less contradictory regulatory environment that would motivate companies to integrate new technologies with sustainability goals (Rathnasiri et al., 2024). Such policies reveal weaknesses characteristic of the region, including the lack of consistency of policies across jurisdictions, which remains unhelpful for investors and innovators. Policies at the national and regional levels must be aligned to afford sustainability projects with little interference from bureaucracies and other bureaucratic undertakings that slow progress. Governments should also regulate sandbox environments, which provide an exclusive area for organisations experimenting with releasing new technologies without triggering regulatory frameworks. Such an approach would enable emerging innovative players, including startups, to try out new solutions for sustainability in a manner that sustains basic safety and environmental requirements.

Finally, policy frameworks must emphasise affordable solutions for organisations interested in sustainability innovations and offer specific financial motivators (Rathnasiri et al., 2024). Governments should provide subsidies, tax exemptions, and cheap credit facilities to organisations that seek to invest in green technologies. This would in any way assist in lessening the financial burdens, especially for SMEs, as most of them cannot afford to delve into expensive sustainability ventures. Suppose the governments can provide structures for companies to foresee what changes they should expect concerning sustainability initiatives in the long term. In that case, organisations will be encouraged to improve in the right direction.

3.5.3 Promoting Public–Private Partnerships (PPPs) to Tackle Financial and Infrastructural Constraints

Availing finances and infrastructure, PPPs have assumed the role of a strategic weapon in various social projects, especially renewable energy, waste management, and sustainable agriculture (Hossin et al., 2024). The governments of the South Asian region should encourage public–private partnership funding structures that employ public funds and private finance to establish significant sustainability initiatives (Kang et al., 2019). For instance, successful PPPs in the energy sector have realised the establishment of the region's solar and wind power plants. These partnerships also help balance risks and rewards more fairly, enabling governments to harness private sector effectiveness in project delivery.

Sometimes, PPPs can provide solutions for problems associated with technological infrastructure complementary to financial partnerships. Namely, governments can enlist private technology companies to upgrade national grids, modify transportation networks, enhance waste management facilities, and incorporate relatively novel sustainability technologies. PPPs can also provide for creating needed digital environments and applications for process optimisation of energy efficiency, resource use, and coordination with the key participants. In addition, governments must encourage infrastructure access where private businesses enjoy an efficient technological system that can be made available to struggling businesses at lower prices to help lower the penetration expenses for new entrant businesses in the region.

3.5.4 Enhancing Capacity-Building Efforts and Developing Innovation Ecosystems

Another approach to eliminating barriers is developing the relevant competencies to implement and foster sustainable innovations (Aleixo et al., 2018). Therefore, governments and organisations should invest more in training and education to prepare the workforce to operate, maintain, and deploy new technologies. In South Asia, because local clients may barely grasp the basics of digital media or lack adequate technical skills to implement these innovations, capacity building should come first.

Moreover, the building of the so-called innovation ecosystems: structures uniting universities, research centres, startups, and established companies is crucial for the development of an innovation culture. That is why governments can help by creating startup spaces or technology parks where people can work and get benefits such as tax relief, subsidies, or grants, as well as provide fundamental requirements for research. They contribute to establishing the continuity of worldwide generations of new ideas, technologies, and solutions for sustainability issues. Promoting such conditions can help South Asia develop innovations that suit the region's environmental and social niches.

3.5.5 Promoting Leadership and Strategic Initiatives for Driving Sustainability

Leadership is essential for enhancing sustainable practices within organisations (Boeske, 2023). Today's firms and governments require influential leaders who can transform compliance into mainstream practice and a sustainable growth advantage. More leadership development programmes, especially in sustainability, should be deployed in public and private organisations to help foster a culture of innovation and environmental and long-term thinking. The significant finding discussed in this chapter is that organisations should consider investing in leaders who support sustainable development since they will likely face internal and external barriers when pursuing innovative solutions.

Besides leadership, organisations should implement strategic management focusing on their businesses' sustainability objectives (Bellucci et al., 2020). Such measures may cover activities such as adopting ambitious targets for cutting emissions, making purchasing decisions based on the principle of sustainability, and promoting innovations through specialised development activities for green projects. Sustaining innovation involves directing the organisational planning frameworks with sustainability, which includes a durable mechanism for measuring outcomes and stock of organisational innovation.

3.6 Future Outlook

3.6.1 Trends in Technological Innovations for Sustainability in South Asia Based on the Review Findings

The systematic literature review identifies several emerging technological innovation trends towards sustainability in South Asia. Indeed, one prominent area of continued development is the growth and popularity of renewable energy systems such as solar and wind. India and Bangladesh are among the rising solar farm investments and wind energy projects due to lower technological costs and favourable government policies. Innovations in power storage also make it easier to incorporate renewable sources into current grid systems because battery technologies improve energy stability and efficacy. Therefore, there is a need to encourage the use of renewable energy to satisfy the increasing energy needs in the region without further degrading the environment.

Another strong trend is the trend towards agricultural technologies that enhance sustainability. The use of precision agriculture technologies such as drones and the IoT and the application of artificial intelligence to farm management enable farmers to apply appropriate inputs appropriately, reducing wastage while at the same time increasing production. However, shifting to data-driven precision farming generates efficiency and helps conserve water and chemicals used in agriculture. Furthermore, new agricultural methods like vertical culture, aquaculture, and hydroponics are being developed in urban places, making producing foods with few environmental impacts possible.

Last, there is a gradual shift to circular economy models that can become an essential concept of everyday sustainability (Singh et al., 2023). Some organisations are gradually waking up to the reality that products and services should be designed with a focus on recycling, reuse, and resource efficiency. The activities that sought to curb the use of plastics encourage the use of environmentally sound packaging materials, and encourage people to adopt sustainable consumption patterns are becoming essential strategies for sustainable development in the region. The shift towards circularity is one of the exposures of consciousness to address sustainability concerns by detaching growth from resource use, which is crucial for sustainable development in Asia.

3.6.2 Potential for Overcoming Key Constraints Through Evolving Policies and Global Trends

The changing policy environment observed in South Asia today is a significant advantage in addressing the limitations of advanced technologies for sustainability. The world government acknowledges the need to confront climate change and environmental deterioration and, therefore, develop better-coordinated policies on sustainability. In this regard, ratifying the Paris Agreement and other international accords is creating pressure on South Asian countries to set higher climate goals that can set directives for the necessary legal changes to advance green technologies.

This study identifies various factors, including policy change, developing trends like the low-carbon economy, and the changing corporate social responsibility discourse that there is a shift in organisational practices in South Asia. Many MNCs and local firms are integrating their business strategies with the objectives of global sustainability strategies because corporations realise that active involvement in sustainability is going to improve their corporate image and their competitiveness. This shift leads to increased organisations practising sustainability and investing in environmentally friendly and socially responsible technologies.

In addition, more advanced international funding sources such as climate finance and green bonds are beginning to be recognised as helpful funding tools by organisations intending to undertake sustainability projects. The lack of funding from global institutions and investors can hamper the economic barriers organisations face, and thus, they can adopt new technologies that were once considered costly. While these trends are in their infancy, the sustainability-oriented dynamics of development offer South Asian countries great potential for utilising global momentum in the face of local constraints by aligning their policies with international best practices and engaging in regional and global cooperation.

3.6.3 Vision for Sustainable Futures Through Enhanced Innovation Efforts by Key Organisations

This chapter argues that South Asian organisations' increased innovation efforts are critical to the long-term vision of a sustainable future. When governments, multinational corporations, startups, and NGOs work harmoniously, they also lay the foundations of an ecosystem that supports technological advancement and simultaneously solves prevalent regional sustainability problems. Organisations can play significant roles in enhancing economic resilience, social justice, and the protection of the environment, with sustainability as a strategic business objective.

As for the future, which needs to be built, further development of cooperation between sectors seems most promising. For example, collaboration between governments and business companies increases information sharing. It encourages the development of new technologies and the investment in physical capital needed for sustainable development. Furthermore, grassroots projects enabled by civil society organisations and lower tiers of government can engage the populace and owners of small enterprises in proactive sustainability that underpins community and regional security and ecological responsibility.

Ultimately, it will take political will to bring about visions to help South Asia usher in a more sustainable future through innovation and partnership. Organisations must be flexible to adapt to the emerging threats and opportunities concerning sustainability. In other words, South Asian countries can integrate the enhancement of the quality of life through available technologies, employ diverse expertise from around the globe, and extend sustainable development goals worldwide for people and the planet.

3.7 Conclusions and Implications

This chapter addresses the need for more substantial technology applications to enhance sustainability in South Asia. The findings underscore an increasing concern with renewable energy, sustainable agriculture, and the circular economy as critical avenues for responding to the region's environmental problems. They are all essential factors in tackling climate change and sustaining economic and social growth within a context of growing urbanisation and resource competition.

However, reviewing the literature on sustainable technologies reveals considerable organisational barriers to increasing sustainability. The main challenges include regulatory factors, lack of capital, a poor technological foundation, and a culture that presents formidable hurdles to the efficient implementation of sustainability. Overcoming these constraints is essential for realising the potential of innovations for sustainable regional development. The chapter also presents measures that can be taken to address such barriers: the development of PPPs, the more effective implementation of capacity-building measures, and the creation of favourable regulatory conditions. Building organisational leadership and strategic commitment to sustainability may enable organisations to address them well.

The consequences of this study go beyond the theoretical contributions: A systems approach to sustainability that involves the integration of sustainable policies, the incorporation of sustainable business practices, and the engagement of organisational stakeholders is the key recommendation made in this research. In this way, organisations can successfully enlist their collective knowledge and efforts to share and address their complex issues based on the mutual pursuit of organisational goals and sustainable targets of the specific region and globe. Moreover, developing education and skills is also essential to create demand for technological change in sustainability and an adequate labour force to support it.

Technological developments for a sustainable South Asian future: Organisations' contribution is essential and innovative. Apart from their clients, organisations can significantly help the region and the world change for the better through out-of-the-box politics, effective partnerships, and working towards sustainability. This chapter's findings can be used for further research and practice, stressing further work in decreasing such barriers and intensifying technological advancement in South Asia sustainability advancement.

References

Adomako, S., & Nguyen, N. P. (2024). Digitalisation, inter-organizational collaboration, and technology transfer. *The Journal of Technology Transfer, 49*(4), 1176–1202. https://doi.org/10.1007/s10961-023-10031-z

Ahmed, E. M., & Elfaki, K. E. (2024). Green technological progress has implications for long-run sustainable economic growth. *Journal of the Knowledge Economy, 15*(2), 6860–6877. https://doi.org/10.1007/s13132-023-01268-y

Al-Emran, M., & Griffy-Brown, C. (2023). The role of technology adoption in sustainable development: Overview, opportunities, challenges, and future research agendas. *Technology in Society, 73*, 102240. https://doi.org/10.1016/j.techsoc.2023.102240

Aleixo, A. M., Leal, S., & Azeiteiro, U. M. (2018). Conceptualisation of sustainable higher education institutions, roles, barriers, and challenges for sustainability: An exploratory study in Portugal. *Journal of Cleaner Production, 172*, 1664–1673. https://doi.org/10.1016/j.jclepro.2016.11.010

Bansal, N. (2021). Promoting solar energy in India to meet the country's commitment to climate change and energy security. In M. Asif (Eds.) *Energy and environmental security in developing countries*. Advanced Sciences and Technologies for Security Applications. Springer. https://doi.org/10.1007/978-3-030-63654-8_22

Bashir, S., Dahlman, C., Kanehira, N., & Tilmes, K. (2021). The converging technology revolution and human capital: Potential and implications for South Asia. *World Bank Publications*. https://doi.org/10.1596/978-1-4648-1719-9

Bellucci, M., Bini, L., & Giunta, F. (2020). Implementing environmental sustainability engagement into business: Sustainability management, innovation, and sustainable business models. In *Innovation strategies in environmental science* (pp. 107–143). Elsevier.

Boeske, J. (2023). Leadership towards sustainability: A review of sustainable, sustainability, and environmental leadership. *Sustainability, 15*(16), 12626. https://doi.org/10.3390/su151612626

de Arruda Pollice, F. F., & Scarcelli, M. (2024). Purchasing and supply management journey into unilever's circular economy strategy. In A. R. Ometto, J. Sarkis, & S. Evans (Eds.), *A systemic transition to circular economy*. Greening of Industry Networks Studies, vol 12. Springer. https://doi.org/10.1007/978-3-031-55036-2_8

Dewasiri, N. J. (2024). Sustainable strategies for net zero: An Asia pacific perspective. In R. Singh, & D. Crowther (Eds.), *Transition towards a sustainable future*. Approaches to Global Sustainability, Markets, and Governance. Springer. https://doi.org/10.1007/978-981-97-5756-5_1

Dewasiri, N. J., Baker, H. K., Banda, Y. K. W., & Rathnasiri, M. S. H. (2022). The dividend decision model: A possible solution for the dividend puzzle. In S. Rana, Sakshi, & J. Singh (Eds.), *Exploring the latest trends in management literature* (Review of Management Literature, Vol. 1, pp. 249–267). Emerald Publishing Limited, Leeds. https://doi.org/10.1108/S2754-586520220000001013

Dewasiri, N. J., Karunarathne, K. S. S. N., Menon, S., Jayarathne, P. G. S. A., & Rathnasiri, M. S. H. (2023). Fusion of artificial intelligence and blockchain in the banking industry: Current application, adoption, and future challenges. In A. Saini, & V. Garg (Eds.), *Transformation for sustainable business and management practices: Exploring the spectrum of industry 5.0* (pp. 293–307). Emerald Publishing Limited, Bingley. https://doi.org/10.1108/978-1-80262-277-520231021

Dewasiri, N. J., Dharmarathna, D. G., & Choudhary, M. (2024). Leveraging artificial intelligence for enhanced risk management in banking: A systematic literature review. In R. Singh, S. Khan, A. Kumar, & V. Kumar (Eds.), *Artificial intelligence enabled management: An emerging economy perspective* (pp. 197–213). De Gruyter, Berlin. https://doi.org/10.1515/9783111172408

Domínguez-Escrig, E., Mallén-Broch, F. F., Lapiedra-Alcamí, R., & Chiva-Gómez, R. (2019). The influence of leaders' stewardship behaviour on innovation success: The mediating effect of radical innovation. *Journal of Business Ethics, 159*, 849–862. https://doi.org/10.1007/s10551-018-3833-2

Gamage, T. C., Dewasiri, N. J., Gnanapala, A., & Rathnasiri, M. S. H. (2024). ICT sector development and its contribution to innovation and economic growth in South Asia: A critique. In B. S. Sergi, A. K. Tiwari, & S. Nasreen (Eds.), *Modeling economic growth in contemporary India* (pp. 25–33). Entrepreneurship and Global Economic Growth. Emerald Publishing Limited, Leeds. https://doi.org/10.1108/978-1-80382-751-320241002

Govindarajan, L., Batcha, M. F. B. M., & Abdullah, M. K. B. (2023). Solar energy policies in southeast Asia towards low carbon emission: A Review. *Heliyon, 9*(3), 14294. https://doi.org/10.1016/j.heliyon.2023.e14294

Hossin, M. A., Alemzero, D., Abudu, H., Yin, S., Mu, L., & Panichakarn, B. (2024). Examining public-private partnership investment in energy towards achieving sustainable development goal 7 for ASEAN region. *Scientific Reports, 14*(1), 16398. https://doi.org/10.1038/s41598-024-66800-9

Huang, L., Dong, D., & Dong, X. (2023). Natural resources extraction, financial expansion and remittances: South Asian economies perspective of sustainable development. *Resources Policy, 84*, 103767. https://doi.org/10.1016/j.resourpol.2023.103767

Jayawardena, N. S., & Dewasiri, N. J. (2023). Food acquisition and consumption issues of South Asian countries: A Systematic literature review and future research agenda. *FIIB Business Review*. https://doi.org/10.1177/23197145231194113

Kang, S., Mulaphong, D., Hwang, E., & Chang, C. K. (2019). Public-private partnerships in developing countries: Factors for successful adoption and implementation. *International Journal of Public Sector Management, 32*(4), 334–351. https://doi.org/10.1108/IJPSM-01-2018-0001

Karunarathna, K. S. S. N., Dewasiri, N. J., Singh, R., & Rathnasiri, M. S. H. (2024). What does artificial intelligence–powered ChatGPT bring to academia? A review. In R. Singh, S. Khan, A. Kumar, & V. Kumar (Eds.), *Artificial intelligence enabled management: An emerging economy perspective* (pp. 3–13). De Gruyter, Berlin. https://doi.org/10.1515/9783111172408

Khan, S., Ali, S. S., & Singh, R. (2022). Determinants of remanufacturing adoption for circular economy: A causal relationship evaluation framework. *Applied System Innovation, 5*(4), e62. https://doi.org/10.3390/asi5040062

Khan, M. I., Khan, I. A., & Chang, Y. C. (2020). An overview of global renewable energy trends and Pakistan's current practices—a policy implications perspective. *Journal of Renewable and Sustainable Energy, 12*(5), 056301. https://doi.org/10.1063/5.0005906

Mishra, U., Rathnasiri, M. S. H., & Dewasiri, N. J. (2024). Net-zero economy and energy sustainability: A Nepalese perspective. In R. Singh, S. Khan, A. Kumar, S. Luthra, & H. Chokshi (Eds.), *Net zero economy, corporate social responsibility and sustainable value creation* (pp. 145–161). CSR, Sustainability, Ethics & Governance. Springer. https://doi.org/10.1007/978-3-031-55779-8_9

Rhaman, M. S., Kibria, M. G., & Hoque, A. (2022). Climate change and its adverse impacts on plant growth in South Asia: Current status and upcoming challenges. *Phyton (0031–9457), 91*(4), 695–75. https://doi.org/10.32604/phyton.2022.018898

Rathnasiri, M. S. H., Dewasiri, N. J., & Kumar, A. A. (2024). Policy framework and implementation strategies for Sri Lanka's transition to a net-zero economy. In R. Singh, S. Khan, A. Kumar, S. Luthra, & H. Chokshi (Eds.), *Net zero economy, corporate social responsibility and sustainable value creation* (pp. 43–60). CSR, Sustainability, Ethics & Governance. Springer. https://doi.org/10.1007/978-3-031-55779-8_3

Rathnasiri, M. S. H., & De Silva, V. (2023). The role of shopping values, cognitive and affective evaluations on the relationship between store environment and store patronage intention: A future research agenda, Sri Lanka. *Journal of Management Studies, 5*(1), 100–120. https://doi.org/10.4038/sljms.v5i1.109

Rathnasiri, M. S. H., & Gunathilaka, D. H. P. M. (2024). Net zero policies for sustainable future: Issues and challenges in the Asia pacific region. In R. Singh, & D. Crowther (Eds.), *Transition towards a sustainable future*. Approaches to Global Sustainability, Markets, and Governance. Springer. https://doi.org/10.1007/978-981-97-5756-5_12

Singh, R., Khan, S., & Centobelli, P. (2022). Investigating the interplay between social performance and organisational factors supporting circular economy practices. *Sustainability, 14*(24), e16781. https://doi.org/10.3390/su142416781

Singh, R., Khan, S., & Dsilva, J. (2023). A framework for assessment of critical factor for circular economy practice implementation. *Journal of Modelling in Management, 18*(5), 1476–1497. https://doi.org/10.1108/JM2-06-2021-0145

Song, C., Wu, Z., Dong, R. K., & Dinçer, H. (2023). Greening South Asia: Investing in sustainability and innovation to preserve natural resources and combat environmental pollution. *Resources Policy, 86*, 104239. https://doi.org/10.1016/j.resourpol.2023.104239

Srivastava, R. K. (2020). Manifestations of climate change in South Asia. In *Managing urbanization, climate change and disasters in South Asia*. Disaster Studies and Management. Springer. https://doi.org/10.1007/978-981-15-2410-3_4

Ul-Haq, A., Jalal, M., Sindi, H. F., & Ahmad, S. (2020). Energy scenario in South Asia: Analytical assessment and policy implications. *IEEE Access, 8*, 156190–156207. https://doi.org/10.1109/ACCESS.2020.3019648

Wang, C., Zhou, D., Guo, X., & Kayani, U. N. (2024). Role of natural resource rents, financial development and technological research in achieving sustainable development: A study of South Asian Countries. *Resources Policy, 89*, 104632. https://doi.org/10.1016/j.resourpol.2023.104632

Weeraratne, B. (2019). Transitioning to middle income and beyond: The urbanization challenge in post-conflict Sri Lanka. In D. Weerakoon, & S. Jayasuriya (Eds.), *Managing domestic and international challenges and opportunities in post-conflict development*. South Asia Economic and Policy Studies. Springer. https://doi.org/10.1007/978-981-13-1864-1_14

Chapter 4
Digital Circular Economy: Driving Sustainable Innovation Through Stakeholder Synergy

Oyenuga Michael Oyedele

Abstract The digital circular economy changes how businesses and societies regard sustainability and resource management. It mixes the power and sophistication of digital interfaces with the efficiency that characterizes an open model, while still being founded on circular economy principles which demand collaboration across stakeholders to make sustainable innovation happen. Even though the potential of the digital circular economy is huge, it comes with challenges such as a lack of awareness among citizens and companies of sustainable practices and insufficient sorting infrastructure for packaging waste including policy hurdles, etc. This paper identifies several hurdles in adopting a digital circular economy and presents a strategy that includes education, policy reform, as well as investment into digital infrastructure to help overcome these obstacles. But even more importantly, it underlines the significance of engaging stakeholders in an open innovation process that brings about value and system change. Ultimately, the article advocates a comprehensive model for deploying digital technologies that incorporate stakeholder engagement to enhance sustainable innovation for future resilience.

Keywords Digital circular economy · Stakeholder engagement · Sustainable innovation

4.1 Introduction

The emphasized call for sustainable development caused by the worsening problems related to the natural environment has triggered a novel systemic approach based on a more heuristic, circular and environmentally friendly design premised concerning the interplay between man-made construction features and nature starting from a digital circular economy perspective. An unmistakable call for a departure from the customary linear economic approach (Khan et al., 2022a, b) where essential resources

O. M. Oyedele (✉)
School of Business, Woxsen University, Telangana, India
e-mail: oyenuga.michael@woxsen.edu.in

R. Singh and V. Kumar (eds.), *Sustainable Innovations and Digital Circular Economy*,
https://doi.org/10.1007/978-981-96-1064-8_4

are utilized once and then disposed signalling renewed focus on circular development and sustainability; an agenda that underscores low resource use, smart production techniques as well as aligning with nature—rather than going against it (Singh et al., 2022). The adoption of a digital circular economy symbolizes an important change in the mindset with which societies historically have dealt, and the need to increasingly address sustainability, innovation and economic development. Across the globe, there is a real sense that numerous environmental problems like climate change, resource scarcity and trash disposal need to be addressed now through sustainable means. The digital circular economy uses the developments of technology and digitalization to align different actors in society, also including businesses, governments, as well as civil society towards innovative solutions that can contribute to sustainability goals by rethinking common linear economic models (such as "take-make-dispose") into hundreds of opportunities, builds around resource efficiency, waste mitigation and further material continuous recycling use (Larsson & Larsson, 2020). In this light, multi-stakeholder collaboration comes across as a key ingredient of sustainable innovation. Through the pooling of information, ideas and resources that are required to drive systemic change (Lozano et al., 2021), such partnership provides value-added in good governance by creating cooperation between a variety of actors including businesses, government agencies at different levels from central ministries' department(s) on one end then down through decentralized mechanisms to local governments; civic organizations as well community-based entities within marginalized areas, and even more specialized groups or broader society. In turn, more stakeholders involved in collaborative business relationships can result in greater support for sustainability initiatives and increase the overall impact of innovation (Larsson & Larsson, 2020).

Thus, one cannot overemphasize the need for multi-stakeholder collaboration to drive sustainable innovation. According to Ayuso et al. (2011), stakeholder engagement is critical for an organization's sustainability as it keeps innovation orientation within the organization over time. Cross-functional teams bring different perspectives and skills to help create new products, services or processes that support sustainability objectives. With digitization, resource coordination becomes efficient through better communication with stakeholders; hence this kind of collaboration is even more relevant in a digital circular economy for sharing and pooling data on mobile phones.

Secondly, the move towards a digital circular economy requires organizations to adopt fundamentally different ways of thinking and operating. The message underlined by (Ekins & Zenghelis, 2021) is that sustainability cannot come from new technologies solely; it has also to mean different behaviours and business models. Organizations must adopt resilient practices that are sustainable across the environment, society and economy thus creating value for all its stakeholders. A comprehensive culture of innovation will be necessary to confront the many serious climate change and resource depletion puzzle pieces.

In this transformation process, the Information and Communication Technology (ICT) sector is a key driver of monitoring resource flows; production optimization and channelling desired behaviour among consumers to sustainable practices using ICT technology (Jung et al., 2017). In addition, information communication technologies (ICT) being incorporated into circular economy practices leads to better information flow among stakeholders of the process and improved transparency promoting all parties involved towards more genuinely acting decisions together with other capable partners in that trend through a proper integration method which ultimately fosters material optimization for achieving desirable collaboration (Armenia et al., 2019).

However, while there is a growing acknowledgement of the need for sustainable innovation and multi-stakeholder collaboration simultaneously, there are still barriers. Problems with collaboration arise because stakeholders may have dissimilar priorities, and be at odds on how to use power and resources (Lozano et al., 2021). These barriers cover regulatory, financing and capacity building among other key stakeholders. As pointed out by Widodo et al. (2022), regulations and policies are needed for any innovation effort to survive, including alignment with broader sustainability goals. This in turn is necessary to ensure, that an innovation-friendly environment that enables organizations to achieve environmental sustainability through green innovations is encouraged (Alshammari, 2023). Overcoming these challenges and finding ways to address them accordingly are crucial defining factors that may manifest in the digital circular economy.

4.2 What is a Circular Economy?

A circular economy (CE) is an alternative economic model which, through the principles of sustainability and resource efficiency with minimum waste production involved in both a linear value chain system based on; make-use-dispose seeks instead to create a closed manufacturing loop ecosystem that focuses more specifically at end-of-life products when it considers reuse-recycle-handling. The circular economy is a departure from traditional linear economies that work under a "take-make-dispose" philosophy—this alternative model moves towards an inclusive and regenerative system in which resources are recovered towards their utmost capacity throughout time to generate as minimal waste while inflicting the least harm upon our global ecosystem. It is a model that increasingly, to many international observers appears as imperative in confronting the mounting problems of resource depletion fall-out and climate change.

4.3 Principles of Circularity

The principles of circularity are fundamental ideas that shape the move from a linear economy (extract, produce dispose) to a closed-loop or recycling-centred production system. The following are the key principles of circularity:

1. **Reduce, Reuse, Recycle (3Rs)**: No doubt the 3Rs are at the foundational core of a circular economy. Reduce—This principle involves reducing the consumption of resources and the generation of waste by efficiently sourcing them. Reuse directs

to repurpose, recycle or extend products and materials' lifetime and process bottlings it for new product; this provides a loop back in the production processes (Tambovceva et al., 2021).

2. **Design for Longevity**: Firms should design their products to last for a long time and not break down after a few heavy use. This principle together with the development of higher-quality products can be repaired, upgraded or reconditioned rather than thrown away (Levänen et al., 2022). Intentional design that prioritizes longevity can help to combat waste and create a more sustainable customer culture.
3. **Resource Efficiency**: This applies to how consumed resources, materials and energy are used efficiently with the lowest possible disposal and environmental impact. This principle pushes for using cleaner production techniques and renewable resources (Gomide, 2024). It minimizes the operational costs for companies and helps in environmental sustainability by optimal usage of resources.
4. **Systems Approach**: When developing products for circularity, a holistic systems-level view must be adopted at every stage of the product life cycle (material sourcing → production/manufacturing → distribution and consumption → end-of-life/diversion/disposal). Both dispositive points of view are only possible by assuming a more strategical view of a circular economy that identifies the industry as processes being closed cycles (not lines) including material extraction, manufacturing and usage to disposal. An approach that sees the dynamic relationships among producers, consumers and politicians as a composite plan for creating an economy of sustainability (Kravchenko et al., 2020). Keeping this principle in mind gradually leads to upward alignment of the solutions and creates an organic cross-pollination as various stakeholders contribute towards making the ecosystem a better place.
5. **Business Models**: It probably requires a change in business model, moving from a linear or stratified edge model of operating. As such the organizations at present are advancing towards work around business models like pay-per-use or product-as-a-service to evade this kind of limitation (Hina et al., 2023). As a reminder as well for manufacturers to have great consideration in material and construction that enables things much more repairable or easier on single-material recycling, which is always an imperative with sustainable sourcing.
6. **Stakeholder Engagement**: Wide-ranging collaboration among producers to consumers, involving governments and non-profit organizations together with the community is called for to achieve circularity. Finally, involving multiple actors could enable many different scales and views to be brought into the equation to achieve creativeness within innovation (Fassio & Tecco, 2019), which might result as an outcome of this "joint" process a shared knowledge capital engaging co-created resources can contribute significantly through providing varying best practices for advancing circular economy at high-speed.
7. **Local Solutions**: The principles of circularity are around promoting local place-based solutions that address the needs of communities with as little environmental impact as possible. By pointing to local resources, as well as their capabilities

businesses can create sustainable practices back to context (Hina et al., 2023). That is a resilience and adaptability principle in our connected world.

8. **Continuous Improvement and Learning**: The circular economy is a concept in transition, always learning and never being perfect. However, organizations have to ultimately change to remain informed of new understanding, technologies and market conditions (Testa et al., 2020). This principle cherishes a practice of innovation and checks the practice as we allow businesses to evolve their strategies for sustainability.

4.4 Digital Circular Economy

Digital circular economy is a pioneering approach that combines digital technologies and the principles of the framework to promote cleaner, more resource-efficient economic performance. The category aims at reducing waste, expanding product lives and increasing more efficient use of the earth's sustainable resources by utilizing digital tools and knowledge that is extracted from big data. The digital circular economy utilizes technological advancements to turn linear economic models into regenerative systems that meet the needs of society and ensure environmental stewardship.

The digital circular economy is a new idea aiming to promote sustainability and resource efficiency by applying the most advanced digital technologies with revolutions in practices, moving from the classical circular economy into its next version. The new framework builds on traditional principles of waste prevention (reduce, reuse and recycle), but also stresses the importance of collaboration among stakeholders, technology as an enabler to circular practices through digitalization and a paradigm change in systemic production-consumption patterns.

4.4.1 Key Features of the Digital Circular Economy

1. **Integration of Digital Technologies**: The digital circular economy is centred around the adoption of a suite of digital technologies, including IoT (Internet of Things), big data analytics, AI (Artificial Intelligence) and blockchain to better manage the content of products. They have been designed to enable real-time monitoring, data collection and analytics across the entire product lifecycle (production-consumption end-of-life management) (Walden et al., 2021) which serves as a means of optimizing resource utilization. Thanks to, for example, enhanced transparency and traceability at the level of digital product passports which provide interested parties with more detailed information on materials used as well as their possible recycling alternatives (Walden et al., 2021). This allows real-time monitoring and data-driven decisions that yield improved efficiency, which reduces waste (Mbavarira & Grimm, 2021).

2. **Enhanced Resource Efficiency**: The digital circular economy puts a strong focus on resource efficiency by allowing businesses to make the most of their operations and minimize waste. Scale-up through digitization and other solutions that simplify processes can further strengthen supply chain management by making it much better equipped for making more sustainable decisions (Liu et al., 2022). Firms can track key metrics and then intervene at regular times to reduce inefficiencies so that they focus equipment use away from hoot lad redundant and generate less waste.
3. **Collaboration and Multi-Stakeholder Engagement**: One fundamental principle of the digital circular economy is that it is designed to fit together all stakeholders, manufacturers, consumers, policymakers and researchers. This collaborative distributed generation promotes sharing knowledge and innovation by involving groups of stakeholders working together towards common goals for sustainability (Vaska et al., 2021). The sharing formulation of digital platforms can assist in communications and collaborations for co-creation, which enable shared value solutions to address environmental issues and circular practices. The digital circular economy seeks to address this challenge by promoting partnerships and knowledge sharing to make the sustainability approach more inclusive and efficient (Levänen et al., 2022).
4. **Business Model Innovation**: It motivates us to form new sustainable business models in the area of innovative digital circular economy. The next phase of circular economy leads to new business models which are driven by innovation and sustainability. This may result in product-as-a-service, where firms offer the use of a good to customers without actual ownership being involved. For example, product-as-a-service is based on the payment for a service to use related products instead of buying and owning those assets fully which can increase resource efficiency, thereby reducing wastage (Moggi & Dameri, 2021). This change not only minimizes waste, but it encourages manufacturers to create products that last long and are easily repaired or recyclable (Crovella et al., 2021). It is through changes in conventional business models that the reformation of value creation processes in which geospatial divides can be achieved, leading to fewer pollution-causing outputs.
5. **Closing the Loop:** The circular economy in digital will involve moving to close the loop of product lifecycles by aligning with principles such as recycling, remanufacturing and/or reuse, eventually digitally upcycling obsolete products into resources. Through the use of digital technologies, businesses can monitor materials and products throughout their lifecycle ensuring that valuable resources are being recovered and lower dependence on virgin raw materials (Vetrova & Ivanova, 2021). Not only is this sustainable all-in-one production method good for the environment because it saves resources, but also reduces greenhouse gas emissions related to both producing a product and managing its waste (Fig. 4.1).
6. **Focus on Systemic Change**: Digital circular economy focuses on the need for systemic transformation as a core component enabling sustainability and resource efficiency within current economic models. A systemic focus is necessary for addressing the web of challenges in natural resource depletion, environmental

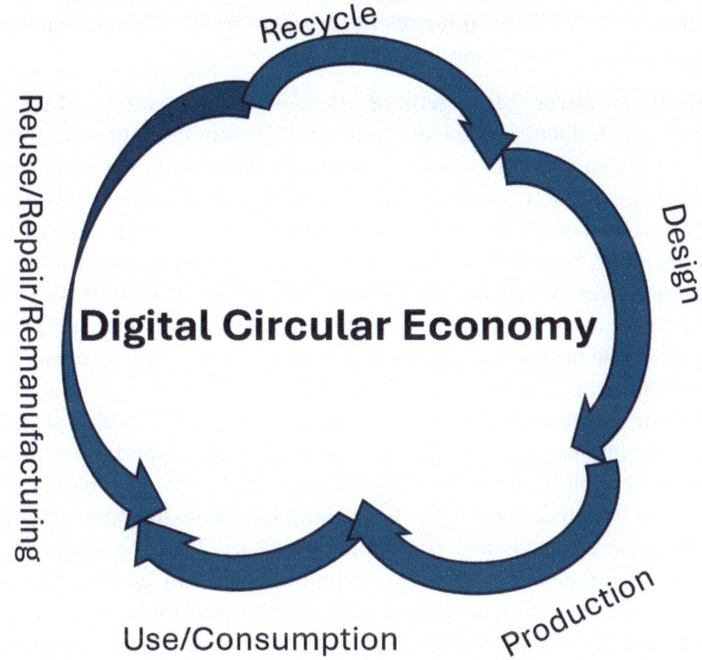

Fig. 4.1 Digital circular economy

degradation and climate change that we face. Systemic change is a deep shift in the way an economic system functions, including changes to its structures, processes and relationships. It necessitates rethinking the production, consumption and life cycle management of resources in every product or service. As defined by Blomsma et al. (2022), circular disruption is a radical transformation in some socio-technical systems—allowing this opportunistic shift from the damaging linear "take-make-use-dispose" to an environmentally regenerative model. Such a transition needs to be implemented with an upstream system thinking approach that takes into account the interdependencies of all actors and the general socio-economic environment. All this is from the design and production of smartphones, not to mention consumption recycling or waste handling. Through the intersection of these stages, firms could establish greener behaviour (Hamam et al., 2021).

4.5 Benefits of the Digital Circular Economy

The digital circular economy moves us towards several benefits that cannot be modelled in standard ways by traditional economics. Combining digital technologies and multi-stakeholder collaboration, it provides an upgrade to sustainability, resource efficiency as well economic resilience. Here are the main advantages of this kind of advanced circular economy model:

1. **Optimized Resource Management**: By favouring practices like recycling, reusing or remanufacturing, the digital circular economy promotes a more effective use of resources. This change can lead to decreased operational costs and environmental impacts, meaning better resource management throughout the entire value chain (Bawono, 2022). And, this efficiency does not only reduce the natural resources but it helps a lot to lower material carbon footprint also during the overall process of production. The design of the digital circular economy allows organizations to save a great portion of costs and reduce their production/consumption environmental footprint by resource use optimization, reducing waste (Millar et al., 2019).
2. **Catalyst for Economic Growth**: A transition towards circular practices can accelerate economic growth through new business opportunities, increased competitiveness and innovation. This is particularly true in low-income countries, where circular economy practices could represent an opportunity to create jobs and achieve sustainable development (Chao, 2023; Siregar, 2023). The circular economy can both increase economic growth by creating new markets and reducing costs, as well as spur innovation. It has also made clear that practices of the circular economy can generate employment and increase competitiveness in different sectors (Gomide, 2024).
3. **Reducing Greenhouse Gas Emissions, Conserving Natural Resources and Minimizing Pollution**: The digital circular economy is based on three principles that drastically cut greenhouse gas emissions while at the time of conserving natural resources and significantly limiting pollution; working towards a sustainable future in the ground of climate change, resource depletion and so on is one of the main points (Redesigning Fashion Industry: A Transformational Circular Approach, 2019). The circular economy contributes to climate protection and natural resource conservation by avoiding waste in the use of resources as much as possible on a large scale, promoting resource efficiency. This is especially crucial given the challenges of climate change and environmental deterioration (Bocken & Coffay, 2022).
4. **Social Benefits**: Moving towards a digital circular economy offers vast social benefits beyond the scope of conventionally structured economic models. The digital circular economy is a new model that embeds digital due diligence and connects the dots between economic, social and environmental factors to drive sustainability, resource efficiency and justice on all levels of our society through multi-stakeholder collaboration reducing inequality. Changing to a digital circular economy is not only good for the planet, but it can contribute to people thriving, and communities living in symbiosis with nature (Cerya, 2023; Suchek et al., 2021). Given that circular economy strategies can be oriented towards providing social equity and working against environmental injustice, it is a way to tackle superficial symptoms of systemic inequalities as well (Linder et al., 2017).

5. **Generate Economic Growth and Jobs**: The implementation of a circular economy 2.0 means that it will provide new business opportunities as well as markets. As (Bucur, 2023) points out the circular economy stimulates innovation and entrepreneurship for sustainability purposes promoting new product or service developments. At the same time, it provides an opportunity to create jobs (Maricuţ & Grădinaru, 2023) in sectors like recycling and repair as well as sustainable product design—among others—which can help sustain local economies (Hashemizadeh, 2023a, b).
6. **Improve Competitive Advantage**: Companies that adopt the digital circular economy should be able to achieve a competitive advantage over the traditional competitor's thanks to the more sustainable products/services they create on cooperatively implemented acquisitions, reuses and ownership concepts, respectively, in market offerings segment at least for certain groups of clients. Circular business practices can serve as a global brand image and build customer and client loyalty through sustainable procurement already on the agenda for consumers (Hao et al., 2020). In a world of increasing awareness (Hashemizadeh, 2023a, b) over environmental-related issues, this competitive advantage plays an essential role in appealing to environmentally minded consumers and investors (Kapoor et al., 2021).
7. **Supply Chain Robustness**: The digital circular economy promotes resilience to supply chain disruptions by localizing sourcing, recovering resources and fostering closed loops: Through reducing reliance on finite materials and external suppliers, firms are likely to become more resilient against economic uncertainty and shocks from global events like pandemics or geopolitical tensions (The Circular Economy Transition in the European Union, 2023).
8. **Innovation and Technological Advancement**: The digital circular economy brings innovation at different levels which are product design, manufacturing process as well as delivery of services. The other new technologies, including IoT, big data analytics and AI, could be used for monitoring resource utilization in companies' operating processes or during recycling as well as an intelligent way of re-designing the business models themselves into more circular (Bag & Pretorius, 2020) thereby fostering creativity and innovation(Geissdoerfer et al., 2017) which births long-term sustainability (Lacko et al., 2021; The Circular Economy Transition in the European Union, 2023).

4.6 Stakeholder Synergy in Driving Sustainable Innovation

At the heart of a digital circular economy lies stakeholder synergy, which is essential in supporting sustainable innovation. Collective stakeholder action refers to the combined measures taken by different stakeholders like businesses, governments, non-governmental organizations (NGOs) and consumers working in a concerted effort towards similar sustainability goals. It is this collaborative spirit that the industry must embrace to spur innovation, both in responding to natural market forces and solving critical environmental and social concerns. Insights on how synergy among key stakeholders can drive sustainable innovation include:

1. **Collaborative Innovation Processes**: Involving the stakeholders can help to improve the innovation process through wide perspectives, knowledge and resources (Goodman et al.(2017)). Based on their review of the literature, identify at least eight stakeholder roles in sustainability-oriented innovation processes: stimulators, initiators and broker/mediator among other types. All of these enable ideas and knowledge exchange among the stakeholders which creates conditions that will make them co-create innovative solutions to sustainability challenges. As previously mentioned (Velter et al., 2021), the involvement of various consortia and stakeholder collaborations can significantly boost realizing sustainable business models by playing strengths into one another. This can include businesses proposing joint technology development projects with research institutes (which ultimately work towards supporting circular practices), or NGOs sharing information about the needs of their communities and carbon impacts. The outcome of this approach, again is more free-market solutions quicker because of the efficiency which can be gained from mass production but also are just generally widely accepted as better and more effective than those created by the government.
2. **Multi-Stakeholder Engagement**: Involving several stakeholders in the innovation procedure creates shared ownership and accountability that is very essential to make any sustainable practices work (Singh et al., 2023). Velter et al. (2021) highlight the requirement of a boundary instrument to intensify multi-stakeholder collaboration for sustainable business model innovation within companies. Engagement in this process not only increases the quality of innovations but also ensures that they are closely suited to serve all stakeholders. This enables the organizations to generate shared value from innovation and benefit all stakeholders including themselves. Tapaninaho and Heikkinen (2022) stress the importance of understanding their subjective value considerations to energize stakeholders into joint value creation. It is in this collision of opposing interests that solutions can be found regarding social, environmental and economic challenges. For instance, a telecommunications firm might work with local communities to devise sustainable communication solutions that would improve connective quality while reducing environmental harm.
3. **Responding to Environmental Challenges**: Performance will be higher among firms that cater to an external stakeholder group. Stakeholders will be able to share their knowledge and invest together in solving major environmental issues, which generates more efficient ways of reducing the environmental loads. This is equally relevant in sectors with enormous sustainability woes like manufacturing and energy. By working together, whether it be public or private sector landowners and users alike, stakeholders can better address these concerns by utilizing the backyard assets we now know more about—our knowledge of local best practices (Zhang et al., 2020). Together, stakeholders can develop integrated solutions that are scalable and better strengthened for future events.

4. **Knowledge Sharing and Capacity Building**: Stakeholder synergy can encourage knowledge sharing between stakeholders, facilitating capacity building which is the requirement for sustainable innovation (Adomako, 2020). Sharing best practices and lessons learned between stakeholders to improve their capacity for innovation and successful implementation of sustainable solutions is very essential.
5. **Facilitating Policy Development**: It can also drive policy development and regulatory frameworks to improve the ecosystem (Mariani et al., 2022). Policy developers can act as catalyst entities capable of fostering collaborations between stakeholders in the pursuit of SDGs. Improving policies to promote stakeholder engagement and innovation will help these organizations foster the broader environment needed for sustainable practices.
6. **Overcoming Barriers to Innovation**: Breaking new ground with stakeholder collaboration can help overcome some of the key innovation challenges, including resource constraints and lack of technology. (Cooper, 2024; Zhang et al., 2020) stress the path of cooperative innovation and its principle application through organizational partnering to gain access needed resources along with knowledge from complementary aspects that deriving competitive advantage firm operations twice in success or occasional manner, contribute new value-added environment product as partnering output.
7. **Strengthening Regulatory Compliance and Support**: Stakeholders working together can also help in getting through the regulatory outcries that come with sustainable innovation. This all can, according to Eisenreich et al. (2021), be achieved by organizations through influencing regulatory bodies as well as cooperating with the policymakers so that they promote "circular innovation supportive policies". Even if a business cannot prevent regulation or remove existing regulations, the industry can guide the crafting of new ones to ensure that they both encourage sustainability and support innovation while still providing for an appropriately benign environment.
8. **Trust and Social Capital**: A successful collaboration follows a collaborative design principle that creates and nurtures trust and social capital between the stakeholders. Marjamaa et al. (2021) suggested that better relationship-building among stakeholders can increase cooperation and the transfer of information as well as resources. Trust is crucial for sustainable innovation as stakeholders (Palafox-Alcantar et al., 2020) must be able and willing to work together, providing one another with information for shared aims.

4.7 Obstacles and Challenges of the Digital Circular Economy

On one hand, the shift to the circular economy comes with plenty of benefits regarding sustainability and resource-savings; on the other hand, companies face several obstacles (Gürsel et al., 2022). For instance, the implementation of the digital circular economy faces significant awareness and understanding barriers among all stakeholders (i.e. businesses, consumers and policymakers). Zemanová (2023) asserted that many organizations do not know the concepts of circular economy at all or they fail to appreciate any output from their operations. The unawareness of consumers and many businesses in terms of circular economy principles may also remain an obstacle to the practice of sustainability (Poponi et al., 2019, Oyedele et al. 2023). Besides, where infrastructure is lacking, it can be difficult to close the loop and ensure that resources are recovered or waste is managed responsibly. Hence, the needed set-up and needs of a digital circular economy might not be supported today by the existing foundations in many regions. In other words, the effective implementation of circular practices can be conditional on a lack of recycling facilities and networks and poor waste management systems (Rizos et al., 2016) which can be very expensive. These costs can be prohibitive for many organizations, especially small and medium-sized enterprises (SMEs) (Assmann, 2023). Digital circular strategies are often perceived as high-cost upfront investments and therefore they remain projects suitable for only those with the financial resources to see them through. Providing financial support and incentives is crucial to reinforce investment in sustainable practices, as access to funding for circular economy initiatives could be restricted especially by small and medium-sized enterprises (SMEs) inevitably challenged with economic aspects that require their resources attention (Alonso-Almeida et al., 2021).

There is also the challenge of cultural resistance inside companies to change to implement the principles of the circular economy. Marrucci et al. (2021) mention that employees, as well as the official management, maybe too close bread absorbing new processes or business layoffs. This cultural resistance may also be a result of the need for specialized knowledge and expertise not present in the workforce. It may be difficult for organizations to find employees with the specific skills needed in circular design resource management and sustainability (Santos et al., 2021). In developing nations, one major challenge some stakeholders face with adopting digital circular economy practices is the regulatory frameworks in place. Munro (2023) suggests that conflicting regulations, a lack of supportive policies and the bureaucratic process can make it difficult for businesses to commit fully to digital circular strategies. As such, the migration to a circular economy calls for enabling frameworks that are coordinated by policymakers (Oladele et al, 2023; Rakesh, 2023). Due to market uncertainties (fluctuating demand for circular products and services), how a consumer thinks (Marcus et al., 2020) and acts (Oyenuga et al., 2021) are hugely influential when it comes to circular adaption. If consumers refuse to adopt circular products or are unable to accept services in a similar light, organizations might fall short of their goals (Toni, 2023). To increase demand for it, we need consumer insight to understand the preferences and push further into promoting circular practice benefits. Issues relating to supply chain, customer requirements and technology are concerns that organizations will face (Khan et al., 2020) perhaps because of a lack of collaboration (García-Quevedo et al., 2020). Without solving these barriers the possible benefits of a digital circular economy will never be achieved.

4.8 Risks to Be Recognized in Digital Circular Economy

In addition to inherent barriers that currently undermine the promise of a digital circular economy, there are other risks or potential deviations which should be addressed to avoid undesirable outcomes and ensure robust frameworks for such an integration. Even though advancing circularity conceivably results in a more sustainable, competitive and climate-neutral economy (Kirchherr et al., 2017), frequent lack of alignment with efforts to reduce emission rates or transform production-consumption schemes abound. To increase the volumes of recycled materials, it is not enough to simply expand recycling—this will only concern quality and demand for secondary raw materials. These include, for example, qualities of advanced durability or recyclability which do not itself lead to less energy-requiring actions and emissions. In another example, tethered caps on plastic bottles would likely increase the global warming of a product by increasing greenhouse gas emissions dependent upon newly required virgin or recycled plastic. Using electric appliances for extended periods may lead to increasing emissions because the greenhouse gas balance used is higher than new ones (Neugebauer et al., 2015). More data and digital products being used do not automatically mean fewer greenhouse gas (GHG) emissions or resource efficiency. It will rather lead to the same positive "fanfare for blossom" which if not properly managed or supervised can backfire. There is no automatic link between data sharing and digital-enabled solutions to be used in production and consumption with greater sustainability. Data must be actionable to initiate change. E-commerce and digital native services could both become part of the alluring excessive consumerism. Secondly, one of the most polluting industries in the world is also digital/ICT. Energy and resource-hungry data centres, digital devices or infrastructures are often running at critical levels (Enerdata, 2018). Now, ICT accounts for 5 to 9% of the total electricity and greenhouse gas emissions category. However, this could go up to 20% by the end of the next decade because demand for data centres and cloud computing as well as energy-hungry new technologies such as blockchain continue to rise (Neugebauer et al., 2015). If immense energy is required for digitalization, it seems equally necessary to change the system with a switch of portfolio from coal and natural gas through more efficiency in production or electricity as well as emphasis on renewable sources. This is increasing due to the rapid development of technology around the globe with a production rate reaching 50 million tonnes, annually. This results in a massive economic loss since large quantities of useful and invaluable materials are disposed of. The last thoughts that come to mind are related to landfilling or informal recycling, disposing of this trash in dumps and places specifically for it; that can also have adverse health effects on people as well as damage the environment. Moreover, the digitalization of our economy and society can lead to negative impacts on safety (e.g. cyberattacks) as well as health risks such as radiation and exposure to hazardous substances (Geissdoerfer et al, 2017).

Digitalization could lead to a digital divide, where only the people who can afford it will be able to use whatever technology is available. Labour markets are becoming polarized, where highly educated specialists skilled in using these new technologies could have practically no competition from low-skilled workers who may be under threat of losing their jobs as a result of automation. The gap in the digital divide may also widen between giant companies and small and medium-sized enterprises (SMEs) that might not have similar financial resources to spend on re-skilling programmes or implementing new technology. Failure to address the seriousness of this situation may lead to a social divide (Liu et al., 2022).

4.9 Implications

There are multiple implications which are essential for the promotion of sustainable practices and efficient generation of value in the circular economy. These lessons have important implications for policymakers, businesses and those seeking to base a more sustainable and resilient economic model. One example is multi-stakeholder collaboration and other symbiotic alliances as key determinants of sustainable innovation in the context of the digital circular economy. Organizations can co-create solutions which focus on environmental challenges and help achieve resource efficiency by eliciting participation from diverse stakeholder groups (businesses, governments, NGOs and consumers). Better collective action not only ensures that the products and services are being developed wisely but also in a way such that it aligns with sustainability grounds for the overall good of stakeholder groups (Walden et al., 2021). For this reason, organizations need to invest in strengthening their digital infrastructure and capabilities so as not only take full advantage of these technologies. This investment will not only support the soaring change to a digital circular economy but also support boosting competitive advantage through operational efficiencies and cost-saving (Khan et al., 2022a, b).

It underscores the need for regulatory frameworks to promote digital circular economy adoption. This can include everything from offering financial benefits for sustainable practices within organizations to creating an environment that supports digital circular business models (Geissdoerfer et al., 2017). Equally, there is a need for educational interventions on sustainability literacy and skills for the implementation of diverse circular practices (Colombi & D'Itria, 2023). This paper has identified several barriers which obstruct the successful implementation of circular economy practices, such as unawareness, availability and legal dispositions. As Vaska et al. (2021) highlight, to overcome these barriers, there is a need for the coordination of efforts through organizations for intervention implementation including infrastructure improvements, engaging stakeholders and supporting change regulation.

4.10 Conclusion

There is no doubt that the shift towards the digital circular economy is a major leap forward in the sustainable, resource-effective transformation of our world. Combining technologies with multi-stakeholder engagement, the digital circular economy expands the opportunities for sustainable innovation in multiple industry sectors. The main learning that has been covered in the rest of this chapter is around the governing force of stakeholder synergy where businesses, governments and communities collaborate to make a regenerative economy. Stakeholder engagement has been demonstrated in the literature as key to mitigating barriers faced when attempting circular practices and promoting an environment supportive of innovation. By adding additional stakeholder diversity and expertise in the mix, organizations are likely to develop radical ideas of what they can do within those resource constraints that will allow for widespread change as well on how we manage waste or approach environmental sustainability. Adopting circular business models can also enhance competitiveness, create jobs and contribute to improving the quality of life in local communities. However, progressing to a digital circular economy is not without its challenges. Some of the barriers that are discussed include a lack of public awareness, underdeveloped infrastructure and regulatory hurdles. To resolve those worries a mix of educational programmes, policies and investments need to come together in one comprehensive solution. Cutting away these barriers and fostering cooperation along value chains can make the digital circular economy part of a positive-sum game that greatly benefits sustainable development. Hence, transitioning into a digital circular economy requires imagination and vision, political buy-in and support from the highest level of leadership. It requires significant changes in policy structure and implementation, as well as collaboration across sectors and value chains. The required steps go far beyond normal environmental and technological challenges and will only be achievable by collaboration between stakeholders.

References

Adomako, S. (2020). Environmental collaboration, sustainable innovation, and small and medium-sized enterprise growth in Sub-Saharan Africa: Evidence from Ghana. *Sustainable Development, 28*(6), 1609–1619. https://doi.org/10.1002/sd.2109

Alonso-Almeida, M., Antón, J., Bagur-Femenías, L., & Perramon, J. (2021). Institutional entrepreneurship enablers to promote circular economy in the European Union: Impacts on transition towards a more circular economy. *Journal of Cleaner Production, 281*, 124841. https://doi.org/10.1016/j.jclepro.2020.124841

Alshammari, K. (2023). Green innovation and its effects on innovation climate and environmental sustainability: The moderating influence of green abilities and strategies. *Sustainability, 15*(22), 15898. https://doi.org/10.3390/su152215898

Armenia, S., Dangelico, R., Nonino, F., & Pompei, A. (2019). Sustainable project management: A conceptualization-oriented review and a framework proposal for future studies. *Sustainability, 11*(9), 2664. https://doi.org/10.3390/su11092664

Assmann, I. (2023). Determinants of circular business model adoption—a systematic literature review. *Business Strategy and the Environment, 32*(8), 6008–6028. https://doi.org/10.1002/bse.3470

Ayuso, S., Rodríguez, M., García-Castro, R., & Ariño, M. (2011). Does stakeholder engagement promote sustainable innovation orientation? *Industrial Management & Data Systems, 111*(9), 1399–1417. https://doi.org/10.1108/02635571111182764

Bag, S., & Pretorius, J. (2020). Relationships between industry 4.0, sustainable manufacturing and circular economy: Proposal of a research framework. *International Journal of Organizational Analysis, 30*(4), 864–898. https://doi.org/10.1108/ijoa-04-2020-2120

Bawono, S. (2022). Circular economy role in industry and environmental sustainability: Qualitative content analysis approach. *Asia Pacific Journal of Management and Education, 5*(3), 1–13. https://doi.org/10.32535/apjme.v5i3.1546

Blomsma, F., Bauwens, T., Weissbrod, I., & Kirchherr, J. (2022). The 'need for speed': Towards circular disruption—what it is, how to make it happen and how to know it's happening. *Business Strategy and the Environment, 32*(3), 1010–1031. https://doi.org/10.1002/bse.3106

Bocken, N., & Coffay, M. (2022). The circular experimentation workbench—a lean and effectual process. *Circular Economy and Sustainability, 3*(3), 1361–1383. https://doi.org/10.1007/s43615-022-00239-w

Bucur, L. (2023). Exploring the role of consumers in promoting a circular economy: Increasing awareness and engagement. In *Proceedings of the international conference on business excellence,* (Vol. 17(1), pp. 38–47). https://doi.org/10.2478/picbe-2023-0006

Cerya, E. (2023). Analysis of SMEs readiness in developing countries to implement the circular economy (pp. 148–162). https://doi.org/10.2991/978-94-6463-158-6_13

Chao, C. (2023). How emotions and green altruism explain consumer purchase intention toward circular economy products: A multi-group analysis on willingness to be environmentally friendly. *Business Strategy and the Environment, 33*(4), 2803–2816. https://doi.org/10.1002/bse.3632

Cooper, M. (2024). Exploring the role of stakeholder collaboration in sustainable supply chain management. *Research Square*. https://doi.org/10.21203/rs.3.rs-4286600/v1

Colombi, C., & D'Itria, E. (2023). Fashion digital transformation: Innovating business models toward circular economy and sustainability. *Sustainability, 15*(6), 4942. https://doi.org/10.3390/su15064942

Crovella, T., Paiano, A., Lagioia, G., Cilardi, A., & Trotta, L. (2021). Modelling digital circular economy framework in the agricultural sector. An application in southern Italy, *45*, 15. https://doi.org/10.3390/engproc2021009015

Eisenreich, A., Füller, J., & Stuchtey, M. (2021). Open circular innovation: How companies can develop circular innovations in collaboration with stakeholders. *Sustainability, 13*(23), 13456. https://doi.org/10.3390/su132313456

Ekins, P., & Zenghelis, D. (2021). The costs and benefits of environmental sustainability. *Sustainability Science, 16*(3), 949–965. https://doi.org/10.1007/s11625-021-00910-5

Enerdata. (2018). Between 10 and 20% of electricity consumption from ICT sector in 2030?.

Fassio, F., & Tecco, N. (2019). Circular economy for food: A systemic interpretation of 40 case histories in the food system in their relationships with SDGs. *Systems, 7*(3), 43. https://doi.org/10.3390/systems7030043

García-Quevedo, J., Jové-Llopis, E., & Martínez-Ros, E. (2020). Barriers to the circular economy in European small and medium-sized firms. *Business Strategy and the Environment, 29*(6), 2450–2464. https://doi.org/10.1002/bse.2513

Geissdoerfer, M., Savaget, P., Bocken, N., & Hultink, E. (2017). The circular economy—a new sustainability paradigm? *Journal of Cleaner Production, 143*, 757–768. https://doi.org/10.1016/j.jclepro.2016.12.048

Gomide, F. (2024). How can the circular economy contribute to resolving social housing challenges? *Applied System Innovation, 7*(2), 21. https://doi.org/10.3390/asi7020021

Goodman, J., Korsunova, A., & Halme, M. (2017). Our collaborative future: Activities and roles of stakeholders in sustainability-oriented innovation. *Business Strategy and the Environment, 26*(6), 731–753. https://doi.org/10.1002/bse.1941

Gürsel, I., Elbersen, B., Meesters, K., & Leeuwen, M. (2022). Defining circular economy principles for biobased products. *Sustainability, 14*(19), 12780. https://doi.org/10.3390/su141912780

Hamam, M., Chinnici, G., Vita, G., Pappalardo, G., Pecorino, B., Maesano, G., D'Amico, M., et al. (2021). Circular economy models in agro-food systems: A review. *Sustainability, 13*(6), 3453. https://doi.org/10.3390/su13063453

Hao, Y., Wang, Y., Wu, Q., Sun, S., Wang, W., & Cui, M. (2020). What affects residents' participation in the circular economy for sustainable development? Evidence from China. *Sustainable Development, 28*(5), 1251–1268. https://doi.org/10.1002/sd.2074

Hashemizadeh, A. (2023a). Analyzing influential indicators in the sustainable development of Guangdong province: The role of circular economy (pp. 299–304). https://doi.org/10.2991/978-94-6463-260-6_39

Hashemizadeh, A. (2023b). Developing a machine learning algorithm to investigate the role of energy consumption in sustainable development: A case study of China (pp. 170–177). https://doi.org/10.2991/978-94-6463-262-0_20

Hina, M., Chauhan, C., Sharma, R., & Dhir, A. (2023). Circular economy business models as pillars of sustainability: Where are we now, and where are we heading? *Business Strategy and the Environment, 32*(8), 6182–6209. https://doi.org/10.1002/bse.3480

Jung, S., Nam, C., Yang, D., & Kim, S. (2017). Does corporate sustainability performance increase corporate financial performance? focusing on the information and communication technology industry in Korea. *Sustainable Development, 26*(3), 243–254. https://doi.org/10.1002/sd.1698

Kapoor, N., Jauhari, D., & Maheshwari, D. (2021). Understanding of circular processes and their impact on Indian economy. *Journal of Contemporary Issues in Business and Government, 27*(3). https://doi.org/10.47750/cibg.2021.27.03.256

Khan, O., Daddi, T., & Iraldo, F. (2020). The role of dynamic capabilities in circular economy implementation and performance of companies. *Corporate Social Responsibility and Environmental Management, 27*(6), 3018–3033. https://doi.org/10.1002/csr.2020

Khan, S., Umar, M., Asadov, A., & Tanveer, M. (2022a). Technological revolution and circular economy practices: A mechanism of green economy. *Sustainability, 14*(8), 4524. https://doi.org/10.3390/su14084524

Khan, S., Ali, S. S., & Singh, R. (2022b). Determinants of remanufacturing adoption for circular economy: A causal relationship evaluation framework. *Applied System Innovation, 5*(4), 62. https://doi.org/10.3390/asi5040062

Kirchherr, J., Reike, D., & Hekkert, M. (2017). Conceptualizing the circular economy: An analysis of 114 definitions. *Resources Conservation and Recycling, 127*, 221–232. https://doi.org/10.1016/j.resconrec.2017.09.005

Kravchenko, M., Pigosso, D., & McAloone, T. (2020). A procedure to support the systematic selection of leading indicators for sustainability performance measurement of circular economy initiatives. *Sustainability, 12*(3), 951. https://doi.org/10.3390/su12030951

Lacko, R., Hajduová, Z., & Zawada, M. (2021). The efficiency of circular economies: A comparison of visegrád group countries. *Energies, 14*(6), 1680. https://doi.org/10.3390/en14061680

Larsson, J., & Larsson, L. (2020). Integration, application and importance of collaboration in sustainable project management. *Sustainability, 12*(2), 585. https://doi.org/10.3390/su12020585

Levänen, J., Park, S., & Rosca, E. (2022). Circular solutions in developing countries: Coping with sustainability tensions using technical functionality and business model relevance. *Business Strategy & Development, 6*(1), 75–94. https://doi.org/10.1002/bsd2.224

Linder, M., Sarasini, S., & Loon, P. (2017). A metric for quantifying product-level circularity. *Journal of Industrial Ecology, 21*(3), 545–558. https://doi.org/10.1111/jiec.12552

Liu, Q., Trevisan, A., Yang, M., & Costa, J. (2022). A framework of digital technologies for the circular economy: Digital functions and mechanisms. *Business Strategy and the Environment, 31*(5), 2171–2192. https://doi.org/10.1002/bse.3015

Lozano, R., Barreiro-Gen, M., & Zafar, A. (2021). Collaboration for organizational sustainability limits to growth: Developing a factors, benefits, and challenges framework. *Sustainable Development, 29*(4), 728–737. https://doi.org/10.1002/sd.2170

Marcus, G. O., Oyenuga, M. O., & Ahungwa, A. I. (2020). Effects of sales promotion on consumer buying behaviour of food seasoning among Nigerian households: A case study of nestle maggi Naijapot. *Budapest International Research and Critics Institute Journal (BIRCI Journal)3*(1),134–139. https://doi.org/10.33258/birci.v3i1.724

Mariani, L., Trivellato, B., Martini, M., & Marafioti, E. (2022). Achieving sustainable development goals through collaborative innovation: Evidence from four european initiatives. *Journal of Business Ethics, 180*(4), 1075–1095. https://doi.org/10.1007/s10551-022-05193-z

Maricuţ, A., & Grădinaru, G. (2023). The impact of circular economy on economic development a review of EU's countries. In *Proceedings of the international conference on business excellence* (Vol. 17(1), pp. 1487–1496). https://doi.org/10.2478/picbe-2023-0134

Marjamaa, M., Salminen, H., Kujala, J., Tapaninaho, R., & Heikkinen, A. (2021). A sustainable circular economy: Exploring stakeholder interests in Finland. *South Asian Journal of Business and Management Cases, 10*(1), 50–62. https://doi.org/10.1177/2277977921991914

Marrucci, L., Daddi, T., & Iraldo, F. (2021). The circular economy, environmental performance and environmental management systems: The role of absorptive capacity. *Journal of Knowledge Management, 26*(8), 2107–2132. https://doi.org/10.1108/jkm-06-2021-0437

Mbavarira, T., & Grimm, C. (2021). A systemic view on circular economy in the water industry: Learnings from a Belgian and Dutch case. *Sustainability, 13*(6), 3313. https://doi.org/10.3390/su13063313

Millar, N., McLaughlin, E., & Börger, T. (2019). The circular economy: Swings and roundabouts? *Ecological Economics, 158*, 11–19. https://doi.org/10.1016/j.ecolecon.2018.12.012

Moggi, S., & Dameri, R. (2021). Circular business model evolution: Stakeholder matters for a self-sufficient ecosystem. *Business Strategy and the Environment, 30*(6), 2830–2842. https://doi.org/10.1002/bse.2716

Munro, M. (2023). Assessing the impact of strategic implementation of circular economy on the competitive advantage of Canadian manufacturing firms. *Journal of Strategic Management, 7*(3), 1–10. https://doi.org/10.53819/81018102t4143

Neugebauer, S., Martinez-Blanco, J., Scheumann, R., & Finkbeiner, M. (2015). Enhancing the practical implementation of life cycle sustainability assessment—proposal of a Tiered approach. *Journal of Cleaner Production, 102*, 165–176. https://doi.org/10.1016/j.jclepro.2015.04.053

Oladele, T. O., Oyenuga, M. O., & Adoga, G. J. (2023). Green management practices and organisational efficiency in higher institution. *Budapest International Research and Critics in Linguistics and Education (BirLE) Journal, 6*(3), 276–290. https://doi.org/10.33258/birle.v6i3.7716276

Oyedele, M. O., Marcus, G. O., & Ahungwa, A. I. (2023). Do consumers care about green marketing practices? Insight from a developing nation. *Budapest International Research and Critics Institute Journal (BIRCI Journal), 6*(3), (1424–1436) https://doi.org/10.33258/birci.v6i3.7665

Oyenuga, M. O., Ahungwa, A. I., & Onoja, E. (2021). Effect of Brand equity on consumer behaviour among students of Veritas university, Nigeria: A study of apple smartphones. *Marketing and Branding Research, 8,* 48–64. https://doi.org/10.33844/mbr.2021.60329

Palafox-Alcantar, P., Hunt, D., & Rogers, C. (2020). A hybrid methodology to study stakeholder cooperation in circular economy waste management of cities. *Energies, 13*(7), 1845. https://doi.org/10.3390/en13071845

Poponi, S., Colantoni, A., Cividino, S., & Mosconi, E. (2019). The stakeholders' perspective within the B corp certification for a circular approach. *Sustainability, 11*(6), 1584. https://doi.org/10.3390/su11061584

Rakesh, C. (2023). Towards a circular economy: Challenges and opportunities for recycling and re-manufacturing of materials and components. In *E3s web of conferences, 430*, 01129. https://doi.org/10.1051/e3sconf/202343001129

Redesigning fashion industry: A transformational circular approach. (2019). *Journal of Applied Business and Economics, 21*(8). https://doi.org/10.33423/jabe.v21i8.2592

Rizos, V., Behrens, A., Gaast, W., Hofman, E., Ιωάννου, A, Kafyeke, T., Topi, C., et al. (2016). Implementation of circular economy business models by small and medium-sized enterprises (SMEs): Barriers and enablers. *Sustainability, 8*(11), 1212. https://doi.org/10.3390/su8111212

Singh, R., Khan, S., & Dsilva, J. (2023). A framework for assessment of critical factor for circular economy practice implementation. *Journal of Modelling in Management, 18*(5), 1476–1497. https://doi.org/10.1108/JM2-06-2021-0145

Singh, R., Khan, S., & Centobelli, P. (2022). Investigating the interplay between social performance and organisational factors supporting circular economy practices. *Sustainability, 14*(24), 16781. https://doi.org/10.3390/su142416781

Siregar, M. (2023). Application of circular economy in manufacturing industry in Indonesia. *AMCA Journal of Community Development, 3*(1), 19–24. https://doi.org/10.51773/ajcd.v3i1.211

Santos, H., Marques, V., & Paschoali, L. (2021). The analysis of barriers for implementing circular economy practices using the analytic hierarchy process (ahp). *Revista Gestão Da Produção Operações E Sistemas, 16*(3), 99. https://doi.org/10.15675/gepros.v16i3.2793

Suchek, N., Fernandes, C., Kraus, S., Filser, M., & Sjögrén, H. (2021). Innovation and the circular economy: A systematic literature review. *Business Strategy and the Environment, 30*(8), 3686–3702. https://doi.org/10.1002/bse.2834

Tambovceva, T., Melnyk, L., Dehtyarova, I., & Nikolaev, S. (2021). Circular economy: Tendencies and development perspectives. In *Mechanism of an economic regulation*, (2) (pp. 33–42). https://doi.org/10.21272/mer.2021.92.04

Tapaninaho, R., & Heikkinen, A. (2022). Value creation in circular economy business for sustainability: A stakeholder relationship perspective. *Business Strategy and the Environment, 31*(6), 2728–2740. https://doi.org/10.1002/bse.3002

Testa, F., Iovino, R., & Iraldo, F. (2020). The circular economy and consumer behaviour: The mediating role of information seeking in buying circular packaging. *Business Strategy and the Environment, 29*(8), 3435–3448. https://doi.org/10.1002/bse.2587

The circular economy transition in the European Union. (2023). *Journal of Management, 39*(2). https://doi.org/10.38104/vadyba.2023.2.12

Toni, M. (2023). Conceptualization of circular economy and sustainability at the business level: Circular economy and sustainable development. *International Journal of Empirical Research Methods, 1*(2), 81–89. https://doi.org/10.59762/ijerm205275791220231205140635

Vaska, S., Massaro, M., Bagarotto, E., & Mas, F. (2021). The digital transformation of business model innovation: A structured literature review. *Frontiers in Psychology, 11*. https://doi.org/10.3389/fpsyg.2020.539363

Velter, M., Bitzer, V., & Bocken, N. (2021). A boundary tool for multi-stakeholder sustainable business model innovation. *Circular Economy and Sustainability, 2*(2), 401–431. https://doi.org/10.1007/s43615-021-00103-3

Vetrova, M., & Ivanova, D. (2021). Closed product life cycle as a basis of the circular economy. *Gatr Journal of Business and Economics Review, 5*(4), 36–50. https://doi.org/10.35609/jber.2021.5.4(4)

Walden, J., Steinbrecher, A., & Marinkovic, M. (2021). Digital product passports as enabler of the circular economy. *Chemie Ingenieur Technik, 93*(11), 1717–1727. https://doi.org/10.1002/cite.202100121

Widodo, S., Marpaung, Z., Santoso, A., Al-fadheeli, M., Sulaiman, N., Hizam, H., Shehab, M., et al. (2022). What makes public sector innovation sustainable?: A case study from Indonesia. *Tem Journal*, 1376–1384. https://doi.org/10.18421/tem113-48

Zemanová, V. (2023). Circular economy implementation from the perspective of benefits and barriers. https://doi.org/10.15240/tul/009/lef-2023-31

Zhang, R., Wang, Z., Tang, Y., & Zhang, Y. (2020). Collaborative innovation for sustainable construction: The case of an industrial construction project network. *IEEE Access, 8*, 41403–41417. https://doi.org/10.1109/access.2020.2976563

Chapter 5
Global SDGs Strategic Policies: Catalyst for Digital Circular Economy

Ananya Pandey, Jipson Joseph, Shahbaz Khan, and Vikas Kumar

Abstract The worldwide environmental movement originated with Rachel Carson's 1962 book *Silent Spring,* which cautioned the world about the adverse impact of chemical pesticides on ecosystem and sustainability. Thereafter, an unprecedent increase in global warming causing severer and harmful climate changes and environmental instabilities led to the establishment of Sustainable Development Goals (SDGs). In 2015, the United Nations accepted them into its global agenda for eradicating poverty, protecting the environment, and promoting peace to all people by 2030. Unsustainable development leads to depletion of natural resources, which is not simply a moral or ecological issue but are pertaining to economics, development, human health, and justice. The interconnected 17 SDGs acknowledge that the decision made in one area will have its repercussions in other areas and insist to strike a balance between 'environmental, social, and economic dimensions of sustainability'. The digital circular economy (DCE) has been well-known in recent years as a tool that answers some of the most urgent cross-cutting sustainable development challenges facing the globe. This model holds great promise for accelerating the 2030 Agenda's implementation with the potential to achieve the various goals of sustainable development like 'SDGs 6 (energy), 8 (economic progress), 11 (sustainable cities), 12 (sustainable consumption and production), 13 (climate change), 14 (oceans), and 15 (life on land)'. This requires significant changes in the organization of societies and businesses to progress towards sustainable future. *Limitations* The study is limited to analyze the global SDGs and related policies that are helpful in catalyzing DCE. *Findings* This chapter analyzes the global strategic plans and

A. Pandey (✉) · J. Joseph
Christ University, Bangalore, India
e-mail: ananya.pandey@res.christuniversity.in

J. Joseph
e-mail: jipson.joseph@res.christuniversity.in

S. Khan
Department of Industrial Engineering, University of Tabuk, Tabuk, Kingdom of Saudi Arabia

V. Kumar
University of Portsmouth, Portsmouth, UK
e-mail: Vikas.Kumar@port.ac.uk

R. Singh and V. Kumar (eds.), *Sustainable Innovations and Digital Circular Economy*,
https://doi.org/10.1007/978-981-96-1064-8_5

policies aimed at enhancing SDGs affirming the significance of digital economy for synergizing with the SDGs.

Keywords Digital circular economy · Sustainable development · Green economy · Climate change · Environmental sustainability · Right to repair

5.1 Introduction

The United Nations (UN) established 'the Sustainable Development Goals (SDGs)' with the objective of setting up a global agenda for sustainability. The UN initiated this global strategic design in September 2015 under the tagline "transforming our world: the 2030 Agenda for Sustainable Development," expanding the groundwork established by the 2012 Rio + 20 Earth Summit, which was framed with the title "The Future We Want" (Giller et al., 2018). 'The UN's 2030 Agenda for Sustainable Development, which was unanimously accepted and proposed by all the member-States in 2015, consists of 17 Sustainable Development Goals (SDGs), 169 targets, and 231 indicators'. 'This UN framework shapes international and national development policies and offers novel possibilities and parameters for bridging the already existing gap between human rights and human development. It functions and a directive for both national and international development initiatives (och Regeringskansliet, 2015). The implementation of the SDGs has been a priority of both international, regional, and national strategic plans and policies. The Paris Agreement, Green Climate Fund (GCF), European Green Deal (EGD), etc. are some of the initiatives that aimed at promoting the realization of SDGs'.

The UN initiated this global agenda of SDGs with a vision to equip the whole globe to confront the interconnected environmental, economic, and social issues that cause serious inconsistencies and injustices in the planet. The aforesaid problems had adverse impacts on the international political scenario. In such a context, the incorporation of the 17 SDGs into the 2030 agenda of sustainable future was laudable as they stood for a sustainable development, which has three pillars in confrontation with the 'environmental, economic, and social issues' of the globe. Each pillar is designed to focus a particular area of issues, so that a comprehensive sustainable development vision will be able to resolve them and promote the "universalism of life claims" (UNDP, 1994). The SDGs are designed in such a way, there is an underlying interconnection between them, which supports mutual promotion and mutual accomplishment. For instance, poverty, health, gender disparity, climate change, etc. are all interrelated issues. 'Managing the threat posed by climate change affects how one manages limited natural resources; attaining gender parity or improved health contributes to the eradication of poverty; and cultivating an inclusive and peaceful society will lessen disparities and boost economic growth' (UNDP, n.d.). The abovementioned achievement of SDGs shows that the achievement of one particular SDGs has its impact towards the achievement of other SDGs.

There has never been a more pressing need for innovative remedies and critical discourses on these subject matters in a world dealing with an expanding global population, the fallout from one of the most devastating pandemics in recent memory, and a plethora of urgent sustainability issues. Threats to the global developmental agenda include resource depletion, extremism, economic disparity, growing political and social intolerance, climate change, and a host of other interrelated problems that make it difficult for the globe to address the pressing and complicated sustainability challenges of modern times (Caradonna, 2022). The convergence of sustainable digital technologies with the digital circular economy (DCE) can establish notable progress in sustainability practices by enhancing environmental impact assessments and maximizing resource efficiency across various industries (Hariyani et al., 2024).

Keeping in view the above, the chapter provides an inclusive understanding of the SDGs including the related strategic global and regional policy initiatives that are actively employed to promote sustainability. It analyzes the role of sustainable innovations and DCE in catalyzing the achievement of SDGs in the global scenario. It explores the need for a synergy between sustainable innovation, SDGs, and DCE. Realizing that the DCE cannot be properly implemented without a substantial regulatory framework, the chapter also examines the current regulations enacted by 'European Union (EU), China, and the United States of America (USA)' for giving a regulatory design of DCE for the whole globe.

5.2 Strategic Policies in Advancing SDGs

The goal-setting is not novel but the setting of goals as a tool for global governance draws a whole new significance. The most recent and renowned example of new forms of "global governance by goals" are the 17 SDGs (Bogers et al., 2022). The setting of an agenda for the whole globe's sustainable future includes the peaceful and prosperous life of both the present and future generations of the planet. It does not exclude any nation from the global policy design affirming that both developed and developing States have equal significance in this inspirational vision. It thus suggests for global partnerships. The UN member-States understand that "eradicating poverty and other forms of deprivation require concerted efforts to combat climate change, protect our seas and forests, enhance health and education, and lower inequality in addition to promoting economic growth" (UN, n.d.). Advancement of SDGs will be impracticable in the absence of international, regional, and national agreements, policies, and plans. The various global partnerships, so far undertaken, affirmed that the collective efforts can make meaningful results in relation to SDGs. In this part, the various global and regional initiatives towards the achievement of SDGs are analyzed.

5.2.1 Global Perspective

Globalization has changed the face of the globe. Interconnected and partnerships have become essential phenomena for sustainability and global vision. Such a scenario also modifies the attempts for the realization of SDGs from the independent initiative to collaborative efforts. International cooperation has become vital in a globalized world, where the interests or policies of any nation is closed connected with others. This typically calls for global partnership and international collaboration among nations to enhance sustainability globally as well. This includes significant agreements, committee reports, and various other strategic plans ratified by several nations in coherence.

5.2.1.1 Paris Agreement on Climate Change

The actualization of the SDGs has been prioritized by the international community in the twenty-first century. The unforeseeable changes in the climate forced the States to frame the Paris Agreement in 2015. Together with other international agreements, it also focused on empowering the attempts towards sustainability. It became effective and a legally binding international treaty of the 195 Parties, including the European Union and 194 States in November 2016. The parties were given long-term objectives in order to lessen the harmful effects and dangers caused by climate change through individual and coordinated efforts. "The agreement stood for taking up positive steps to drastically reduce global greenhouse gas (GHG) emissions in order to keep the rise in global temperature to far below 2 °C over pre-industrial levels and to pursue attempts to restrict it to 1.5 °C over pre-industrial levels (UN Climate Change, n.d.). The agreement's long-term objectives and overall purpose are evaluated on a regular basis. Developing nations get funding to reduce global warming, build resilience, and improve their capacity to adapt to its effects" (Huang & Zhai, 2021). This creative move also designed the idea of Nationally Determined Contributions (NDCs) for outlining the steps that are undertaken by any party to reduce its GHG emissions. It also includes many other goals and priorities that support more broadly the sustainable development aspirations (Dzebo et al., 2017; Moreno et al., 2023). The NDCs unequivocally affirm the interconnectedness with sustainable development, especially in relation to 'SDGs 3, 4, and 5 (health, education, and gender equality respectively)', which are substantially underrepresented in the social SDGs compared to the environmental and economic goals. Numerous climate measures address several 'SDG themes (i.e., socio-economic sectoral categories)', suggesting that there are several possible synergies and chances for policy coherence (Janetschek et al., 2021).

5.2.1.2 Green Climate Fund

The 'United Nations Framework Convention on Climate Change (UNFCCC)" established the 'Green Climate Fund (GCF) in 2010' as one of its flagship programs. It is the largest climate finance fund. GCF functions with the objective of increasing the number of human activities that can make substantial and positive impact on the issue of climate change. "GCF primarily targets the climate change adaptation and mitigation activities of the developing countries and seeks to invest in low-emission and climate-resilient development on the planet by raising large amounts of capital" (Umashankar et al., 2022; Schalatek et al., 2012). GCF has a positive correlation with the realization of SDGs as well. The emphasis on climate action for sustainability (SDG13) necessitates global collective efforts and partnerships (SDG17) that frames certain prerequisites for decent work and economic stability (SDG8). Such an emphasis on collaboration provides a global action plan to confront various issues like climate change. GCF originated in this positive and progressive perspective, which will energize the realization of SDGs. Along with mitigating climate change by granting green funds to the vulnerable nations, through the GCF, inclusive and sustained economic growth can advance society, raise living standards for the most vulnerable members of the population, and provide good jobs for all individuals (Georgieva et al., 2022).

5.2.1.3 Global Environment Facility

The SDGs offer a comprehensive and integrated vision of development that can make extensive influence in global, regional, and local ventures and policies. Within the global partnership for sustainable development, the Global Environment Facility (GEF) holds a special position and is ideally positioned to assist the goals outlined in the SDGs (Carugi & Bryant, 2021). The GEF is in close association with the objective of ensuring sustainability in relation to life below the water (SDG14) by encouraging the cooperative and all-encompassing management of seas and oceans. "In accordance with the United Nations Convention to Combat Desertification (UNCCD), the GEF supports several attempts of the States to stop land degradation by promoting sustainable land management techniques that maintain soil health, backed by suitable policy frameworks (Ulitto & Naidoo, 2019). Concurrently, the GEF promotes international collaboration in the field of freshwater management. In connection with this, it additionally influences SDGs 1 (No Poverty), 5 (Gender Equality), 8 (Decent and Economic Work), 12 (Responsible Consumption and Production) and 13 (Climate Action) through various projects and strategies. In order to finance a wide range of integrated, creative, and catalytic initiatives to increase renewable energy, enhance energy efficiency, lessen vulnerability, and promote technology transfer, the GEF has assisted in raising a total of $46 billion thereby advancing the achievement of SDG13 (Climate Action)" (CB et al., 2021; Ishii, n.d.).

5.2.2 Regional Perspective

Apart from the global perspective, several regions have adopted separate enhanced vision in connection to the advancement of SDGs. State and non-State, regional and subregional, and intergovernmental and interparliamentary actors are the key players in the developmental agenda for a progressive sustainable vision. To guarantee multi-stakeholder ownership, the SDGs must be localized. Governments are helped by UN Women, UN Regional Commissions, and other organizations to incorporate the global sustainable development agenda into their own national development plans (UN ESCAP, 2020). However, the national agenda will have its own limitations, which can be enhanced by regional cooperations. For instance, the European Union (EU) plays on the front foot to address the rising issues and abnormalities in the environment such as increase in global warming and changing climatic conditions that hamper the achievement of SDGs. Several regional collaborations like the 'African Union (AU), and the Association of South East Asian Nations (ASEAN)' also follow a similar pattern to address the ecological issues and enhance sustainability.

5.2.2.1 European Green Deal

Climate neutrality, reducing the GHG emissions to the maximum possible limit and compensating for the unsuccessful attempts, has been the motto of almost all the nations at the verge of climate change related complexities. The EU designed the EGD as a strategy to accomplish climate neutrality by 2050. It addresses a broad range of topics, such as 'transportation, energy and development, agriculture, industry, infrastructure, environment and biodiversity, and research and innovation (Koundauri et al., 2024). The SDGs in relation to clean energy (SDG7), climate action (SDG13), and ecofriendly production and consumption (SDG12) are significantly aligned with EGD policies'. Based on EGD, the DCE strategy has a 39.18% alignment with SDG12 (Responsible Consumption and Production), which makes sense by emphasizing the strategy of waste reduction and resource efficiency (Koundouri et al., 2023). Integrating energy systems for combating climate change is seen in the significant convergence between "EU Energy Integration Strategy" and "EU Hydrogen Strategy" with the objectives of a responsible climate action (SDGs 12 & 13). SDG17 (Partnerships for the Goals) and the Sustainable Growth Strategy 2021 are significantly aligned, indicating a focus on collaboration in promoting sustainable growth (Soergel et al., 2021).

5.2.2.2 African Union Agenda

The AU is putting Agenda 2063 into action at the continental level. This agenda is being implemented in large part by regional intergovernmental and interparliamentary bodies like the 'Southern African Development Community (SADC), Common

Market for Eastern and Southern Africa (COMESA), Intergovernmental Authority on Development (IGAD), and East Africa Community (EAC)' (Women, 2017). Renewable Energy Certificates (RECs) are actively used in AU to ensure green energy targets. Because of their broad-based cross-sectoral mandate, universal coverage, convening authority, intergovernmental nature and strength, and expertise in generating regional consensus on significant international agreements, RECs have a unique role to play in integrating the SDGs and Agenda 2063 (Ademola, 2022).

5.2.2.3 ASEAN Community Vision

ASEAN nations have pledged to execute two concurrent yet interconnected initiatives: "the United Nations 2030 Agenda for Sustainable Development (2030 Agenda) and the 'ASEAN Community Vision 2025' (ASEAN Vision 2025). As part of their first efforts to raise the living standards of the people in the region, ASEAN member-States emphasized the complementary nature of these two agendas" (UN ESCAP, 2017). One of the main points of emphasis in the "ASEAN Socio-Cultural Community Blueprint" of 2025 is the prerequisite for "promoting and ensuring balanced social development and sustainable environment that meet the peoples' needs at all times," which aims at "vibrant, sustainable and highly integrated economies, enhanced ASEAN Connectivity as well as strengthened efforts in narrowing the development gap" (Collins, 2007). Besides, it enlists four priorities of the Blueprint such as "the conservation and sustainable management of biodiversity and natural resources, environmentally sustainable cities, sustainable climate, and sustainable consumption and production." The UN's 2030 agenda furnishes a global framework for the ASEAN to enhance its community-building approach in harmony with the objectives of sustainability paradigm. Additionally, each of the aforementioned four priorities is a manifestation of the SDGs 15, 11, 12, and 13 respectively. The UN and ASEAN have both underlined how integrated approaches to sustainable development are required to address the 'environmental, social, and economic aspects' (Taichang et al., 2022).

5.3 Digital Circular Economy: A Catalyst for Sustainability

The shift to a DCE—"an economic system in which resource use is minimized (refuse and reduce) and material value is maximized (e.g., thorough reuse, repair, refurbishment, remanufacturing, recycling, and energy recovery)"—is supported by digitalization in both scientific literature and public discourse (Piscicelli, 2023; Reike et al., 2018). Achieving net zero and the SDGs will be of utmost importance to businesses and organizations (Bhushan & Singh, 2024). A number of techniques are being created to include and encourage SDGs and net zero-related practices in businesses; one such strategy is the DCE, which is gaining traction because of its documented influence on SDG and net zero components (Singh et al., 2021; Govindan, 2023). In any economy, the societal hurdles associated with DCE ideology frequently prevent it from being adopted and scaled by firms. The move toward improved product lifecycle management is one of the DCE's most important developments. This includes minimizing waste and the demand for new resources by designing products with lifetime, repairability, and recycling in mind. Due to a combination of consumer demand for sustainable products and legal obligations, businesses are implementing these practices at an increasing rate (Roleders et al., 2022). The 'Organization for Economic Cooperation and Development (OECD)' highlights ignorance, unfamiliarity, incompetence, and inexperience among the workforce as barriers for incorporating and implementing the strategies and initiatives of DCE. Thus, the main objective of the transnational and interdisciplinary partnership among organizations should be to establish a center of expertise on corporate ethics that will aid in the adoption of DCE universally by instructing businesses (of which 95% are SMEs), policy makers, researchers, and higher education institutions (Singh et al., 2022, 2023).

5.4 Global Governance Towards Digital Circular Economy

The Global concerns Report 2024 from the World Economic Forum predicts that environmental concerns will remain prominent in the upcoming decades. In terms of long-term severity, the top four global risks are extreme weather occurrences, abrupt changes to Earth systems, loss of biodiversity and ecosystem collapse, and scarcity of natural resources. The tenth-ranked major risk is pollution. By putting its principles into practice, the DCE has the ability to help reduce these dangers (Panwar & Niesten, 2020). Several regions and nations adopted and regulated DCE through various rules, regulations, and different strategic approaches.

5.4.1 European Union (EU)

In 2022, the EU unveiled the DCE Action Plan, which advanced the shift towards DCE as an effective tool for its 2050 aspirations for climate neutrality. Increasing the use of sustainable products, educating customers about the green shift, reviewing the building product legislation, and developing a plan for sustainable textiles are some of its recommendations (Almeida et al., 2023). The European Council (EC) established new eco-design guidelines for sustainable products on 27 May 2024. The new regulations seek to establish standards of environmental sustainability for nearly all products sold in the EU. It attempts to create a digital product passport that will not only forbid the destruction of some unsold consumer products but also provide information about a product's environmental sustainability (EC, 2024). A provisional agreement was made by the EU on a more robust right of repair for customers (PRODlaw.eu, 2024; Repp et al., 2021).

5.4.2 China

DCE principles were first included into Chinese policies in 2002, and since then, they have progressively become key components of the country's economic strategy. Examining 'China's Five-Year Development Plan (FYDP)' and the National Congress agenda enables researchers to keep track of the evolution of DCE policies (Koty, 2021). It also enables the scholars to understand the stringent institutional structure and operational framework that the implementation of the DCE policies adheres to, which produces tangible actions and quantifiable results. The development of China's DCE is still being hampered by problems like poor coordination, unequal development, unsustainable success, and a dearth of basic research, despite the efforts and results being worth embracing (Imperatives, 1987). Many efforts are made to find solutions for addressing the aforementioned problems. The 'Ellen MacArthur Foundation' (2020) has released a report stating that there are substantial opportunities for them in industrial or business domains. This assertion is supported by a set of recommendations for development pertaining to the built 'environment, mobility, nutrition, textiles, and electronics' (Chen, 2023; Feng & Lam, 2021). In China for the digital DCE to flourish, 'The Made in China 2025' strategy underlines the creation and implementation of smart and green manufacturing systems. These systems have the potential to remove technological barriers in the manufacturing sector that limit industrial production, energy use, and waste emissions while also enabling cities to run more sustainably (Zenglein & Holzmann, 2019). Furthermore, the Made in China 2025 program sets forth standards for boosting technical innovation to enhance environmental protection and energy efficiency, increasing-rate of green energy use, modernizing conventional sectors, and fortifying environmental oversight. The Made in China 2025 pilot program aims to steadily raise the percentage of low-carbon and green energy utilization while lowering energy consumption (Agarwal & Choudhary, 2021).

5.4.3 United States of America

Almost '80 million tons of packaging waste are annually generated in the USA' and only less than half of this huge waste is recycled. In order to solve the issues such as environmental degradation and change in climatic conditions related to industrial innovation, Extended Producer Responsibility (EPR) is a viable option and may be extremely important (Tumu et al., 2023). The EPR is defined as "an environmental policy approach in which a producer's responsibility for a product is extended to the post-consumer stage of a product's life cycle" (Diggle et al., 2023). This definition identifies the following three primary advantages of EPR. Firstly, the local institutions bear "the burden of end-of-life" management; producers bear the burden (the "polluter pays" principle). Secondly, there is an increase in recycling and material recovery rates. Lastly, the manufacturers are encouraged to use more environment friendly product designs (design for environment, DfE) (Mayanti & Helo, 2024). Businesses are held responsible for the whole lifecycle of their products under rules known as EPR. Its objectives are to aid in the DCE and lessen waste generated by packaging (Joltreau, 2022).

5.5 Synergy Between Sustainable Innovation and Digital Circular Economy

At the present time where there is a dire need for enhancement of sustainability, it is becoming more crucial to comprehend how the DCE affects the SDGs as it rises in importance within business, policy, and research agendas. The techniques of "reduce" and "repair, refurbish and remanufacture" hold significant promise in relation to SDGs. It showed potential in furthering the campaign against 'climate change (Goal 13)'. Additionally, the "reduce" technique has a significant impact on 'clean, cheap energy (Goal 7)'. "This can be effectuated by reducing the amount of energy consumption and resources exploitation, including the non-renewables. Moreover, industry, innovation, and infrastructure can all be supported by reducing the complexity of a product's design and offering cascading opportunities, particularly in infrastructure (Goal 9)" (Ortiz-de-Montellano et al., 2023). 'SDG 12, along with resource efficiency, are essential components of sustainable development and they constitute the foundation of DCE. Enhancing the utilization of resources in the processes of production and consumption while lowering externalities in the input to output process is known as resource efficiency (Polasky et al., 2019). SDG 12 is the use of resources and services that enables people to meet their basic needs and improve their life's quality with the prerequisite to minimize the use of natural resources and toxic materials and reduce the emissions of waste and pollutants for not jeopardizing the needs of future generations'. SDG 12 on "Ensuring Sustainable Consumption and Production patterns" is one of the key goals that is especially pertinent to a DCE since it focuses on recycling waste materials through closed-loop processes that minimize negative externalities (Robertson, 2005). DCE requires achieving objectives like SDG 8.4 that targets to "improve progressively, through 2030, global resource efficiency in consumption and production and endeavor to decouple economic growth from environmental degradation". Using 'SDG 8's target 4 and SDG 12 as general stand-ins for a DCE', it is possible to identify a network of pertinent trade-offs and synergies with the majority of SDGs and their corresponding targets (Pearce et al., 2013).

5.6 Role of Distinct Bodies in SDG Alignment with the Digital Circular Economy

No single organization is capable of addressing today's global development issues, necessitating collaborative efforts. The ability of governments, private sector, and international institutions to collaborate effectively will determine whether or not the world can handle some of the most urgent development concerns. This requires national initiatives and regional cooperations. A robust institutional framework balancing the three essentialities of sustainable development is a vital precondition for such efforts. This framework must also address present and future challenges and fill in any gaps in the agenda's implementation (Iyer et al., 2017).

5.6.1 *International Organizations*

Balancing of economic growth, environmental protection, and societal welfare are indispensable for the global transition towards sustainability. "Within such a strategic balancing, international organizations play crucial functions that range from data collection to policy advancement, from mediating disputes between nations to encouraging collective action" (Cheng, 2014). More broadly, these actions also translate into intentionally tackling the many issues arising from a more multipolar and integrated international order. The OECD takes the lead in implementing the 2030 Agenda. The OECD has been using its resources as a key organization for the promotion of sustainable development, to which it has a strong institutional commitment, particularly to its own distinctive membership and wide range of policy expertise (Silva, 2021). Further, the 2030 Agenda for SDGs, which establishes goals to be met by 2030 in sectors including poverty alleviation, health, education, and the environment, rely heavily on the 'World Trade Organization (WTO) (Sampson, 2021). From its inception in 1948, the World Health Organization (WHO)' has prioritized the health and basic needs of all people, especially the marginalized sections of the society. The nine principles that form the foundation of the WHO Constitution remain just as pertinent to the sustainable development objective nearly seven decades later. By adopting the 'Alma-Ata Declaration in 1978, the Global Strategy for Health for All in 1981, and the findings of the Commission on Social Determinants of Health in 2008, WHO and its member-States have continually reaffirmed their commitment to improving the health of all peoples over the intervening years' (Rifkin, 2018). In addition to directly supporting 'SDG 1 (Ending poverty in all its forms everywhere), the World Bank Group's (WBG) goal of ending extreme poverty—reducing the percentage of the world's population living in extreme poverty to 3% by 2030—also contributes to the 2030 Agenda' in a number of complementary and mutually reinforcing ways (Kanbur et al., 2018).

5.6.2 Private Sector

Private sector plays a more significant part in the mechanism of economic growth and employment. It creates tax revenues, offers goods and services, finances essential social and economic infrastructure, and comes up with innovative and different ways to address development issues. It is crucial in the fight against climate change as well (Frey & Sabbatino, 2018). 'The private sector contributes to development going beyond the conventional strategy of fostering economic growth to include the creation of more intricate and all-encompassing public–private partnerships. The importance of the private sector is not just restricted to investment and financing' (Garcia, 1997). Its other potential contributions include the ability to introduce highly influential innovations, technical information, technological inputs, and prudent maintenance of human resources. The private sector contributes significantly to development by fostering opportunities for value creation and sustainable economic growth (Khan et al., 2022). Furthermore, it is explicit that the initiatives for sustainable development will enhance the marketplace and business climate. Funds from the public and commercial sectors totaling trillions of dollars will be transferred to the achievement of SDGs (Rashad & Shah, 2021).

5.6.3 Non-governmental Organizations

NGOs, as acknowledged by the 2030 Agenda, are essential to achieving the SDGs. NGOs can provide substantial contributions to stimulate the pro-people initiatives of government and the actualization of SDGs. These groups are committed to bringing about positive social and environmental change and run without interference from governments. Since they have a strong foundation in the community, NGOs are better able to comprehend the unique demands and issues that marginalized and vulnerable groups experience. Their ability to operate at the grassroots level allows them to carry out focused, situation-specific initiatives that directly support the SDGs (Nikkhah & Redzuan, 2010; Harangozó & Zilahy, 2015).

5.6.4 Public–Private Partnerships

The 'World Bank defines Public–Private Partnerships (PPPs)' as "a long-term contract between a private party and a government institution, for providing a public asset or service, in which the private party carries major risk and management responsibilities, and remuneration is tied to performance and meets performance standards" (The World Bank, 2024). The 'United Nations Economic Commission for Europe (UNECE)' offers a broader definition to PPPs as "the mechanism for financing, designing, implementing and operating public sector facilities and services" (Ning et al., 2023). In addition, the UNECE describes PPPs as the "creative methods used by the public sector to contract with the private sector, who bring their capital and their ability to deliver projects on time and within budget, while the public sector retains the responsibility to provide these services to the public in a way that benefits the public and delivers economic development and an improvement in the quality of life" (Singh et al., 2021). PPPs are key components of social inclusivity, environmental sustainability, and economic growth. A fresh emphasis on accelerating the advancement of SDGs, particularly the digital and DCE, is imperative given the current circumstances and swift escalation of fiscal spending (Rejeb et al., 2023).

5.7 Integrating Digital Circular Economy for SDGs: Need for Policy Framework

The DCE's fundamentals have been included into the plans of numerous nations. For instance, in "China, India, and the Republic of Korea, the DCE is pushed as a top-down national political goal; in contrast, in other regions and nations, like the USA, Japan, and the EU, it is a vehicle for creating bottom-up waste management and environmental laws" (McDowell et al., 2017). The implementation of the DCE system necessitates the cooperation of various stakeholders, including businesses, academic institutions, plant operators, local government agencies, NGOs, and government departments, in addition to government support through laws and policies (Ghosh & Eduljee, 2023). The technologies that are currently on hand can assist in increasing efficiency, managing energy consumption more effectively, and promoting low-carbon economic growth and job creation. However, a number of obstacles stand in the way of the effective deployment of these technologies (Wang et al., 2021). These include non-technical ones like lack of policy support, perceived and actual project risk, lack of funding, lack of low-cost capital availability, lengthy permitting processes, community backlash, rebound effects, and supply chain difficulties. The realization of sustainable development and the 2030 agenda requires the support of intermediaries, policymakers, and innovative designers due to its complexity.

5.7.1 Policy Instruments Supporting Sustainability Through Circular Economy

Policymakers, producers, and consumers face significant issues with regard to sustainability and DCE. DCE methods stand for reducing waste generation and thereby they can become effective tools to mitigate the harms of climate change. As such, they place less pressure on the environment. Furthermore, public policies that promote DCE are created at the lowest levels of public administration using various policy instruments in accordance with the principle of subsidiarity (Arauzo-Carod et al., 2022). This empowers regional governments with the opportunities "to create, carry out, and oversee these policies."

5.7.1.1 Carbon Pricing

One important instrument for combating climate change and promoting climate action is carbon pricing (SDG 13) (Sen & Sahoo, 2024). 'The most successful and economical tools for encouraging investments in clean technology are carbon pricing laws, such as carbon taxes and emissions trading systems' (Parry, n.d.). The main mitigation and adaptation approach for sustainable development is carbon price in the form of climate action (SDG 13). Carbon taxes are the most essential policy tool for promoting transformation pathways. Other policy tools include energy efficiency, renewable energy subsidies, urban green infrastructure, and more. Carbon taxes are perhaps the most important way to achieve sustainable growth (Baranzini et al., 2017). According to the Polluter Pays Principle, the carbon pricing can make significant alterations in the relative costs of all commodities and services. "Thus, businesses, consumers, and investors take into account not only their own costs and advantages while making decisions that result in GHG emissions, but also estimates the social costs related to both direct and indirect emissions generated in every stage of product life, from the exploitation of resources to the disposal of the product. When all producers and consumers base their decisions on prices adjusted for the climate externality, the entire economy becomes less carbon intensive" (Kohlin et al., 2021).

5.7.1.2 Green Finance

Green finance is one of the most effective policy instruments that encompasses 'all sustainable development criteria established by the UN SDGs' and promotes transparency and long-term thinking of investments flowing into environmental objectives (Wang et al., 2022). A key tool in reaching carbon neutrality and greening 'sustainable economic growth is green finance, which is centered on eco-friendly investments and activities' (Bhatnagar et al., 2022; Fu et al., 2023). Climate finance initiatives develop through the analysis of various advantages of carbon–neutral processes in green finance and the active contributions of organizations like GCF. By analyzing and disclosing climate risks and political uncertainty, financial institutions and industry can foresee and reduce possible losses associated with environmental variables. This strategy helps in reducing the negative impact on sustainability (Volz, 2018). Green finance operates through its various agents like green bonds and green insurance for achieving the SDGs. Growing public awareness is crucial for protecting the environment and increased funding for green initiatives have created a plethora of opportunities in the field of green finance (Clark et al., 2018). The goal of climate action is a cross-cutting theme that supports several SDGs, including SDG 13, but green finance can designate other categories for the application of tools like green bonds and insurance for achieving goals, such as 'Socioeconomic Advancement and Empowerment (SDG 1), Terrestrial and Aquatic Biodiversity Conservation (SDG 14 and SDG 15), Affordable Basic Infrastructure (SDG 3), or Climate Change Adaptation (SDG 13)' (Barua, 2020).

5.8 Conclusion and Future Scope

The present economy's interfaces still operate in an open linear system discarding millions of tons of waste into the nature. This confirms that one of the main reasons for environmental deterioration and unsustainable production strategies is industrialization. But it is crucial to note that in the context of sustainable development, the DCE has recently drawn more attention as a tool for an integrative and multidisciplinary approach, offering itself as a response to some of the most pressing global challenges (Rosa et al., 2023). Over the next 40 years, it is anticipated that material consumption would more than double, with notable variations among geographies and material types. The demand for materials, particularly metals and non-metallic minerals, is predicted to intensify throughout the developing world, whereas the OECD and BRICS countries are predicted to reduce. Global demand for these two commodities is expected to nearly quadruple. Environmental burdens such as GHG emissions, local pollution, and trash generation will increase in tandem with the rising usage of materials. The shift to renewable energy and materials is the foundation of a DCE. It is a durable and powerful system that can equip the industry, people, and the environment with sustainable benefits by disconnecting the link between economic activity and the exploitation of limited resources. In the absence of a progressive DCE, the globe cannot ensure sustainable future for its inhabitants. A close analysis of the strategies of SDGs affirms the 2030 agenda's close connection and aspiration for DCE.

References

Ademola, A. A. (2022). Real deal or an unworkable framework? Agenda 2063, SDG goals and the challenges of development in 21st century Africa. Office of the Special Adviser on Africa.

Agarwala, N., & Chaudhary, R. D. (2021). 'Made in China 2025': Poised for success? *India Quarterly, 77*(3), 424–461. https://doi.org/10.1177/09749284211027250

Almeida, D. V., Kolinjivadi, V., Ferrando, T., Roy, B., Herrera, H., Gonçalves, M. V., & Van Hecken, G. (2023). The "greening" of the empire: The EGD as the EU first agenda. *Political Geography, 105*, 102925. https://doi.org/10.1016/j.polgeo.2023.102925

Arauzo-Carod, J. M., Kostakis, I. & Tsagarakis, K. P. (2022). Policies for supporting the regional circular economy and sustainability. *The Annals of Regional Science, 68*, 255–262. https://doi.org/10.1007/s00168-022-01124-y.

Baranzini, A., Van den Bergh, J. C., Carattini, S., Howarth, R. B., Padilla, E., & Roca, J. (2017). Carbon pricing in climate policy: Seven reasons, complementary instruments, and political economy considerations. *Wiley Interdisciplinary Reviews: Climate Change, 8*(4), e462. https://doi.org/10.1002/wcc.462

Barua, S. (2020). Financing sustainable development goals: A review of challenges and mitigation strategies. *Business Strategy & Development, 3*(3), 277–293. https://doi.org/10.1002/bsd2.94

Bhatnagar, M., Taneja, S., & Özen, E. (2022). A wave of green start-ups in India—The study of green finance as a support system for sustainable entrepreneurship. *Green Finance, 4*(2), 253–273.

Bhushan, M., & Singh, R. (2024). Talent management facilitates net-zero transition through employee green behavior. In *Net zero economy, corporate social responsibility and sustainable value creation: Exploring strategies, drivers, and challenges* (pp. 117–129). Springer Nature Switzerland. https://doi.org/10.1007/978-3-031-55779-8_7.

Bogers, M., Biermann, F., Kalfagianni, A., & Kim, R. E. (2022). Sustainable development goals fail to advance policy integration: A large-n text analysis of 159 international organizations. *Environmental Science & Policy, 138*, 134–145. https://doi.org/10.1016/j.envsci.2022.10.002

Caradonna, J. L. (2022). *Sustainability: A history*. Oxford University Press.

Carugi, C., & Bryant, H. (2021). From evaluation of joint programmes to joint evaluation of SDGs-ready interventions: Lessons from the Joint GEF-UNDP Evaluation of the Small Grants Programme. In *Evaluating environment in international development* (pp. 180–196). Routledge. https://doi.org/10.1177/1098214019865936

CB, P., Ravikumar, K., & Das, V. (2021). Significance of Climate finance and Global Environment Facility in the climate changes and its implications at SAARC countries.

Chen, R. (2023). *China's circular economy policies: Review and reflection*. Circular Innovation Lab.

Cheng, D. S. (2014). Contributions of international organizations towards globalization in the 21st century. *Journal of Global Business Management, 10*(2), 94. https://doi.org/10.36475/9.4.4.

Clark, R., Reed, J., & Sunderland, T. (2018). Bridging funding gaps for climate and sustainable development: Pitfalls, progress and potential of private finance. *Land Use Policy, 71*, 335–346. https://doi.org/10.1016/j.landusepol.2017.12.013

Collins, A. (2007). Forming a security community: Lessons from ASEAN. *International Relations of the Asia-Pacific, 7*(2), 203–225. https://doi.org/10.1093/irap/lcl007

Diggle, A., Walker, T. R., & Adams, M. (2023). Examining potential business impacts from the implementation of an extended producer responsibility program for printed paper and packaging waste in Nova Scotia, Canada. *Circular Economy, 2*(2), 100039. https://doi.org/10.1016/j.cec.2023.100039

Dzebo, A., Brandi, C., Janetschek, H., Savvidou, G., Adams, K., Chan, S., & Lambert, C. (2017). *Exploring connections between the Paris agreement and the 2030 agenda for sustainable development*. Stockholm Environment Institute.

Feng, K., & Lam, C. Y. (2021). An overview of the circular economy in China: How the current challenges shape the plans for the future. *The Chinese Economy, 54*(5), 355–371.

Frey, M., & Sabbatino, A. (2018). The role of the private sector in global sustainable development: The UN 2030 agenda. In *Corporate responsibility and digital communities: An international perspective towards sustainability* (pp. 187–204).

Fu, C., Lu, L., & Pirabi, M. (2023). Advancing green finance: A review of sustainable development. *DESD, 1*, 20. https://doi.org/10.1007/s44265-023-00020-3

Garcia, L. E. (1997). Sustainable development and the private sector: A financial institution perspective. In *Bridges to sustainability: Business and government working together for a better environment* (pp. 130–141). https://doi.org/10.1007/978-3-030-02006-4_381-1

Georgieva, K., Gaspar, V., & Pazarbasioglu, C. (2022). *Poor and vulnerable countries need support to adapt to climate change*. International Monetary Fund. https://www.imf.org/en/Blogs/Articles/2022/03/23/blog032322-poor-andvulnerable-countris-need-support-to-adapt-to-climate-change

Ghosh, S. K., & Eduljee, G. (Eds.). (2023). *The circular economy: Meeting sustainable development goals* (Vol. 51). Royal Society of Chemistry. https://doi.org/10.1039/9781837671984.

Giller, K. E., Drupady, I. M., Fontana, L. B., & Oldekop, J. A. (2018). Editorial overview: The SDGs–aspirations or inspirations for global sustainability. *Current Opinion in Environmental Sustainability, 34*, 1–2. https://doi.org/10.1016/j.cosust.2019.02.002

Govindan, K. (2023). How digitization transforms the traditional circular economy to a smart circular economy for achieving SDGs and net zero. *Transportation Research Part E: Logistics and Transportation Review, 177*, 103147. https://doi.org/10.1016/j.tre.2023.103147

Harangozó, G., & Zilahy, G. (2015). Cooperation between business and non-governmental organizations to promote sustainable development. *Journal of Cleaner Production, 89*, 18–31. https://doi.org/10.1016/j.jclepro.2014.10.092

Hariyani, D., Hariyani, P., Mishra, S., & Sharma, M. K. (2024). Leveraging digital technologies for advancing circular economy practices and enhancing life cycle analysis: A systematic literature review. *Waste Management Bulletin*. https://doi.org/10.1016/j.wmb.2024.06.007

Huang, M. T., & Zhai, P. M. (2021). Achieving Paris Agreement temperature goals requires carbon neutrality by mid-century with far-reaching transitions in the whole society. *Advances in Climate Change Research, 12*(2), 281–286. https://doi.org/10.1016/j.accre.2021.03.004

Imperatives, S. (1987). *Report of the world commission on environment and development: Our common future* (42,427).

Ishii, N. (n.d.). *The green environment facility and SDGs*. GEF.

Iyer, D., Rubian, R., & Pizarro, G. (2017). *Institutional and coordination mechanisms: Guidance note on facilitating integration and coherence for SDG implementation*. UNDP.

Janetschek, H., Brandi, C., Dzebo, A., & Hackmann, B. (2021). The 2030 Agenda and the Paris Agreement: voluntary contributions towards thematic policy coherence. In *Making climate action more effective* (pp. 26–38). Routledge. https://doi.org/10.1080/14693062.2019.1677549

Joltreau, E. (2022). Extended producer responsibility, packaging waste reduction and eco-design. *Environmental and Resource Economics, 83*(3), 527–578. https://doi.org/10.1007/s10640-022-00696-9

Kanbur, R., Patel, E., & Stiglitz, J. (2018). Sustainable development goals and measurement of economic and social progress. In *For good measure: Advancing research on well-being metrics beyond GDP* (p. 33).

Khan, S., Ali, S. S., & Singh, R. (2022). Determinants of remanufacturing adoption for circular economy: A causal relationship evaluation framework. *Applied System Innovation, 5*(4), 62. https://doi.org/10.3390/asi5040062

Köhlin, G., Dorband, I., Christoph, J., & Sterner, T. (2021). Recipe for a green recovery: Carbon taxes. *World Development, 115*, 246–257.

Koty, A. C. (2021). China's circular economy: Understanding the new five-year plan. *China Briefing, Dezan Shira & Associates, 16*.

Koundouri, P., Alamanos, A., Plataniotis, A., Stavridis, C., Perifanos, K., & Devves, S. (2024). Assessing the sustainability of the EGD and its interlinkages with the SDGs. *npj Climate Action, 3*(1), 23. https://doi.org/10.1038/s44168-024-00104-6

Koundouri, P., Theodossiou, N., Stavridis, C., Devves, S., & Plataniotis, A. (2023). The EGD and the 17 SDGs: Uncovering their connection with a ML-based approach. *Decision making using AI in energy and sustainability: Methods and models for policy and practice* (pp. 21–48). Springer International Publishing.

Mayanti, B., & Helo, P. (2024). Circular economy through waste reverse logistics under extended producer responsibility in Finland. *Waste Management & Research, 42*(1), 59–73. https://doi.org/10.1177/0734242X231168801

McDowall, W., Geng, Y., Huang, B., Barteková, E., Bleischwitz, R., Türkeli, S., Doménech, T., et al. (2017). Circular economy policies in China and Europe. *Journal of Industrial Ecology, 21*(3), 651–661. https://doi.org/10.1111/jiec.12597

Moreno, J., Van de Ven, D. J., Sampedro, J., Gambhir, A., Woods, J., & Gonzalez-Eguino, M. (2023). Assessing synergies and trade-offs of diverging Paris-compliant mitigation strategies with long-term SDG objectives. *Global Environmental Change, 78*, 102624. https://doi.org/10.1016/j.gloenvcha.2022.102624

Nikkhah, H. A., & Redzuan, M. R. B. (2010). The role of NGOs in promoting empowerment for sustainable community development. *Journal of Human Ecology, 30*(2), 85–92. https://doi.org/10.1080/09709274.2010.11906276

Ning, L., Abbasi, K. R., Hussain, K., Alvarado, R., & Ramzan, M. (2023). Analyzing the role of green innovation and public-private partnerships in achieving sustainable development goals: A novel policy framework. *Environmental Science and Pollution Research,* 1–17. https://doi.org/10.1007/s11356-023-26414-6

och Regeringskansliet, R. (2015). *The global goals and the 2030 agenda for sustainable development*. Government Offices of Sweden.

Ortiz-de-Montellano, C. G. S., Samani, P., & van der Meer, Y. (2023). How can the circular economy support the advancement of the sustainable development goals (SDGs)? A comprehensive analysis. *Sustainable Production and Consumption, 40*, 352–362. https://doi.org/10.1016/j.spc.2023.07.003

Panwar, R., & Niesten, E. (2020). Advancing a circular economy. *Business Strategy and the Environment, 29*(6), 2890–2892. https://doi.org/10.1002/bse.2602

Parry, I. W. (n.d.). What is the best policy instrument for reducing Co_2 emissions? In *Fiscal policy to mitigate climate change*. International Monetary Fund.

Pearce, D., Barbier, E., & Markandya, A. (2013). *Sustainable development: Economics and environment in the third world*. Routledge.

Polasky, S., Kling, C. L., Levin, S. A., Carpenter, S. R., Daily, G. C., Ehrlich, P. R., Lubchenco, J., et al. (2019). Role of economics in analyzing the environment and sustainable development. *Proceedings of the National Academy of Sciences, 116*(12), 5233–5238. https://doi.org/10.1073/pnas.1901616116

Piscicelli, L. (2023). The sustainability impact of a digital circular economy. *Current Opinion in Environmental Sustainability, 61*, 101251. https://doi.org/10.1016/j.cosust.2022.101251

PRODlaw.eu. (2024). The right to repair explained to consumers and manufacturers. *PRODlae.eu.*

Rashed, A. H., & Shah, A. (2021). The role of the private sector in the implementation of sustainable development goals. *Environment, Development and Sustainability, 23*(3), 2931–2948. https://doi.org/10.1007/s10668-020-00718-w

Reike, D., Vermeulen, W. J., & Witjes, S. (2018). The circular economy: New or refurbished as CE 3.0? exploring controversies in the conceptualization of the circular economy through a focus on history and resource value retention options. *Resources, Conservation and Recycling, 135*, 246–264. https://doi.org/10.1016/j.resconrec.2017.08.027

Rejeb, A., Rejeb, K., Appolloni, A., Treiblmaier, H., & Iranmanesh, M. (2023). Circular economy research in the COVID-19 era: A review and the road ahead. *Circular Economy and Sustainability, 3*(4), 2127–2157. https://doi.org/10.1007/s43615-023-00265-2

Repp, L., Hekkert, M., & Kirchherr, J. (2021). Circular economy-induced global employment shifts in apparel value chains: Job reduction in apparel production activities, job growth in reuse and recycling activities. *Resources, Conservation and Recycling, 171*, 105621. https://doi.org/10.1016/j.resconrec.2021.105621

Rifkin, S. B. (2018). Alma Ata after 40 years: Primary health care and health for all—from consensus to complexity. *BMJ Global Health, 3*(Suppl 3), e001188. https://doi.org/10.1136/bmjgh-2018-001188

Robertson, J. (2005). *New economics of sustainable development.*

Roleders, V., Oriekhova, T., & Sysoieva, I. (2022). Trends in a global circular economy. *Management Theory and Studies for Rural Business and Infrastructure Development, 44*(2), 176–184. https://doi.org/10.15544/mts.2022.18

Rosa, L. A. B. D., Cohen, M., Campos, W. Y. Y. Z., Ávila, L. V., & Rodrigues, M. C. M. (2023). Circular economy and sustainable development goals: Main research trends. *Revista De Administração Da UFSM, 16*(1), e9. https://doi.org/10.5902/1983465971448

Sampson, G. P. (2021). The WTO, trade agreements, and sustainable trade. *Asia policy, 16*(4), 7–21.

Sen, S., & Sahoo, P. (2024). Carbon pricing for sustainable transition in India. *World Development Perspectives, 34*, 100586. https://doi.org/10.1016/j.wdp.2024.100586

Singh, R., Dsilva, J., Centobelli, P., & Tripathi, V. (2021). A review on sustainable HRM: A study evaluating sustainability for organisational development. *The Empirical Economics Letters, 20*(2), 1–10.

Singh, R., Khan, S., & Centobelli, P. (2022). Investigating the Interplay between social performance and organisational factors supporting circular economy practices. *Sustainability, 14*(24), 16781. https://doi.org/10.3390/su142416781

Singh, R., Khan, S., & Dsilva, J. (2023). A framework for assessment of critical factors for circular economy practice implementation. *Journal of Modelling in Management, 18*(5), 1476–1497. https://doi.org/10.1108/JM2-06-2021-0145

Silva, A. L. (2021). Innovation in development cooperation: Emerging trajectories and implications for inclusive sustainable development in the 21st century. *Innovation and Development, 11*(1), 151–171. https://doi.org/10.1080/2157930X.2020.1807100

Soergel, B., Kriegler, E., Weindl, I., Rauner, S., Dirnaichner, A., Ruhe, C., Popp, A., et al. (2021). A sustainable development pathway for climate action within the UN 2030 agenda. *Nature Climate Change, 11*(8), 656–664. https://doi.org/10.1038/s41558-021-01098-3

Schalatek, L., Nakhooda, S., & Watson, C. (2012). The green climate fund. In *Overseas development institute and Heinrich Böll Stiftung North America* (pp. 1–8).

Taichang, Z., Yunle, W., & Xiaochen, X. (2022). Enhancing China-Asean regional connectivity: Synergising the master plan on Asean connectivity (Mpac) 2025 and the belt and road initiative (Bri). *Вестник науки, 3*(48), 18–40.

The World Bank. (2024). PPPs. The World Bank.

Tumu, K., Vorst, K., & Curtzwiler, G. (2023). Global plastic waste recycling and extended producer responsibility laws. *Journal of Environmental Management, 348*, 119242. https://doi.org/10.1016/j.jenvman.2023.119242

Umasankar, M., Boopathy, S., Padmavathy, S, Leena, N. F., & Rubee, S. (2022). Disconnect to reconnect: Employee wellbeing through digital detoxing. *Journal of Positive School Psychology, 6*(2), 4663–4673.

United Nations (UN). (n.d.). *SDGs*. UN.

UN Climate Change. (n.d.). *Key aspects of the Paris agreement*. UN Climate Change.

UN ESCAP. (2017). *Complementarities between the ASEAN community vision 2015 and the UN 2030 Agenda for sustainable development: a framework for action*. ESCAP.

UN ESCAP. (2020). *Asia-Pacific regional guidelines on voluntary local reviews: Reviewing local progress to accelerate action for the sustainable development goals*. ESCAP.

United Nations Development Program (UNDP). (n.d.). *SDGs: Background on the goals*. UNDP.

UNDP. (1994). *Human development report*. UNDP.

Uitto, J. I., & Naidoo, I. A. (2019). Local, national, global: Evaluative evidence of scaling up the SDGs. In *American association of geographers annual meeting* (vol. 3).

Volz, U. (2018). Fostering green finance for sustainable development in Asia. In *Routledge handbook of banking and finance in Asia* (pp. 488–504). Routledge. https://doi.org/10.2139/ssrn.3198680

Wang, F., Harindintwali, J. D., Yuan, Z., Wang, M., Wang, F., Li, S., Chen, J. M., et al. (2021). Technologies and perspectives for achieving carbon neutrality. *The Innovation, 2*(4). https://doi.org/10.1016/j.xinn.2021.100180

Wang, K. H., Zhao, Y. X., Jiang, C. F., & Li, Z. Z. (2022). Does green finance inspire sustainable development? Evidence from a global perspective. *Economic Analysis and Policy, 75*, 412–426. https://doi.org/10.1016/j.eap.2022.06.002

Women, U. N. (2017). The sustainable development goals (SDGs) and Africa's agenda 2063′. *Sustainable Development Goals & Africa's Agenda, 2063*, 39.

Zenglein, M. J., & Holzmann, A. (2019). Evolving made in China 2025. *MERICS Papers on China, 8*, 78.

Chapter 6
Fostering Sustainable Innovation Through Multistakeholder Collaboration in the Digital Circular Economy

Azamat Ali, Jitendra Singh, and Mikhail Tyurkin

Abstract In this chapter, we explore into the role of multistakeholder collaboration in propelling sustainable innovation for the digital circular economy. As the world transforms towards sustainable and resource-efficient systems, collaboration is key. The interplay between governments, businesses, academia, and civil society becomes crucial. Understanding what motivates the adoption of technologies like artificial intelligence (AI), blockchain, and the Internet of Things (IoT) is essential. These technologies can drive the implementation of circularity. It deals with how such collaborations can yield innovations, decrease consumption of resources, and help reduce waste by digital solutions to improve circular processes. A whole discussion pertaining to challenges like regulatory, technological, and funding challenges has been used to highlight how those challenges could be precluded. Collective action leading to new business models, sustainable patterns in the production and consumption pattern, and the efficient management of digital resources are discussed here. This chapter will examine how collaborative approaches can unlock the digital CE within broad long-term sustainability and economic resilience, by exploring through policy frameworks innovation ecosystems and stakeholder engagement.

Keywords Sustainable innovation · Digital circular economy · Multistakeholder collaboration · Resource efficiency · Digital technologies · Circular systems

---

A. Ali (✉) · J. Singh
School of Liberal and Creative Arts (Journalism and Mass Communication), Lovely Professional University, Phagwara, Punjab, India
e-mail: azamatcug@gmail.com

M. Tyurkin
School of Arts and Humanities, Department of Media, HSE University, St. Petersburg, Russia

R. Singh and V. Kumar (eds.), *Sustainable Innovations and Digital Circular Economy*,
https://doi.org/10.1007/978-981-96-1064-8_6

6.1 Introduction

Sustainable innovation is the generation of new or improved products, processes, or practices that respond to environmental, economic, and social challenges toward the needs of more-than-one generation. It involves embedding sustainability principles directly into innovation activities in a manner of delivery as reducing negative environmental impacts, enhancing resource efficiency, and promoting social equity (Bocken et al., 2016). Sustainable innovation differs from traditional innovation. It aims to improve economic or technological conditions while balancing multiple objectives. Unlike traditional approaches, it ensures that economic gains do not compromise environmental or social well-being.

The digital circular economy has been defined as the thoughtful amalgamation of the principles of a circular economy with digital technologies to make resource usage optimized and waste reduced. Unlike the linear economy's "take-make-dispose" model, the circular economy aims to reduce waste by focusing on resource efficiency, recycling, and extended product lifecycles achieved through repair, refurbishment, and remanufacturing (MacArthur, 2015). The digital aspect of this approach utilizes technologies like Internet of Things, artificial intelligence, and blockchain in tracing, managing, and optimizing resources throughout the entire products' lifecycle (Lacy & Rutqvist, 2015). These technologies improve the better implementation of circular practices by providing real-time monitoring tools and promoting predictive maintenance and material flows that are transparent and traceable (Geissdoerfer et al., 2017).

The joint collaboration of governments, businesses, academia, and civil society will be key in making progress in the digital circular economy. Government will act as a facilitator for setting standards and creating incentives for circular economy activities, hence shaping the regulatory and policy environment consonant with circular practices. They are now involved in facilitating the development of the required infrastructure as well as driving research and development in sustainable technologies.

On the other hand, businesses play a core role in the actualization of the circular economy through various circular business models like product-as-a-service, take-back schemes, and closed-loop supply chains (Bocken et al., 2016). Businesses will be able to establish benchmark business standards by investing in and creating new technologies that enhance circular processes and will enable industry leaders to spur companies to start using more circular practices. Academia contributes through high-quality research and theoretical underpinnings of the circular economy, new knowledge, and developments in technologies, as well as theory–practice bridging (Geissdoerfer et al., 2017).

It is through civil society, incorporating consumers, NGOs and community groups, which sustainability and circular practice will be accomplished. Civil society organizations raise the awareness, alter consumer behaviour and also ensure responsible business and government practices that help in environment-friendly and socially commendable impacts (Gibbs, 2021). Involving civil society may lead to widespread public engagement and support in circular economy initiatives.

Overcoming regulatory challenges, technological shortcomings, and financial constraints is crucial for the transition to a digital circular economy. Lacy and Rutqvist (2015) emphasized the importance of effective multistakeholder collaboration. This involves knowledge sharing, mobilizing resources, and coordinated problem-solving on collaborative platforms. Such efforts can accelerate the shift to sustainable and circular economic models. By leveraging the strengths of diverse groups, the benefits for the environment and society will be significant in the long term.

6.2 The Concept of Digital Circular Economy

The digital CE therefore has an advanced economic model that combines circular economy ideas with digital technologies to make this economy's sustainability better and more efficient in terms of resources. It breaks away from the traditional "take-make-dispose" of the linear economy to a more regenerative one that keeps resources in use for as long as possible and reduces waste. The general idea of the digital circular economy encompasses quite a few critical principles and objectives, with an aim to correct flaws that are present in linear approaches, while at the same time leveraging opportunities presented by digital technologies.

6.2.1 Key Principles and Objectives

Efficient use of resources, waste prevention, and extending product life are the core principles of the digital circular economy. This principle of resource efficiency implies maximizing material and energy inputs with no adverse environmental impact through the product cycle. It designates the products and systems with fewer in taken resources and waste generated, thereby lowering the total environmental footprint.

Another principle of the digital circular economy is waste reduction. This kind of digital circular economy no longer considers final wastes but instead treats it as a recovered resource used again (Murray et al., 2017). Products of a circular economy will be designed for easy disassembly and repair, and recycling is undertaken at the end of their life cycle. This approach closes the loops for product lifecycles by remanufacturing, refurbishment, and recycling strategies.

This forms one of the key objectives in the digital circular economy through practices that improve product durability and reparation (Bocken et al., 2016). Here, it deals with product design that is easy to maintain and repair while minimizing replacement requirements and, as a result, the overall consumption of resources.

6.2.2 Role of Digital Technologies in Circular Economy

Digital technologies play a critical role in enabling and advancing ideas about the digital circular economy. However, for instance, the IoT, AI, and blockchain, among others are playing a very influential role in improving efficient resource management, enhancing product tracking, and optimizing circular processes (Lacy & Rutqvist, 2015).

The Internet of Things is defined as the integration of connected devices through sensors and the sharing of data in real time with several types of communication. In the light of circular economy, IoT devices enable the online monitoring and management in real time of resources, products, and processes (Rusch et al., 2023). For example, this IoT sensor can determine the condition and usage of a product and, thus, provide useful data in predicting maintenance requirements, optimizing supply chains, and optimizing the use of resources. All of these pieces of information help a business in informed decisions regarding the lifespan of a product and resource recovery.

AI offers advanced analytics and decision-making support in the digital circular economy. It enhances capabilities by processing large volumes of data to find patterns, predict outcomes, and optimize processes. Lee suggests that AI is crucial for improving recycling processes and supply chain management in the circular economy. It can also lead to the development of new business models. AI helps design products that are easier to disassemble and recycle. It predicts the most efficient methods for material separation and processing, making the circular economy more effective.

This attribute would have major benefits in terms of traceability and transparency in a digital circular economy. Blockchain is a decentralized digital ledger in which transactions may be recorded across a network of multiple nodes in a safe and unalterable manner (Tapscott & Tapscott, 2016). In a circular economy context, blockchain can be used to track origin and movement of the material, such that those materials are handled responsibly throughout their life cycle. Such transparency will actually help to validate claims made about recycling and resource recovery, while creating a healthy supply chain with integrity, thus building stakeholder trust.

6.2.3 Benefits of Adopting Digital Circular Models

The benefits of adopting digital circular models are minimal environmental impact. They maximize the use of resources and avoid wastes with extended product lifecycles to minimize raw materials utilization and waste generation at source (Geissdoerfer et al., 2017). With this reduction in environmental impact, greenhouse gas emissions decline with a corresponding reduction in the energy usage and reduced levels of pollution. This helps to work towards the global fight against climate change and degradation of the environment.

Economic benefits are also significant. Digital circular models result in cost savings for the companies through increased resource efficiency and lowered waste management costs (Bocken et al., 2016). Product design for durability, and to be reparable, helps to reduce the number of replacements and thereby extends revenue generations by businesses. Circular models may also develop ancillary lines of revenue, for example by leasing of products, product take-back programs, and remanufacturing; these lines can stimulate business growth and innovation opportunities (Lacy & Rutqvist, 2015).

Promoting a circular economy offers various social benefits. It stimulates sustainable consumption patterns and creates employment in new sectors linked to the circular economy. The digital circular economy fosters a culture of sustainability and resource awareness. It encourages consumers to repair goods instead of discarding them (Murray et al., 2017). Industries in circular economies, such as recycling and remanufacturing, also contribute to job creation and economic growth.

6.3 Stakeholders in the Digital Circular Economy

The digital circular economy is an idea of wide applicability toward sustainability because it strongly relies on the involvement of all stakeholders: governments, businesses, academia, and civil society. This focus, importance, and scope shape, implement, and advance the principles of a digital circular economy. Their input is therefore essential for resource efficiency, waste reduction, and extended product lifecycles brought through the application of digital technologies.

6.3.1 Governments: Policy Making and Regulation

Governments need regulations since they enhance a digital circular economy by governing. The legal and regulatory framework the government establishes provides an enabling environment for circular practices. Comprehensive policies and regulations cause catalysts to facilitate the adoption of principles of circular economy at all industry and sectoral levels.

Supporting governments in the realization of regulations focusing on efficient use of resources, waste reduction, and recycling can help many developments in the circular economy. For example, extended producer responsibility, where manufacturers can be made to take full responsibility for the entire life cycle of the product, up to how it is disposed of at the end and recycled. Regulations encourage companies to design their products with a view to recycling and reusing in the pursuance of circular economy objectives.

Other ways through which the government can facilitate businesses embracing circular business models include financial incentives and subsidies. The other ways in which the government can influence the shift toward a circular economy are through grants and tax incentives concerning research and development of sustainable technologies or financing infrastructure projects favourable toward recycling and waste management. For example, the governments may advocate for the adoption of circular economy principles through public procurement policies to encourage market demand in the market for sustainable products and services (Kirchherr et al., 2017).

6.3.2 Businesses: Industry Practices and Innovation

Business is driving and innovative the digital circular economy. Majorly, it has a role to implement and work with circular principles on operations and develop new models of business in accordance with circular economy goals (Bocken et al., 2016). Today, companies increasingly apply the circular business models of the product-as-a-service, take-back scheme, and closed-loop supply chains in pursuance of the principles of waste reduction and resource efficiency (Lacy & Rutqvist, 2015).

Translating traditional linear processes into circular ones requires a change in product design and production methods. For instance, it becomes possible to have products whose modules are used. These modules can easily be repaired, upgraded, or recycled (MacArthur, 2015). The design leads to the extension of product lifespans while minimizing newly acquired raw materials. Besides, companies utilize IoT, AI, and blockchain technologies to improve resource tracking and management. It is considered the pre-requisites of circular economy practices.

More than just innovation in the product design itself, circular business practices now stretch into innovation for how a company can think of engaging with its customers and creating value through services that enable circularity. It incorporates such strategies as companies adopting the subscription model in which customers pay to use a product rather than fully owning it with the possibility of leading to higher sustainability consumption patterns (Bocken et al., 2016). Adoption of leading-edge practices will save businesses cost on environmental impact, reduces their cost of operation, and opens up new avenues for income generation.

6.3.3 Academia: Research and Development

Academia plays an important role in advancing the digital circular economy by mainly using its research and development entities. Innovations in universities and institutions lead to new knowledge input, develop new technologies, and provide the theoretical frameworks used in ending the linear economy cycle, such as that reviewed by Geissdoerfer et al. (2017). Contributions to new circular solutions and technologies are provided by researching material science, sustainable design, and systems thinking.

Academia also educates and trains the future professionals who are going to practice circular economy. The curriculum development and specific program of academic institutions equip students to understand and implement circular economy concepts in their next working life (Murray et al., 2017). Additionally, academia's findings can be utilized for policy formulation and business practices by developing evidence-based recommendations and insights as claims by Ghisellini et al. (2016).

The key collaborations between academia and industry translate research into practical applications. Industry partnership, joint research projects, and technology transfer initiatives really bridge the gap between academic research and actual implementation (Priyadarshini & Abhilash, 2020). These partnerships might bring about new technologies and business models, which will work well toward the 'digital circular economy goals.

6.3.4 Civil Society: Advocacy and Consumer Behaviour

Consumers, NGOs, as well as other civil society elements participate in the industry advocacy and the conduct of consumers to popularize the digital circular economy. Organizations in civil society spread awareness concerning the advantages of practices involving the circular economy and engage in advocacy for policies that are encouraging of sustainability and resource efficiency (Gibbs, 2021). Civil society often impacts the opinion of the public; therefore, they can generate demand for sustainable products and services.

One of the main critical success factors of the digital circular economy is consumer behaviour. As consumers grow in concern with environmental issues and the advantages brought on by circular practices, they are more inclined towards decisions that serve the sustainability goals (Kirchherr et al., 2017). Such examples include consumers demanding products that have been designed for durability, repairability, and recyclability. They are also supporting companies which display a seriousness of commitment to circular economy principles and sustainable practices.

The other role that civil society plays in accountability is making businesses and governments accountable in the proper use of the environmental resource and the manner in which their ventures impact the social consciousness (Thrassou et al., 2024). These organizations can cause paradigm shifts and make stakeholders respect the principles of the circular economy through advocacy and public persuasion through campaigns. Civil society promotes the culture of sustainability and motivates responsible consumption towards the general success of the digital circular economy.

6.4 The Role of Digital Technologies in Circular Economy

Such technology is transformative and allows the restoration of the CE by unlocking a transformation from less linear, traditional models of producing and consuming to more sustainable and circular ones. Among these, AI, blockchain technology, IoT, and data analytics combined with digital twins are some of the most important technologies driving this change process. These can create many paths to improving resource efficiency, optimizing supply chains, and developing new business models that permeate the principles of the circular economy.

6.4.1 Artificial Intelligence

Artificial intelligence, in general, is referred as AI. It has proven very imperative for improving the circular economy with smarter decisions and better optimizations of resources. Machine learning and predictive analytics are a few of the AI technologies that can identify patterns and insights in driving circular practices. For example, AI can be used to predict equipment failure and optimize maintenance schedules to extend the lifecycle of the assets while reducing waste (Rojek et al., 2023). Predictive maintenance thus not only minimizes the downtime but also reduces costs in operations and consumption of resources. All these aspects are in line with circular economy principles.

AI also enhances better recycling processes. With image recognition, sophisticated AI algorithms can enhance the efficiency of waste sorting systems to identify and separate different materials that are more efficient than the human eye for manual sorting purposes. It increases the purity of recyclable materials and contaminates less; this is important for closing the loop in recycling processes. For example, AI-based optimization may allow for a closed supply chain due to demand forecast and subsequent scheduling adjustment to minimize waste and accumulation of reserves (Turskis & Šniokienė, 2024).

6.4.2 Blockchain Technology

In this respect, blockchain technology provides a decentralized and transparent framework that will apply to the case of circular economy. The properties inherent in blockchain—immutability, transparency, and traceability—are integral to upgrade supply chain management and protect the integrity of circular practice (Singh et al., 2022). In the circular economy, blockchain traces from production to the end-of-life product stages, assuring proper recycling or reusing of materials.

For instance, blockchain allows product passports that include in detail the information that contains the material and parts of the product coupled with recycling instruction (Singh et al., 2022). This will allow consumers and companies to choose what their products do after use and recycle them. This blockchain technology helps create decentralized trading or swapping platforms for second-hand products where reuse instead of further demand for new products creates lesser use of products in areas such as industry.

6.4.3 *Internet of Things (IoT)*

The Internet of Things supports a circular economy by enabling seamless real-time tracking and resource management (Jum'a et al., 2024). IoT devices, such as sensors and smart meters, monitor product and infrastructure usage and condition. This information is vital for circular practices like optimizing energy efficiency and improving waste management. Additionally, IoT data provides insights into the product life cycle.

For example, the applications of IoT upgrade smart grids that have more efficient monitoring and controlling energy usage with decreased wastage and increased energy efficiencies. While in the manufacturing line, IoT sensors can monitor the performance and wear to ensure optimal predictive maintenance when reducing the downtime of the sensors and equipment. The utilization of IoT devices enables companies to design and provide product-as-a-service models whereby these devices monitor product usage and performance for the clients to provide maintenance and upgrade services rather than selling the products outright; it further ensures product efficiency during usage.

6.4.4 *Data Analytics and Digital Twins*

Digital twins and data analytics are the most potent tools or solutions for improving the execution of circular economy practices in terms of offering in-depth insights and simulations of real-world systems. Data analytics is the analysis of big data using advanced statistical and computational techniques to reveal patterns and trends that are used to inform circular strategies (Lee, 2021). This data supports decision-making in such areas as supply chain optimization, resource management, and waste reduction.

Digital twins are a virtual replica of a physical asset, system, or process that massively improves simulation and optimization opportunities within the circular economy. It helps organizations understand through simulations how processes might be affected by several decisions that could be actualized in the real world before they could be actualized. This capability helps ensure that designs for products are optimized to increase their durability and recycling capabilities. In turn, this will improve maintenance strategies to increase asset lifespan (Lattanzi et al., 2021).

Through data analytics and digital twins, closed-loop supply chains are enabled: through understanding how the products perform and throughout their lifecycle, which helps streamline resource efficiency, waste reduction, and recycling processes (Lee, 2021). Through these technologies, companies can realize much greater alignment with the circular economy principles and businesses move much closer toward even more sustainable practices.

6.5 Mechanisms for Multistakeholder Collaboration

By combining disparate strengths and unique perspectives of different actors, cooperation between stakeholders is the central enabler of the Digital CE, driving innovation and delivering sustainable outcomes. More importantly, the effective collaboration of governments, businesses, academia, and civil society solves the complexity of a new economy transition. This section elaborates on how workable partnerships are built through collaboration platforms, networks shared goals and objectives, and best practice for successful collaboration.

6.5.1 Building Effective Partnerships

Among the more fundamental mechanisms of effective multistakeholder collaboration in the circular economy, building effective partnerships stands forth. Effective partnerships include aspects such as trust, clear communication, and a shared vision that sums up the achievement of circular economy goals. Eweje et al (2021) suggest that each stakeholder should be clearly defined with their role and input in the project with mechanisms to solve conflicts or meet expectations. Such partnerships also need to be configured to facilitate aspects of information sharing, resource pooling, and problems being solved jointly.

The main action taken during the development of successful partnerships is entering into formal agreements that clearly mention roles, responsibilities, and commitments. Agreements have the effect of making the collaboration more formal and ensuring that parties are on the same page when it comes to goals and expectations (Cramer, 2022). Take PPPs in the circular economy: this will entail bringing together governmental bodies, businesses, and NGOs to address complex issues like waste management and resource use as a collective effort. Partnerships can pull together the best of each stakeholder to achieve common goals by defining roles and establishing governance structures (Cramer, 2022).

6.5.2 Collaborative Platforms and Networks

These are channels through which facilitating and coordinating multistakeholder engagement in the circular economy occur. They are conduits through which various stakeholders may interact, share knowledge, and collaborate on both projects and initiatives. There exist various forms that such collaborative platforms and networks may take, such as digital platforms, industry forums, and research consortia. Among such ones are the digital-based platforms that considerably influence real-time communication and collaboration of geographically dispersed stakeholders (Suchek et al., 2021).

Examples of such collaborative platforms are the Ellen MacArthur Foundation's Circular Economy 100, which collaborates together businesses, governments, and academia to effect change in the pathway towards a circular economy. Under the CE100, members are allowed space for sharing good practice, access to research, and collaboration on initiatives on the topic of the circular economy. Similarly, the European Circular Economy Stakeholder Platform is a platform in which ideas are shared, best practices are showcased, and collaborative projects are shared as well (European Commission, 2021).

6.5.3 Shared Goals and Objectives

The common goals and objectives must be aligned for successful and focused multi-stakeholder collaborations. In general, since the stakeholders share a common understanding of what should be achieved through collaboration, it increases the potential for stakeholders to work cohesively and more efficiently. Shared goals provide meaning and direction and spur motivation among involved stakeholders as well as channel their efforts (Singh et al., 2020).

To have the goals and objectives in line, there is a need for stakeholders to be engaged on an early collaborating platform and getting them involved in the goal-setting process. Participatory approaches such as workshops and consultations can be effective in identifying common interests and priorities among stakeholders (Singh et al., 2020). For instance, it may form strategies for circular economies by engaging different groups of stakeholders, the resultant views and needs into the plans to eventually lead to more effective and inclusive solutions.

6.5.4 Best Practices for Collaboration

Implementation of the best practices of collaboration will boost the productivity of multistakeholder partnerships and networks. These best practices are open communication, trust building, and accountability processes. This helps to ensure that open communication opens avenues through which all stakeholders will be informed about progress in efforts at collaboration and are able to input their insights and feedback (Purvis et al., 2023). The demanded meetings, updates, and reports will maintain the transparency thereof to continue engaging the concerned stakeholders.

Another best practice is building trust between stakeholders. This is a vital element of successful teamwork because it develops an atmosphere wherein stakeholders can feel safe and work together and share resources as well as information (Purvis et al., 2023). Trust is attained through a course of consistent and transparent actions plus proving commitment to mutual benefits in terms of accomplishing shared goals.

Mechanisms of accountability are equally pivotal in ensuring that stakeholders live up to their responsibilities and, therefore, provide the collaborative activity with adequate inputs. Some of these mechanisms include performance metrics, reporting of progress, and the assessment period on what is being done in collaborative activities (Eweje et al., 2021). Defined expectations and monitoring of progress by stakeholders can ensure that the collaborative work is goal-oriented and productive.

6.6 Overcoming Challenges in Multistakeholder Collaboration

Despite the importance to the progress of the Digital CE, there are still several obstacles to be solved through multistakeholder collaboration so that the results are successful. In this sense, regulatory barriers, technological gaps, financial constraints, and differences in culture and organization, among others, are some of the main obstacles found to date. Therefore, knowing them very well is necessary, as well as developing particular strategies to overcome them.

6.6.1 Regulatory Barriers

Regulatory barriers represent an important stumbling block for effective multistakeholder collaboration in the circular economy. Various jurisdictions have sharply contrasting regulatory regimes, which are highly complicated and hinder the collaboration of enterprises by standards and requirements (Hartley et al., 2023). This includes different waste management regulations, recycling standards, and design of various products for conflicting harmonization among related practices and processes across borders.

These can be overcome through advocacy and policy dialogue by stakeholders aimed at changing the regulatory framework to include amendments of regulations aligned with the objectives of the circular economy. Coordination in developing and promoting common standards and guidelines can address issues of regulatory fragmentation. Partnerships across sectors and participating in forums on policies help to harmonize regulations and establish a more enabling environment for the circular economy initiatives (Hartley et al., 2023).

6.6.2 Technological Gaps

Technological mismatches represent the other significant challenge in multistakeholder collaboration. To a great extent, advanced technologies and AI, IoT, or blockchain are used to create the perfect end-to-end implementation of circular economy practices. However, differences in adoption and capacities for technology between the stakeholders can then become a barrier to effective collaborations (Rejeb et al., 2023).

This gap can be addressed through the promotion of the transfer of technology as well as capacity building. Technological gaps can be achieved through collaborative research and development where there is a sharing of knowledge and existing resources to develop and deploy new technologies. Education and training programs will bridge technological gaps as they will improve stakeholders' ability to participate in circular economy activities effectively. It can also encourage innovation, allow for the easier adoption of new technologies, and prove the value of these technologies to potential collaborators by supporting pilot projects.

6.6.3 Financial Constraints

Financial restriction could become one of the most prominent barriers in most multistakeholder collaborations. Research, technology, and infrastructure investments will be very high for circular economy initiatives. For most stakeholders, especially for SMEs and non-profit organizations, it will be challenging to find enough funding for such initiatives (Gonçalves et al., 2022).

An important strategy in overcoming the dilemma of funds with stakeholders would be to come up with different funding avenues in public–private partnership investments, grants, and subsidies. In this regard, PPPs are a great avenue to tap into the resources and expertise from both sectors so that the circular economy activities can be implemented and the financial risk shared with both public and private entities (Gonçalves et al., 2022). Funding opportunities can also be sought through government programs, international organizations, and other global agencies whose main goal is to give out financial aid and assistance to sustainability projects. The other area could involve collaborative efforts towards developing business models that can integrate the circular principle, thereby attracting investment and outlining the financial viability of circular economy practices (Lacy & Rutqvist, 2015).

6.6.4 Cultural and Organizational Differences

Cultural and organizational differences can be a major barrier to effective multistakeholder collaboration. The array of stakeholders with variable values and prioritization toward sustainability can be destructive in the sense that this friction may even lower the collaboration. For instance, cultural differences regarding attitudes toward waste, resource efficiency, and environmental accountability impact the will of the stakeholders to implement circular economy practices.

These differences need to be made audible before all interested stakeholders through open and inclusive dialogue so that mutual understanding and respect can evolve. Common vision and a set of objectives help to align diverse perspectives toward a collaborative environment (Singh et al., 2022). It would comprise approach strategies on intercultural communication and collaboration such as workshops and joint projects, with teamwork encouraged in an effective form, as suggested by Purvis et al. (2023). A coherent framework of collaboration can be found in making internal policies and practices according to the principles of the circular economy for proper development.

6.7 Strategies for Promoting Sustainable Innovation

Critical to continuing the progression of the Digital CE and long-term environmental and economic sustainability is the promotion of sustainable innovation. Key areas of contributions to this process include the development of a strong policy framework and incentives, the diversification of funding mechanisms and models of investment, building of innovation ecosystems and incubators, and education and capacity building. In fact, this section keeps track of such strategies in detail by indicating how they are important to driving forward sustainable innovation.

6.7.1 Policy Frameworks and Incentives

At the policy level, there will be frameworks and incentives making a conducive environment for the development and adoption of circular economy practices. An example would be the ways governments can encourage investments in sustainable technologies and practices through incentives like tax concessions or subsidies and regulatory support. For instance, investment tax credits on renewable energy technologies or R&D grants in innovations for the circular economy can energize private sector involvement and quicken the pace of transition to a circular economy (Singh et al., 2021).

In addition to this, ambitious and clear sustainability goals from policy frameworks trigger innovation. Regulations such as EPR and eco-design requirements push companies to think about the lifecycle impacts of their products and encourage circular practices. Such policies create market demand for sustainable innovations and incentivize companies to develop solutions that minimize waste and enhance resource efficiency.

6.7.2 Funding Mechanisms and Investment Models

This sustainable innovation requires proper funding to obtain a swing. Alternative funding models and investment schemes enable the development and exploitation of circular economy technologies and business models. Venture capital and private equity are a traditional source of funding, which provides capital to new ventures and established corporations as they produce innovative products and services (Bonini & Capizzi, 2019). Still, such funding sources are restrained in focusing on sustainability-related innovation due to factors such as risk and uncertain returns.

There are other alternative funding sources besides impact investing and green bonds, which include public–private partnerships, where the role of investment sometimes becomes very key. Impact investments are those that involve investing in projects or firms pursuing both financial return as well as positive social or environmental impacts. Green bonds refer to those issued by governments or corporations for environmentally friendly projects, such as initiatives related to the circular economy. Through PPPs, large-scale projects and infrastructure for a circular economy can mobilize resources and expertise from both the public and private sectors (Gonçalves et al., 2022).

Innovation-focused funding models, such as innovation prizes and competitions, stimulate creativity and attract attention to emerging sustainable technologies. Funding models reward the breakthrough innovations and deliver further development as well as commercialization financial support.

6.7.3 Innovation Ecosystems and Incubators

Innovation ecosystems and incubators are very significant in encouraging cooperation and hastening the pace of developing sustainable innovations. Innovation ecosystems are interactions and collaborations of various organizations: businesses, research institutions, and governmental agencies, combined to promote innovation and economic growth (Adner, 2017). Innovation ecosystems offer stakeholders a ground for sharing ideas, resources, and challenges in the circular economy.

Incubators and accelerators serve as important parts in the ecosystems, providing the required resources, mentorship, and support to help scale innovation within early-stage companies. They essentially provide funding opportunities, working space, technical expertise, and networking opportunities. For example, the Circular Economy Incubator founded by the Ellen MacArthur Foundation is on supporting start-ups developing circular economy solutions through comprehensive support as well as facilitation by industry leaders (Singh et al., 2023).

Innovation ecosystems and incubators can offer fertile ground for developing sustainable technologies by making the environment conductive to quick development and dissemination. Innovation ecosystems and incubators bridge the innovation gap by connecting innovative ideas with market uptake, which then translates into progress toward a circular economy.

6.7.4 Education and Capacity Building

Education and capacity building become considerably important for equipping not only the individuals but also the organizations with knowledge and skills sufficient for the opportunities for sustainable innovation. The basic idea here is that educational programs, training initiatives, and awareness programs focused on principles and practices of circular economy enhance the understanding and build the expertise of sustainability (Kirchherr et al., 2017). Universities, vocational training institutions, and online platforms can offer courses and certifications in the areas of sustainable design, resource efficiency, and circular business models.

It also encompasses the development of an innovative and sustainable culture for organizations. Internal company training and workshop sessions on circular economy practices can be undertaken by companies to raise awareness among the workforce about innovative working methodologies. Academia-industry partnerships can further enhance the capacity building process, as it helps in the transmission of knowledge from academia to the industry and education as per the requirements of the industries in institutes (Mazzucato, 2018).

Public awareness campaigns and community engagement programs can enhance the knowledge of sustainable innovation as well as support by the consumers and the community members within and from the society at large (Lacy & Rutqvist, 2015). Knowledge regarding the aware capacity building enhances the capability of providing the required knowledgeable and motivated workforce in progress towards the attainment of circular economy goals.

6.8 Future Directions

As the CE develops, these emerging trends, potential areas for further collaboration and visionary concepts for sustainable innovation and circularity define its future directions. That is to say, in this section, we look at these aspects and elaborate on the most important developments and strategic directions that would influence advancement in circular economy practice.

6.8.1 Emerging Trends in Digital Circular Economy

Over the next decade, the digital circular economy is set to see multitudes of transformative trends brought about by technological evolution and changes within society. One important trend relating to the above is the movement towards incorporating artificial intelligence and machine learning into circular economy practices. AI has enabled optimization of resources, facilitated and predicted the maintenance of machines, and automated recycling processes that usually contribute to much more efficient and powerful circular systems (Chui et al., 2018). For instance, some algorithms of machine learning can be used to filter huge quantities of data for enhanced accuracy in waste sorting and novel recoverable resources.

A related, emerging trend is the increased use of blockchain technology to leverage greater transparency and traceability in circular supply chains. Blockchain can provide an auditable and tamper-evident record that enables stakeholders to track the lifecycle of products and materials with precision (Khan et al., 2022). Increased visibility helps verify the authenticity of recycled materials, enforce compliance with sustainability standards, and earn the trust of both consumers and businesses.

The Internet of Things is equally important in the digital circular economy because it aids in the real-time monitoring and management of resources. This way, IoT devices can collect and send data on product usage, consumption of energy, and waste generation among other things that can improve decision-making and proactive management of resources. Firms with such information data may optimize their operations, reduce waste, and implement circular practices more effectively.

Digital twins are another innovation that has just recently emerged. These can be virtual replicas of physical assets that can be used to simulate and optimize the entire processes that will be involved in the circular economy. Simulations of products and systems' working and behaviour allow the insight into environmental impact and what opportunities could be taken up for its improvement, thereby allowing for scenario analysis-thereby helping in designing more sustainable products and processes.

6.8.2 Potential Areas for Further Collaboration

This shows some significant progress in the development of the digital circular economy. However, several areas were suggested by the participants to be targeted for further cooperation and focus between stakeholders: for instance, the need for greater emphasis on standardizing metrics and methodologies for the evaluation of circularity. That standards and frameworks will help in standardizing the measures used to assess circular practices. This, therefore, shall make it easy for the industries to compare their progress with others' (Sassanelli et al., 2019). Development and general agreement of the standards require close cooperation among industry leaders, researchers, and policymakers.

Another region of co-operation involves circulatory business model improvement. Most of the circular economy initiatives have required the collaboration of various sectors such as manufacturing, retail, and logistics. Through that co-operation, it is possible for stakeholders to devise well-integrated solutions cutting across several aspects of circularity-from design through to end-of-life management (Lacy & Rutqvist, 2015). For example, relationships between manufacturing companies for products and recycling companies improve the recovery and recycling of materials.

The other promising space for collaboration is public–private partnerships (PPPs). Governments, businesses, and NGOs can share their resources to push forward high-impact circular economy projects and infrastructure development. In fact, PPPs enable the exploitation of a wide range of skills and resources in order to address systemic problems and work towards a deeper change in the system (Gonçalves et al., 2022). Projects such as collaborative innovation labs and pilot projects may support the development and scaling of new circular solutions; they may be grounds for learning, attracting broader adoption.

International cooperation can further facilitate the globalization of the circular economy. Cross-border partnerships, in particular, would address challenges related to waste management and resource scarcity across a country's borders, facilitate sharing best practices as well as diffusion of relevant technologies (Ghisellini et al., 2016). International organizations and international agreements can promote mutual knowledge and coordination toward global sustainability goals.

6.8.3 *Vision for Sustainable Innovation and Circularity*

The vision for sustainable innovation and circularity focuses on building a resilient and regenerative economy. This economy is based on resource efficiency, waste reduction, and resource stewardship. Achieving this requires a shift in how products are designed, used, and disposed of. Circularity must be integrated into every stage of the product life cycle (Geissdoerfer et al., 2017).

An important feature of that vision is closed-loop systems with minimal waste and complete resource recovery. The principles of the circular economy propose that products would be designed to achieve maximum useful life, opportunities for repair, and in the ultimate form, recyclability (Korhonen et al., 2018). New ideas, such as modular design and take back, remanufacturing can support this shift towards a more substantial economy.

One of the other components of the vision is through embedding circular economy principles in corporate strategies and business models. Organizations increasingly realize that doing business should be more circular by being the engine for competitive advantage and sustainability. This prompts businesses to benefit from cost-saving opportunities and enhance brand reputation through attracting environmentally conscious consumers through adoption of circular business models (Economics, 2019). It redefines value propositions, optimizes supply chain, and engages with stakeholders to generate value in a circular economy setting.

Further, the sight of sustainable innovation looks into the social and economic perspectives of circularity. Circular economy initiatives should generate not only environmental impacts but also social impacts in the form of employment opportunities and community development (Murray et al., 2017). In this way, the circular economy can lead to more equitably balanced and resilient communities.

6.9 Conclusion

Exploring sustainable innovation within the framework of the digital circular economy reveals many critical insights. Among the most important is how digital technologies like AI, blockchain, and IoT enable a profound transformation of circularity. These technologies are key to driving the shift toward more efficient and sustainable practices. Another crucial takeaway here is the need for strong multistakeholder collaboration to achieve that degree of transformation. These technologies support resource optimization and transparency but also push efficiency and innovation in circular systems.

Continued inter-governmental cooperation by businesses, academia, and civil society is required to drive circular economy goals forward. Positive partnerships and mutual goals are what can help overcome these barriers and improve the scope of full-circle models. Together, by negating regulatory, technological, and financial challenges, the speed at which society moves toward a more sustainable and resilient economy can be increased.

Only through a commitment to apply circular principles to all sectors of economic and social activities will accelerate sustainable innovation take place. Only then can stakeholders contribute to that future characterized by efficient resource use and reduction of impacts on the environment as well as enhancement of economic opportunities through fostering an environment that is productive for collaboration and innovation.

References

Adner, R. (2017). Ecosystem as structure: An actionable construct for strategy. *Journal of Management, 43*(1), 39–58. https://doi.org/10.1177/0149206316678451

Bocken, N. M., De Pauw, I., Bakker, C., & Van Der Grinten, B. (2016). Product design and business model strategies for a circular economy. *Journal of Industrial and Production Engineering, 33*(5), 308–320. https://doi.org/10.1080/21681015.2016.1172124

Bonini, S., & Capizzi, V. (2019). The role of venture capital in the emerging entrepreneurial finance ecosystem: Future threats and opportunities. *Venture Capital, 21*(2–3), 137–175. https://doi.org/10.1080/13691066.2019.1608697

Chui, M., Manyika, J., Miremadi, M., Henke, N., Chung, R., Nel, P., & Malhotra, S. (2018). Notes from the AI frontier: Insights from hundreds of use cases. *McKinsey Global Institute, 2*, 267.

Cramer, J. (2022). Effective governance of circular economies: An international comparison. *Journal of Cleaner Production, 343*, 130874. https://doi.org/10.1016/j.jclepro.2022.130874

Economics, M. (2019). Completing the picture how the circular economy tackles climate change. *Ellen Macarthur Foundation.*

European Commission. (2021). European Circular Stakeholder Platform. https://circular-cities-and-regions.ec.europa.eu/associated-partners/european-circular-stakeholder-platform

Eweje, G., Sajjad, A., Nath, S. D., & Kobayashi, K. (2021). Multi-stakeholder partnerships: A catalyst to achieve sustainable development goals. *Marketing Intelligence & Planning, 39*(2), 186–212. https://doi.org/10.1108/MIP-04-2020-0135

Geissdoerfer, M., Savaget, P., Bocken, N. M., & Hultink, E. J. (2017). The circular economy–a new sustainability paradigm? *Journal of Cleaner Production, 143*, 757–768. https://doi.org/10.1016/j.jclepro.2016.12.048

Ghisellini, P., Cialani, C., & Ulgiati, S. (2016). A review on circular economy: The expected transition to a balanced interplay of environmental and economic systems. *Journal of Cleaner Production, 114*, 11–32. https://doi.org/10.1016/j.jclepro.2015.09.007

Gibbs, D. (2021). The circular economy and planned sustainability. *The Palgrave handbook of global sustainability* (pp. 1–18). Springer International Publishing.

Gonçalves, B. D. S. M., Carvalho, F. L. D., & Fiorini, P. D. C. (2022). Circular economy and financial aspects: A systematic review of the literature. *Sustainability, 14*(5), 3023. https://doi.org/10.3390/su14053023

Hartley, K., Schülzchen, S., Bakker, C. A., & Kirchherr, J. (2023). A policy framework for the circular economy: Lessons from the EU. *Journal of Cleaner Production, 412*, 137176. https://doi.org/10.1016/j.jclepro.2023.137176

Jum'a, L., Ikram, M., & Jabbour, C. J. C. (2024). Towards circular economy: A IoT enabled framework for circular supply chain integration. *Computers & Industrial Engineering, 192*, 110194. https://doi.org/10.1016/j.cie.2024.110194

Khan, S., Ali, S. S., & Singh, R. (2022). Determinants of remanufacturing adoption for circular economy: A causal relationship evaluation framework. *Applied System Innovation, 5*, 62. https://doi.org/10.3390/asi5040062

Kirchherr, J., Reike, D., & Hekkert, M. (2017). Conceptualizing the circular economy: An analysis of 114 definitions. *Resources, Conservation and Recycling, 127*, 221–232. https://doi.org/10.1016/j.resconrec.2017.09.005

Korhonen, J., Honkasalo, A., & Seppälä, J. (2018). Circular economy: The concept and its limitations. *Ecological Economics, 143*, 37–46. https://doi.org/10.1016/j.ecolecon.2017.06.041

Lacy, P., & Rutqvist, J. (2015). *Waste to wealth: The circular economy advantage* (Vol. 91). Palgrave Macmillan.

Lattanzi, L., Raffaeli, R., Peruzzini, M., & Pellicciari, M. (2021). Digital twin for smart manufacturing: A review of concepts towards a practical industrial implementation. *International Journal of Computer Integrated Manufacturing, 34*(6), 567–597. https://doi.org/10.1080/0951192X.2021.1911003

Lee, K. (2021). A systematic review on social sustainability of artificial intelligence in product design. *Sustainability, 13*(5), 2668. https://doi.org/10.3390/su13052668

MacArthur, E. (2015). Towards a circular economy: business rationale for an accelerated transition. *Greener Manag International, 20.* https://www.greenindustryplatform.org/research/towards-circular-economy-business-rationale-accelerated-transition

Mazzucato, M. (2018). *The value of everything: Making and taking in the global economy*. Hachette UK.

Murray, A., Skene, K., & Haynes, K. (2017). The circular economy: An interdisciplinary exploration of the concept and application in a global context. *Journal of Business Ethics, 140*, 369–380. https://doi.org/10.1007/s10551-015-2693-2

Priyadarshini, P., & Abhilash, P. C. (2020). Circular economy practices within energy and waste management sectors of India: A meta-analysis. *Bioresource Technology, 304*, 123018. https://doi.org/10.1016/j.biortech.2020.123018

Purvis, B., Celebi, D., & Pansera, M. (2023). A framework for a responsible circular economy. *Journal of Cleaner Production, 400*, 136679. https://doi.org/10.1016/j.jclepro.2023.136679

Rejeb, A., Appolloni, A., Rejeb, K., Treiblmaier, H., Iranmanesh, M., & Keogh, J. G. (2023). The role of blockchain technology in the transition toward the circular economy: Findings from a systematic literature review. *Resources, Conservation & Recycling Advances, 17*, 200126. https://doi.org/10.1016/j.rcradv.2022.200126

Rojek, I., Jasiulewicz-Kaczmarek, M., Piechowski, M., & Mikołajewski, D. (2023). An artificial intelligence approach for improving maintenance to supervise machine failures and support their repair. *Applied Sciences, 13*(8), 4971. https://doi.org/10.3390/app13084971

Rusch, M., Schöggl, J. P., & Baumgartner, R. J. (2023). Application of digital technologies for sustainable product management in a circular economy: A review. *Business Strategy and the Environment, 32*(3), 1159–1174. https://doi.org/10.1002/bse.3099

Sassanelli, C., Rosa, P., Rocca, R., & Terzi, S. (2019). Circular economy performance assessment methods: A systematic literature review. *Journal of Cleaner Production, 229*, 440–453. https://doi.org/10.1016/j.jclepro.2019.05.019

Singh, N. K., Pandey, S., Sharma, H., & Goel, S. (Eds.). (2020). *Green innovation, sustainable development, and circular economy*. CRC Press.

Singh, R., Dsilva, J., Centobelli, P., & Tripathi, V. (2021). A review on sustainable HRM: A study evaluating sustainability for organisational development. *The Empirical Economics Letters, 20*(2), 1–10.

Singh, R., Khan, S., & Centobelli, P. (2022). Investigating the interplay between social performance and organisational factors supporting circular economy practices. *Sustainability, 14*, 16781. https://doi.org/10.3390/su142416781

Singh, R., Khan, S., & Dsilva, J. (2023). A framework for assessment of critical factor for circular economy practice implementation. *Journal of Modelling in Management, 18*(5), 1476–1497. https://doi.org/10.1108/JM2-06-2021-0145

Suchek, N., Fernandes, C. I., Kraus, S., Filser, M., & Sjögrén, H. (2021). Innovation and the circular economy: A systematic literature review. *Business Strategy and the Environment, 30*(8), 3686–3702. https://doi.org/10.1002/bse.2834

Tapscott, D., & Tapscott, A. (2016). *Blockchain revolution: How the technology behind bitcoin is changing money, business, and the world*. Penguin.

Thrassou, A., Vrontis, D., Efthymiou, L., & Figurek, A. (2024). Delineating non-profit organizations: Society, sustainability and accountability. In *Non-profit organisations: Society, sustainability and accountability* (vol. III, pp. 1–24). Springer Nature Switzerland.

Turskis, Z., & Šniokienė, V. (2024). IoT-Driven transformation of circular economy efficiency: An overview. *Mathematical and Computational Applications, 29*(4), 49. https://doi.org/10.3390/mca29040049

Part II
Digital Circular Economy and Sustainability for Future Generations

Chapter 7
Creating Sustainable Workplace: Enhancing HRM Strategies in the Digital Circular Economy

Rubee Singh, Shahid Mushtaq, Kishore Kumar, and Obaidur Rahman

Abstract The digital circular economy provides a comprehensive framework that aims at achieving optimum resource efficiency by incorporating sustainability principles and advanced digital technologies. It has become imperative for every organization to adopt digital transformation to contribute towards sustainable development goal. Human resource management serves as an important linkage between digital circular economy, framework, adoption, and sustainability goals of the organization. This chapter provides important insights into the critical role of human resource management in helping organizations to navigate the pathway toward digital circular economy. This study outlines the various human resource management strategies required to achieve the organizational sustainability goals by focusing on imparting digital skill sets to the workforce. This chapter also highlights the importance of fostering a sustainability innovation culture in the workplace. HRM strategies should include providing technological proficiency, sustainability, skill sets, change management, and adoption of AI-based tools. This chapter suggests that human resource management must play an active role in developing circular economy framework that focuses on developing sustainability mindset among the employees and improving the sustainability performance of the organization. The relationship between HRM and the digital circular economy is critical for the business to achieve sustainability goals and improve the environmental and social performance of the business. This chapter should be useful for practitioners and policymakers to drive the

R. Singh (✉)
Institute of Business Management, GLA University, Mathura, India
e-mail: dr.rubeerajput@gmail.com

Department of Management Studies, Kumaun University Nainital, Nainital, India

S. Mushtaq
School of Business Studies, Central University of Jammu, Jammu, India
e-mail: shahid@cujammu.ac.in

K. Kumar
School of Management, Bennett University, Greater Noida, India

O. Rahman
Sharda University, Greater Noida, India
e-mail: obaidur.rahman@sharda.ac.in

R. Singh and V. Kumar (eds.), *Sustainable Innovations and Digital Circular Economy*,
https://doi.org/10.1007/978-981-96-1064-8_7

transition towards a digital circular economy and integrate digital circular economy principles across the different sectors.

Keywords Digital circular economy · Digital technologies · Human resource management · Sustainable development · Social responsibility · Industry 4.0

7.1 Introduction

Sustainable development is the new imperative for the world in the present era. It is important due to a renewed focus on inclusive growth, ecological protection, and societal welfare. The notion of circular economy has significantly propagated sustainable development practices (Geissdoerfer et al., 2017; Kathpal et al., 2021). The circular economy approach minimizes environmental waste and enhances resource efficiency through the concept of the 3Rs (recycle, reduce, reuse) (Sánchez-García et al., 2024). With rapid advancement of circular economy practices, the notion of digital circular economy has emerged, which focuses on incorporating digital technologies for achieving efficient resource management and improve circular economy practices (Lampropoulos et al., 2024; Medaglia et al., 2024). These digital technologies are cloud computing, artificial intelligence, IoT, etc. Through digitization in the operations, organizations can develop and implement innovative practices for achieving sustainable development goals (Dantas et al., 2021; Singh et al., 2024a, b, c). Digital technologies like AI, cloud computing, and IoT not only improve the process in the business but also help to ensure transparency in the supply chain and sustainable innovation opportunities in the business (Pagoropoulo et al., 2017; Cagno et al., 2021). The need for digital circular economy is increasing every day due to the tremendous sustainability challenges faced by the world. Rapid exploitation of natural resources and environmental degradation due to the rise in the global population has become a common norm in the past century (Rodriguez-Anton et al., 2019). The traditional economy which is based on the linear economy concept only focuses on the usage of natural resources and the generation of waste. Traditional linear economy has resulted in exploitation of resources and deterioration of the ecosystem (Korhonen et al., 2018; Sariatli, 2017). The traditional linear economy is no longer a feasible option for making this world more sustainable. The concept of digital circular economy provides a unique opportunity to transform toward positive environmental impact and sustainable growth. Digital circular economy integrates a sustainability perspective into its approach to offer several important benefits for the sustainable development of the economy (Piscicelli, 2023; Kristoffersen et al., 2020; Liu et al., 2022). It minimizes waste through waste management by creating closed-loop cycle where the raw material is reused or retained as much as possible. With advanced digital technologies such as AI-based manufacturing, businesses can oversee the life cycle of the material which in turn helps companies to adopt strategies related to reuse, recycle, repair, and refurbishment of the product (Khan et al.,

2022). This helps businesses to ensure resource optimization and results in a reduction in carbon emissions. For instance, advanced technology-based sensors at the industrial facility may track the operations, estimate the maintenance, and forecast the likelihood of possible shutdowns (Ding et al., 2023; Luo et al., 2024).

The digital circular economy also plays an important role in improving transparency and monitoring in supply chain management (Liu et al., 2023). Advanced technologies like AI blockchain could be the way to track the logistics and transportation resulting in sustainable sourcing of the products. The transparency also ensures that organizations make rational decisions about the various suppliers and environmental compliances (Centobelli et al., 2022). These technologies not only help the business to monitor carbon footprints and ensure transparency but also help the consumers to know the adverse environmental impact of the business activities (Singh et al., 2024a, b, c; Trihinas et al., 2022). Human resource management functions have an instrumental role that guiding the organization concerning digital transformation and sustainable development issues (Aust et al., 2024; Brewster & Brookes, 2024; Mishra et al., 2023). In the contemporary era, sustainable development issues are no longer a voluntary approach for the businesses but have become integral part of business models (Chams & García-Blandón, 2019; Kumar et al., 2024). HRM function has become an important instrument in propagating sustainable development practices within the organization. With the increasing pressure on businesses from various stakeholders about the adverse impact on society and environment, the role of HRM has emerged as an important factor (Agrawal & Bansal, 2022; Kumar & Singh, 2024; Sukalova et al., 2022). Today, the HRM system and sustainability goals of the organizations are intertwined HRM strategies help in fostering a sustainability culture and address the challenges faced by the organization in achieving sustainability goals (Zubair & Khan, 2019). HRM strategies also the business to understand the implications of business operation operations, on the environment and society. There is increasing pressure on organizations to adopt sustainable business models by adopting advanced digital technologies (Santana & Lopez-Cabrales, 2019). HRM systems are the key players in driving digital transformation as they facilitate workforce, engagement and alignment with the organizational objectives. HRM function not only ensures employer relations but also builds an environment for fostering sustainable innovation and ethical practices at the workplace. HRM function is a driving force of sustainability practices at the workplace through integration of sustainability goals and concerns in the organizational culture (Mishra, 2017; Olateju et al., 2020). In the present time, sustainability issues revolving around business and not only limited to environmental degradation but also incorporate social development and governance-related issues (Sharma & Kohli, 2024). HRM systems should integrate all the sustainability dimensions with the recruitment, compensation, training and development activities in the organization. recruitment and selection HR ensure that prospective candidates must understand sustainability practices and sustainability values (Sova et al., 2023; Mukhuty et al., 2022; Campos-García et al., 2024). Further training and development should also aim at building sustainability competencies among the employees to contribute towards the sustainability goals of the organization. HRM function also play a critical part in developing sustainability culture in the organization (Ali et al., 2024). This includes propagating, sustainability values like environmental protection, reducing carbon footprints, resource optimization, and social values at the workplace. HRM function encourages green behaviour in the form of waste, minimization, and sustainability practices among the employees. Further, it also incentivizes sustainability performance among the employees by connecting the sustainability practices of the employees with the sustainability goals of the organizations (Khan et al., 2024; Kumar & Tarkar, 2024). It aims to provide monetary and non-monetary rewards to the workforce, which actively adopt sustainability practices at the workplace. HRM function is a critical pillar in the digital transformation of the organization. Digital transformation requires adoption and implementation of new technologies in the business operations. HR systems play a crucial role in this transition by preparing employees for digitization through upskilling practices (Abid et al., 2024). They identify the requirements of skills for digitization and then provide training programs to the employees to facilitate digitization. The digital transformation often results in extensive overhaul in the form of change in the organization, structure, job positions, and workflow (Zhang & Chen, 2024). Thus, HR facilitates these transitions in a smooth manner which may not lead to friction or anxiety among the employees. To this end, HR adopts effective change management in the form of proper and timely communication, and participative decision-making at the workplace (Theres & Strohmeier, 2023; Yusuf et al., 2023).

The circular economy business model has garnered lot of interest in the recent years. It aims to focus on balancing the revenue streams of the business, keeping in mind the responsibility towards the environment and society. It involves development and implementation of new processes and systems that not only result in economic value but also improve the environmental and social performance of the business (Marrucci et al., 2021). Human resource management has a crucial role in developing this kind of organization culture that promotes sustainable innovation. Human resource managers need to align organizational sustainability goals with the organization, vision. Islam et al. (2021) highlighted that HRM significantly impacts the organization, sustainability culture, ethical behaviour and capabilities to foster sustainable innovation at the workplace. HR managers should support sustainability innovation and motivate the employees to adopt green behaviour at the workplace. HR has a crucial role inculcating sustainability into the business model through the training and development, which in turn helps the employees to understand the relevance of sustainable practises at the workplace (Burrichter et al., 2022; Klimovskikh et al., 2023). This chapter focusses on the impact of human resource management strategies to promote a digital circular economy framework in the organization. It presents significant insights about the various HRM strategies that play an important role in fostering sustainability culture and circular economy practises at the workplace (Raza et al., 2022). This should be useful for the policymakers, practitioners, and other stakeholders to implement circular economy principles to create sustainable workplace.

The following chapter is structured as follows. The next section discusses the HRM function in the digital area. Section 7.3 presents discussion on human resource management in the context of digital transformation and sustainability. Section 7.4 links the digital circular economy with a human resource management function, and Sect. 7.5 provides a discussion on employee development and green skills followed by challenges in adopting digital circular economy principles in HRM. Final Sect. 7.6 provides concluding remark and implications of this chapter.

7.2 HRM Function in Digital Era

In this digital era, business organizations need to leverage technological expertise for resource optimization and improving the overall performance of the businesses. Just like other functions of the organization, the human resource management function is also witnessing a significant overhaul due to digital transformation. Development of advanced technologies like bigdata, cloud computing, AI, and other automation tools is rapidly redefining HRM systems (Pan & Froese, 2023). It is not only reshaping the role of HR but also helps in propagating sustainability goals through digital circular practices (Budhwar et al., 2022). These advanced technologies have a profound influence on HRM function through resource optimization and improved decision-making competencies. Previously, recruitment and selection were more labour-intensive and required a lot of documentation and manpower (Alsaif & Sabih Aksoy, 2023). The digitization process has made the recruitment function more efficient with the usage of digital tools, advanced technologies such as AI-based tools, machine learning, and big data facilitate HR professionals to shortlist the candidates based on stipulated criteria. Consequently, it improves the hiring process speedier and more accurate (Radonjić et al., 2022). Digital platform usage allows HR to run through numerous prospective applications and select the best-suited applications. Further, digitization also allows HR professionals to use digital media for recruitment through social media advertisement, and online evaluation of the candidate via video interviews. This results in talent acquisition in the more efficient and cost-effective manner (Vardarlier, 2020). To some extent, usage of digital technologies also mitigates bias chances in the recruitment process. Advanced digital technologies have also revolutionized the training and development functions of human resource management. The rise in online learning platforms, online simulation sessions, and virtual learning technologies offers a wide array of choices to HR to provide learning opportunities to its employees (Thite, 2018). Employees could enroll in online training sessions, courses, and interactive learning platforms, which help in upscaling at their own convenience. Moreover, these digital tools also provide learning management systems that help to evaluate an analyse the employee's performance. The integration of AI-based learning courses provides immersive learnings. HR plays an important role in designing and assigning digital learning solutions based on the employee's requirement of skill sets and organizational goals (Barykin et al., 2021). Digitization has also transformed the performance management system of HRM. The traditional performance management system based on annual assessment has transformed into a real-time assessment system. human resource management function now utilizes a number of performance management software that facilitate HR to evaluate and monitor the employee's performance and progress against their key performance indicators (Curzi et al., 2019).

The performance management systems apply data-driven approach for performance measurement, which enables managers to provide timely support and assessment to the employees. These measurement systems and software could even estimate the potential performance lag, thereby allowing timely interventions to address the issues. Furthermore, performance management, digital tools also propagate transparency and accountability at the workplace. The people at the workplace can have real-time access and can track their progress. HR professionals have also use big data to make informed decisions regarding training and development remuneration reward for the employees. Digital transformation in HRM system also significantly improves employee engagement (Wang et al., 2022). Digital HR portals for employee engagement surveys, skill set requirement, survey, and collaboration needs help the HR to evaluate and examine the satisfaction level of the employees at the workplace. Remote working has also emerged as a new important phenomenon in the HRM functions. Technological advancements have also enabled remote working which has tremendously enhanced employee engagement (Halid et al., 2022; Tripathi et al., 2022). It has fostered employee engagement, particularly during COVID-19 pandemic and since then the HRM function has drastically changed. Big data solutions, video calling, and other digital platforms have enabled businesses to maintain productivity even in the case of remote working. Flexible working hours and remote job locations has become the new norm in recent years without compromising productivity. This transformation at the workplace is a big shift from the traditional, HRM function to a new hybrid-based HRM function, integrating inclusivity and engagement (Chapano et al., 2023; Syarief et al., 2022).

7.3 Human Resource Management in the Context of Digitalization and Sustainability

With the adoption of a circular economy and sustainable business model, the role of human resource management has changed drastically and moved beyond traditional functions. Human resource managers foster organizations' sustainability culture. They designed digital sustainability practices by embracing circular economy principles like sustainable consumption, regeneration of resources, and reuse (Gierlich-Joas et al., 2020). Through sustainability training, courses, social responsibility, practices, and waste management programs, HR encourage employees towards environmental and social concerns. The human resource manager also plays an instrumental role in integrating the employee's role with the organizational goals through sustainability practices and green behaviour. HRM also propagate a long-term sustainability culture within the organization that faster circular business environment at the workplace (Faisal, 2023; Mandip, 2012). It also plays an important role in talent acquisition from sustainability perspective, who possesses the required skill set and mindset based on sustainability principles. This also play an important role in driving sustainability culture in the early orientation of the employees. HRM based on circular principles, not only hires with a sustainability perspective but also develop expertise in the existing employees to achieve the organizational sustainability vision (Alyahya et al., 2023). The circular economy business model aims to integrate human resource management and organization business operations to minimize waste and optimize resource utilization. This could be possible through agile workforce management in the form of remote work location project-based jobs and skill-based contract. Through dynamic HRM practices in the form of remote working and agile workforce, HRM facilitate reduction in carbon emissions and improves workforce well-being (Straus et al., 2023). Digital HRM also promotes skillet-sharing practices wearing the employees are shifted across functions to different roles (Zhou & Cen, 2023). This not only ensures optimum utilization of workforce but also facilitates knowledge sharing and sustainable innovation, thereby contributing to the overall sustainability goals of the organization (Dabić et al., 2023). Digital human resource management also helps in the development and implementation of circular workforce life-cycle management. The principle of circular economy practices could also be applied to the employee's life cycle management where in HRM encourages circularity and sustainability at every phase of employees, life cycle from hiring to retirement. It may involve orientation of employees at the workplace with a sustainability perspective, designing sustainability-oriented virtual training sessions, and a sustainable workplace with minimal waste and energy consumption (Singh et al., 2024a, b, c). The HRM could facilitate life-cycle transition in the employees' careers through upskilling programs and ex-employee engagement programme. This also helped the businesses to maintain long-lasting relationships with all the employees and serve as brand ambassadors of the business in the society. They may also be used by HR in future business projects and contribute to the growth of the organization. The HR managers are leveraging digital technologies and integrating circular principles into HRM to improve the competitiveness and achieve the sustainability goals of the business (Chen et al., 2024; Kumar & Singh, 2024). Advanced technologies like AI big data, machine learning is revolutionizing the human resource management strategies. These technologies have emerged as important tool to ensure resource optimization, mitigate waste, improve life-cycle management, and thereby inform decision-making (Singh et al., 2023). Human resource management functions are increasing adopting AI, big databases, and digital tools for transforming human resource functions, such as talent acquisition, performance, management, processes, training, and development, etc. These digital platforms are crucial in HRM as they could process a vast amount of data to identify the needs and match the job roles and skill set as per the requirements of the organization (Dash, 2023). Moreover, these digital platforms could also provide personalized career development plans for employees. This may help the organization to build more flexible, adaptable, satisfied employees and a sustainable workplace. It is important for human resource manager to leverage these AI, machine, and learning-based tools to foster innovation, collaboration, and creativity among the employees.

7.4 Linking Digital Circular Economy and Human Resource Management Function

The convergence between digital advancement and sustainability has provided unique opportunities and challenges to business organizations across the world. This convergence has not only developed the concept of digital circular economy which integrates logical tools with circular business model for resource optimization, waste, minimization, and sustainable consumption. The success of a digital circular business model largely depends upon the effectiveness of the HRM function how the HRM functions manage employees and implement practises ensures the success of digital transformation with sustainability perspective in the organization (Kumar & Singh, 2024). Thus, human resource management has become a significant function of the organization. It equips employees. The required skill set and motivation to adopt digital circular business practises. Prabhu et al. (2024) emphasizes that HRM strategies is based on digital circular economy principles are essential for fostering green behaviour and sustainable innovation culture in the organization. noted that circular business model should integrate digital technological tools to resource optimization, resource regeneration and refurbishment of material. Sharma et al. (2023) highlighted that digital circular business model practises should also give priority to the talent management function. To promote circular business practises human resource manager must identify the prospective candidate with the sustainability mindset and further cultivate the sustainability expertise and skill set in the employees (Chen, 2023). This implies executing hiring process hiring of employees through the usage of digital tools and the assessment of candidates through digital platforms.

HRM should adopt a strategy to incentivise sustainability behaviour in line with the circular economy principle at the workplace. It may involve a reward system to appreciate the contribution of the employees towards sustainability. This sustainability performance of the employees in the form of resource optimization, reduction in waste and sustainable innovation at the workplace should be rewarded. HR managers need to align the performance metrics of the employees with the organization's sustainability goals. HRM strategies should also focus on encouraging employees to think creatively and sustainability practices at the workplace (Agarwal et al., 2023). HRM needs to integrate sustainability and circular economy principles into the vision and values of the organization. This will foster an environment where the employee will be motivated to actively integrate sustainability perspective into their routine work.

7.4.1 Employee Development and Green Skills

An important factor that links the digital circular economy principle with the human resource management is the development of workforce having sustainability skills, essential to fostering a sustainability culture at the workplace. Sustainability skillsets imply the required competencies and capabilities that help the employees to contribute to the sustainability goals of the organization through knowledge sharing, sustainable consumption, waste mitigation, and reduction in carbon footprints (Din et al., 2024). From the perspective of a digital circular economy, these skills also involve the understanding of digital tools like sustainable manufacturing, waste management etc. The employees should be given the learning to use digital tools to monitor negative environmental impact and track their sustainability performance. Agrawal et al. (2023) emphasize AI and blockchain integration into the supply chain management through imparting digital skills to the employees that can measure, monitor, and interpret the concerned data to make informed decision-making in the business. Human resource management develops learning management programs to foster green technical skills. Development and implementation of virtual learning programs are through training, courses, workshops, and other collaborative initiatives with educational universities. It also develops career progression pathways to encourage the employees to acquire new and advanced skills (Kumar & Tarkar, 2024; Nyathi & Kekwaletswe, 2024). Human resource managers should align the skill development programs in line with the digital circular economy principles to ensure that organization may have competitiveness in this contemporary sustainability-driven business environment.

HR strategies need to focus on employee well-being and engagement, especially in times of rapid technological advancements. They need to promote virtual learnings for the employees, opportunities for reskilling and skilling and encourage collaboration among peers for knowledge sharing. This will help in creating a sustainable workplace where the employees will be adaptable, resilient, and aligned with the sustainability goals of the organization. Chand et al. (2023) noted that the growth of the business organization rests on the HRM strategies to integrate digital technologies with the sustainability goals and building a workforce with the sustainability mindset. Human resource management functions in the organization also need to ensure that employees are equipped with the required technological skill sets to navigate the challenges posed transition towards a digital circular business model (Singh et al., 2024a, b, c).

7.5 Challenges in Adopting Digital Circular Economy in HRM

Setyaningrum and Muafi (2023) noted that to adopt digital circular economy practices, the organization must foster sustainable innovation in the workplace. It requires a fundamental shift in the organizational culture from sustainability perspective. Wikhamn (2019) highlighted that sustainability innovation culture in the organization encourages employees to think out of the box concerning how to ensure resource optimization and create value from the sustainability perspective. Business core strategy should be based on sustainability and circular economy by integrating sustainable development in the vision, mission, values, and purpose of the organization. HRM also promotes circular economy practices through employee engagement practices such as sustainability leaders, surveys, sustainability, and champion programs within the organization to recognize the contribution of best sustainability practices of the employees. Digital circular economy is based on sustainability innovation, which requires businesses to invest in new business models, operations, and products. Human resource management plays an important role to foster sustainable innovation by promoting sustainability perspective in the mindset of the employees and imparting necessary learnings. This results in creating a culture of sustainable innovation to accomplish the sustainability goals of the organization (Duvnjak & Kohont, 2021).

Kumar and Singh (2024) noted that digital circular economy presents significant workforce issues for human resource management. With the increased emphasis on digital technology, automation, and AI-based tools, displacement of jobs has become the common norm and primary challenge that HRM should address. These advanced technologies are primarily for circular economy practices which may turn a few jobs obsolete and result in job losses. Human resource management has a responsibility to prepare the workforce for this transition by developing HRM strategies of reskilling and reallocating the rules to the employees. These HRM strategies may also include the development and implementation of digital technologies based on circular economy practices. Dabić et al. (2023) highlighted that human resource manager also has a responsibility to ensure that the benefits of digital circular practises should lead to inclusive growth and environment, social development, and environmental considerations. Setyaningrum and Muafi emphasized that successful integration of digital circular economy practices in the organization not only depends upon the adoption of digital tools but also implementing human resource management strategies in line with the circular economy principles. Another important challenge that often-business face in the process of digital transition towards circular economy practices is change management at the workplace (Vasilev et al., 2023). The workforce is often hesitant to adopt new digital tools as there not well versed with the digital circular business practices. The HRM function has an important role in addressing this resistance to change by clear communication, required training, and emphasizing benefits of the digital technologies to the employees. They must also support the employees in every possible manner throughout the transition process. The human resource management strategies should focus on building the competencies and capabilities of the employees which involves agility, resilience, and willingness to adapt to the new digital circular business model (Hee & Shanmugam, 2019; Husen et al., 2024; Scholkmann, 2021).

Human resource management should identify the required training and development programs to address the issue of skill gap in the workforce in the context of digital circular economy principles. It is often observed that existing learning programs fails to address the need for required competencies to work in the digital circular economy-based business models. Delmas and Pekovic (2018) Highlighted that training and development programs, often focus on motivation, team management, innovation, but do not including the learnings for advanced digital technologies in the field and sustainable innovation. The HRM strategies must integrate digital circular economy principles and rework screening and development strategies, keeping in mind the skill gaps and digital circular economy framework. The lack of technological infrastructure and resources is another critical factor to integrate digital circular economy principles into HRM practices. Bressanelli et al. (2022) highlighted that for development and implementation of a digital circular economy, business model requires strong technological infrastructure. However, many medium and small enterprises lack technological capabilities, especially in the developing economies. The investments in the digital infrastructure and resources may not be financially feasible for these organizations. They may find it costly to invest in new technologies and systems to measure and monitor sustainability practices. The lack of funds to invest in these technologies is the major issue (Trujillo-Gallego et al., 2022). The human resource manager need understand and identify the benefits of technological transformation to achieve the sustainability goals of the organization. With increased emphasis on digital circular economy principles, labour policies also need to be evolved to support HRM in fostering sustainability. Public policies should provide supporting environment to businesses to bolster sustainability practices. Public policy should emphasize on the development of skills required to implement a digital circular economy. The government could also support human resource management by incentivizing businesses to invest in digital sustainability training programs, focused on waste, minimization resource efficiency, renewable energy, and sustainable production (Agrawal & Kumar, 2024; Aust et al., 2020). These policies must also promote the use of advanced technologies and digital training like AI tools, machine, learning, blockchain, and IOT which are important for implementing digital circular practices. The government can also support organizations by building digital infrastructure and resources on mass scale promoting digital circular economy (Singh et al., 2024a, b, c).

7.6 Conclusions, Limitations, and Future Research Direction

With the increasing emphasis on building, the digital circular economy has drastically transformed the way businesses operate, undertake economic activities, and deal with the environment. The transition not only requires the development and implementation of digital tools and techniques, but also a fundamental shift in the relationship between employees, organizations, and social responsibility. This chapter outlines the importance of the role of human resource management strategies in driving sustainability, the performance of the organization, and integrating digital circular economy framework in the business. The convergence between digital circular economy framework and human resource management systems is the key for sustainable innovation and growth of the business. Human resource management plays an instrumental role in achieving the sustainability goals of the business by fostering a culture of sustainable innovation, reskilling employees, and digital performance management systems. The digital transformation in the form of AI-based tools, blockchain, big data analytics, and cloud computing has drastically changed the human resource management function in terms of manpower, planning, recruitment and selection, and employee relations. Integration of digital technologies into HRM function help the HR managers to predict the requirement of technological skill sets by the employees, future manpower. It also helps HR managers to identify and assess the employees, knowledge, and orientation towards sustainability values and practices. As the digital circular economy focus on waste, minimization, and refurbishment and regeneration of resources, HRM strategies need to focus on building, workforce, competencies, and capabilities in the digital circular economy perspective. HRM strategy should be aimed at inculcating green behaviour, developing sustainability competencies, and aligning performance metrics with a sustainability goal of the organization. HR managers should focus on implementing AI-based tools and learning programs on sustainability practices and cross-department knowledge sharing. This will help the organization to adapt to the new changing requirements of the digital circular business model. Furthermore, fostering a sustainability culture helps organizations to promote sustainability innovation in the product, and life cycle and induce a positive impact on the environment. HRM strategies also include employee engagement for driving sustainability initiatives at the workplace by linking the task with organizational sustainability values. Employees' performance metrics are designed to measure sustainability contributions and further reward system can be established to recognize the sustainability efforts of the employees in terms of resource optimization and waste minimization. These strategies are extremely important for driving employees' performance towards sustainability and circular economy principles. Human resource management strategies aligned with the digital circular economy perspective, not only help the business to grow. It also enhances competitiveness in the long run and helps in building a sustainable competitive advantage for the business.

This chapter offers several implications for policymaker practitioners and HR managers. human resource management strategies could be extremely useful for organizations, aiming for a transition toward a digital circular economy. it will also help companies to inculcate sustainability innovation culture in the organization to achieve sustainability goals and a positive impact on the environment. Performance management, digital tools, and AI-based training programs could significantly improve the digital circular economy practices at the workplace. It is important to integrate HRM with sustainability goals to establish digital circular economy for a sustainable future. It will also help the human resource management department to build sustainable workplace based on the transition towards digital circular economy principles. This chapter also has certain limitations. An empirical study based on primary data about the effectiveness of HRM strategies for achieving a digital circular economy would provide valuable insights in this area. Moreover, it would be interesting to examine the impact of how human resource management practices are driving sustainability performance and digital circular economy in the context of small and medium enterprises across the industries.

References

Abid, U., Faisal, M. N., Al-Esmael, B., Farooq, Z. H., & Nassour, S. (2024). Exploring the moderating role of technological competence and artificial intelligence in green HRM. *Polish Journal of Management Studies, 29*.

Agarwal, S., Gupta, A., & Roshani, P. (2023). Redefining HRM with artificial intelligence and machine learning. In *The adoption and effect of artificial intelligence on human resources management* (pp. 1–13). Emerald Publishing Limited.

Agrawal, D., & Kumar, K. (2024). Achieving SDGs through MSMEs: An empirical assessment of environmental consideration initiatives in India. In *Green transition impacts on the economy, society, and environment* (pp. 358–376). IGI Global.

Agrawal, N., & Bansal, R. (2022). Green HRM: A medium of social responsibility and sustainable development. *International Journal of Innovation and Sustainable Development, 16*(2), 172–185.

Ali, M., Shujahat, M., Fatima, N., Lopes de Sousa Jabbour, A. B., Vo-Thanh, T., Salam, M. A., & Latan, H. (2024). Green HRM practices and corporate sustainability performance. *Management Decision*.

Alsaif, A., & Sabih Aksoy, M. (2023). AI-HRM: Artificial intelligence in human resource management: A literature review. *Journal of Computing and Communication, 2*(2), 1–7.

Alyahya, M., Aliedan, M., Agag, G., & Abdelmoety, Z. H. (2023). The antecedents of hotels' green creativity: The role of green HRM, environmentally specific servant leadership, and psychological green climate. *Sustainability, 15*(3), 2629.

Aust, I., Cooke, F. L., Muller-Camen, M., & Wood, G. (2024). Achieving sustainable development goals through common-good HRM: Context, approach and practice. *German Journal of Human Resource Management, 38*(2), 93–110.

Aust, I., Matthews, B., & Muller-Camen, M. (2020). Common good HRM: A paradigm shift in sustainable HRM? *Human Resource Management Review, 30*(3), 100705.

Barykin, S. Y., Kapustina, I. V., Valebnikova, O. A., Valebnikova, N. V., Kalinina, O. V., Sergeev, S. M., & Volkova, L. (2021). Digital technologies for personnel management: Implications for open innovations. *Academy of Strategic Management Journal, 20*, 1–14.

Bressanelli, G., Adrodegari, F., Pigosso, D. C., & Parida, V. (2022). Circular economy in the digital age. *Sustainability, 14*(9), 5565.

Brewster, C., & Brookes, M. (2024). Sustainable development goals and new approaches to HRM: Why HRM specialists will not reach the sustainable development goals and why it matters. *German Journal of Human Resource Management, 38*(2), 183–201.

Budhwar, P., Malik, A., De Silva, M. T., & Thevisuthan, P. (2022). Artificial intelligence–challenges and opportunities for international HRM: A review and research agenda. *The International Journal of Human Resource Management, 33*(6), 1065–1097.

Burrichter, K., Chen, B., & Marco, G. (2022). Evaluation of modern technology on human resources management and sustainable development in pharma industries. *Journal of Commercial Biotechnology, 27*(2).

Cagno, E., Neri, A., Negri, M., Bassani, C. A., & Lampertico, T. (2021). The role of digital technologies in operationalizing the circular economy transition: A systematic literature review. *Applied Sciences, 11*(8), 3328.

Campos-García, I., Alonso-Muñoz, S., González-Sánchez, R., & Medina-Salgado, M. S. (2024). Human resource management and sustainability: Bridging the 2030 agenda. *Corporate Social Responsibility and Environmental Management, 31*(3), 2033–2053.

Centobelli, P., Cerchione, R., Del Vecchio, P., Oropallo, E., & Secundo, G. (2022). Blockchain technology for bridging trust, traceability and transparency in circular supply chain. *Information & Management, 59*(7), 103508.
Chams, N., & García-Blandón, J. (2019). On the importance of sustainable human resource management for the adoption of sustainable development goals. *Resources, Conservation and Recycling, 141*, 109–122.
Chand, R., Narula, G. S., Nijjer, S., & Jandwani, A. (2023). Utilizing AI in sustaining green HRM practices—a digital initiative towards socially responsible and environment sustainability. In *5th International conference on advances in computing, communication control and networking (ICAC3N)* (pp. 541–544). IEEE.
Chapano, M., Mey, M. R., & Werner, A. (2023). Perceived challenges: Unfounded reasons for not forging ahead with digital human resource management practices. *SA Journal of Human Resource Management, 21*, 2085.
Chen, Q., Gong, Z., Wu, J., & Wang, T. (2024). Does digital transformation affect carbon performance through talent? The moderating role of employee structure. *Journal of Cleaner Production, 435*, 140581.
Chen, Z. (2023). Revolutionising HRM practice with blockchain technology: unleashing disruptive paradigms of work and overcoming management challenges. *Technology Analysis & Strategic Management,* 1–14.
Curzi, Y., Fabbri, T., Scapolan, A. C., & Boscolo, S. (2019). Performance appraisal and innovative behavior in the digital era. *Frontiers in Psychology, 10*, 1659.
Dabić, M., Maley, J. F., Švarc, J., & Poček, J. (2023). Future of digital work: Challenges for sustainable human resources management. *Journal of Innovation and Knowledge, 8*(2), 100353.
Dantas, T. E. T., de-Souza, E. D., Destro, I. R., Hammes, G., Rodriguez, C. M. T., & Soares, S. R. (2021). How the combination of circular economy and industry 4.0 can contribute towards achieving the sustainable development goals. *Sustainable Production and Consumption, 26*, 213–227.
Dash, S. P. (2023). HR digital transformation: Blockchain for business. *Recent Advances in blockchain technology: Real-world applications* (pp. 59–87). Springer International Publishing.
Delmas, M. A., & Pekovic, S. (2018). Corporate sustainable innovation and employee behavior. *Journal of Business Ethics, 150*, 1071–1088.
Din, A. U., Yang, Y., Yan, R., Wei, A., & Ali, M. (2024). Growing success with sustainability: The influence of green HRM, innovation, and competitive advantage on environmental performance in the manufacturing industry. *Heliyon, 10*(10).
Ding, T., Li, J., Shi, X., Li, X., & Chen, Y. (2023). Is artificial intelligence associated with carbon emissions reduction? Case of China. *Resources Policy, 85*, 103892.
Duvnjak, B., & Kohont, A. (2021). The role of sustainable HRM in sustainable development. *Sustainability, 13*(19), 10668.
Faisal, S. (2023). Green human resource management—a synthesis. *Sustainability, 15*(3), 2259.
Geissdoerfer, M., Savaget, P., Bocken, N. M., & Hultink, E. J. (2017). The circular economy–a new sustainability paradigm? *Journal of Cleaner Production, 143*, 757–768.
Gierlich-Joas, M., Hess, T., & Neuburger, R. (2020). More self-organization, more control-or even both? Inverse transparency as a digital leadership concept. *Business Research, 13*(3), 921–947.
Halid, H., Halim, S. N. A., & Ravesangar, K. (2022). Human resource management practices in the digital era. *Technological challenges: The human side of the digital age* (pp. 109–158). Springer International Publishing.
Hee, O. C., & Shanmugam, N. (2019). A review of human resource change management strategies in the digital era. *International Journal of Academic Research in Business and Social Sciences, 9*(3), 521–531.
Husen, S., Wahidah, R. N., & Mustajab, D. (2024). Evolution of HRM strategies in the digital age: A comprehensive review. *Amkop Management Accounting Review (AMAR), 4*(1), 42–58.

Islam, T., Khan, M. M., Ahmed, I., & Mahmood, K. (2021). Promoting in-role and extra-role green behavior through ethical leadership: Mediating role of green HRM and moderating role of individual green values. *International Journal of Manpower, 42*(6), 1102–1123.

Kathpal, S., Kumar, K., Kumar, P., & Mishra, S. (2021). Building a low carbon indian economy. In *IOP conference series: Earth and environmental science* (Vol. 795, No. 1, p. 012023). IOP Publishing.

Khan, K., Khan, Q., Jalaly, S., Shams, M. S., & Shah, T. A. (2024). Green human resource management and green service behaviour in organisations: Examining underlying mechanism. *South Asian Journal of Human Resources Management, 11*(1), 113–129.

Khan, S. A. R., Piprani, A. Z., & Yu, Z. (2022). Digital technology and circular economy practices: Future of supply chains. *Operations Management Research, 15*(3), 676–688.

Klimovskikh, N., Sekerin, V., Makushkin, S., Kuzmicheva, A., Leontev, M., & Kochetkov, E. (2023). Impact of human resource management on improving the innovation potential of an enterprise to achieve the principles of sustainable development. *Journal of Law and Sustainable Development, 11*(1), e0274–e0274.

Korhonen, J., Honkasalo, A., & Seppälä, J. (2018). Circular economy: The concept and its limitations. *Ecological Economics, 143*, 37–46.

Kristoffersen, E., Blomsma, F., Mikalef, P., & Li, J. (2020). The smart circular economy: A digital-enabled circular strategies framework for manufacturing companies. *Journal of Business Research, 120*, 241–261.

Kumar, K., & Singh, R. (2024). *Sustainable human resource management*. Elsevier.

Kumar, K., & Tarkar, P. (2024). Influence of sustainable human resource management on sustainability performance of the organization: Exploring the mediating role of organizational citizenship behaviour towards sustainability. *Journal of the Knowledge Economy*, 1–28.

Kumar, K., Singh, R., Sharma, A., & Lehri, B. P. S. (2024). Climate Change adaptation strategies for achieving net-zero economy. In *Transition towards a sustainable future* (pp. 163–183). Springer.

Lampropoulos, G., Garzón, J., Misra, S., & Siakas, K. (2024). The role of artificial intelligence of things in achieving sustainable development goals: State of the art. *Sensors, 24*(4), 1091.

Liu, L., Song, W., & Liu, Y. (2023). Leveraging digital capabilities toward a circular economy: Reinforcing sustainable supply chain management with industry 4.0 technologies. *Computers & Industrial Engineering, 178*, 109113.

Liu, Q., Trevisan, A. H., Yang, M., & Mascarenhas, J. (2022). A framework of digital technologies for the circular economy: Digital functions and mechanisms. *Business Strategy and the Environment, 31*(5), 2171–2192.

Luo, J., Zhuo, W., Liu, S., & Xu, B. (2024). The optimization of carbon emission prediction in low carbon energy economy under big data. *IEEE Access*.

Mandip, G. (2012). Green HRM: People management commitment to environmental sustainability. *Research Journal of Recent Sciences*.

Marrucci, L., Daddi, T., & Iraldo, F. (2021). The contribution of green human resource management to the circular economy and performance of environmental certified organisations. *Journal of Cleaner Production, 319*, 128859.

Medaglia, R., Rukanova, B., & Zhang, Z. (2024). Digital government and the circular economy transition: An analytical framework and a research agenda. *Government Information Quarterly, 41*(1), 101904.

Mishra, P. (2017). Green human resource management: A framework for sustainable organizational development in an emerging economy. *International Journal of Organizational Analysis, 25*(5), 762–788.

Mishra, P., Kumar, K., Fouzdar, A. S., & Singh, A. (2023). Sustainability disclosure and green finance: Driving the transition towards a sustainable future. In *The sustainable Fintech revolution: Building a greener future for finance* (pp. 20–36). IGI Global.

Mukhuty, S., Upadhyay, A., & Rothwell, H. (2022). Strategic sustainable development of industry 4.0 through the lens of social responsibility: The role of human resource practices. *Business Strategy and the Environment, 31*(5), 2068–2081.

Nyathi, M., & Kekwaletswe, R. (2024). Electronic human resource management (e-HRM) configuration for organizational success: Inclusion of employee outcomes as contextual variables. *Journal of Organizational Effectiveness: People and Performance, 11*(1), 196–212.

Olateju, A. O., Aminu, A. W., & Danmola, R. A. (2020). Green human resources management (green HRM) and sustainable development: Prospects and challenges. *International Journal of Social Sciences and Humanities Review, 10*(3).

Pagoropoulos, A., Pigosso, D. C., & McAloone, T. C. (2017). The emergent role of digital technologies in the circular economy: A review. *Procedia Cirp, 64*, 19–24.

Pan, Y., & Froese, F. J. (2023). An interdisciplinary review of AI and HRM: Challenges and future directions. *Human Resource Management Review, 33*(1), 100924.

Piscicelli, L. (2023). The sustainability impact of a digital circular economy. *Current Opinion in Environmental Sustainability, 61*, 101251.

Prabhu, U., Fathima, Y., Sathyanarayana, N., Srikumar, M. S., & Cavaliere, L. P. L. (2024). Emerging digitalization trends in human resource management. In *Data-driven intelligent business sustainability* (pp. 214–234). IGI Global.

Radonjić, A., Duarte, H., & Pereira, N. (2022). Artificial intelligence and HRM: HR managers' perspective on decisiveness and challenges. *European Management Journal.*

Raza, A., Shaikh, E., Tursoy, T., & Almashaqbeh, H. A. (2022). Economics and business perspectives of sustainable HRM. In *Sustainable development of human resources in a globalization period* (pp. 36–48). IGI Global.

Rodriguez-Anton, J. M., Rubio-Andrada, L., Celemín-Pedroche, M. S., & Alonso-Almeida, M. D. M. (2019). Analysis of the relations between circular economy and sustainable development goals. *International Journal of Sustainable Development & World Ecology, 26*(8), 708–720.

Sánchez-García, E., Martínez-Falcó, J., Marco-Lajara, B., & Manresa-Marhuenda, E. (2024). Revolutionizing the circular economy through new technologies: A new era of sustainable progress. *Environmental Technology & Innovation, 33*, 103509.

Santana, M., & Lopez-Cabrales, A. (2019). Sustainable development and human resource management: A science mapping approach. *Corporate Social Responsibility and Environmental Management, 26*(6), 1171–1183.

Sariatli, F. (2017). Linear economy versus circular economy: A comparative and analyzer study for optimization of economy for sustainability. *Journal on Bioeconomy and Sustainable Development, 6*(1), 31–34.

Scholkmann, A. B. (2021). Resistance to (digital) change: Individual, systemic and learning-related perspectives. In *Digital transformation of learning organizations* (pp. 219–236).

Setyaningrum, R., & Muafi, M. (2023). Green human resource management, green supply chain management, green lifestyle: Their effect on business sustainability mediated by digital skills. *Journal of Industrial Engineering and Management, 16*(1), 1–26.

Sharma, A., Sharma, D., & Bansal, R. (2023). Challenges of adopting artificial technology in human resource management practices. In *The adoption and effect of artificial intelligence on human resources management* (pp. 111–126). Emerald Publishing Limited.

Sharma, R. C., & Kohli, D. (2024). Embracing digital transformation for sustainable development: The synergy between learning organizations, human resource management, and service quality. *Journal of Chinese Human Resources Management, 14*(3), 3–17.

Singh, A., Kanaujia, A., Singh, V. K., & Vinuesa, R. (2024a). Artificial intelligence for sustainable development goals: Bibliometric patterns and concept evolution trajectories. *Sustainable Development, 32*(1), 724–754.

Singh, R., Kumar, K., & Khan, S. (2024). A Comprehensive view of artificial intelligence (AI)–based technologies for sustainable development goals (SDGs). In *Artificial intelligence enabled management: An emerging economy perspective* (p. 183).

Singh, R., Ravesangar, K., Narayanan, S., & Singh, K. K. H. (2024). HRM practices in creating net zero: Supporting organizational sustainability. In *Net zero economy, corporate social responsibility and sustainable value creation: Exploring strategies, drivers, and challenge*s (pp. 1–18). Springer Nature Switzerland.

Singh, S., Thakur, P., & Singh, S. (2023). How does the use of AI in HRM contribute to improved business performance?: A systematic review. *Managing Technology Integration for Human Resources in Industry, 5*, 131–139.

Sova, O., Bieliaieva, N., Antypenko, N., & Drozd, N. (2023). Impact of artificial intelligence and digital HRM on the resource consumption within sustainable development perspective. In *E3S web of conferences* (Vol. 408, p. 01006). EDP Sciences.

Straus, E., Uhlig, L., Kühnel, J., & Korunka, C. (2023). Remote workers' well-being, perceived productivity, and engagement: Which resources should HRM improve during COVID-19? A longitudinal diary study. *The International Journal of Human Resource Management, 34*(15), 2960–2990.

Sukalova, V., Stofkova, Z., & Stofkova, J. (2022). Human resource management in sustainable development. *Sustainability, 14*(21), 14258.

Syarief, F., Nindiasari, H., Martanto, M., Febriani, B., & Wujarso, R. (2022). e-HRM: Changes in business and labor culture in the digital paradigm. *e-HRM: Changes in Business and Labor Culture in the Digital Paradigm, 6*(1).

Theres, C., & Strohmeier, S. (2023). Met the expectations? A meta-analysis of the performance consequences of digital HRM. *The International Journal of Human Resource Management, 34*(20), 3857–3892.

Thite, M. (2018). *E-HRM: Digital approaches, directions & applications*. Routledge.

Trihinas, D., Thamsen, L., Beilharz, J., & Symeonides, M. (2022). Towards energy consumption and carbon footprint testing for AI-driven IoT services. In *IEEE international conference on cloud engineering (IC2E)* (pp. 29–35). IEEE.

Tripathi, M. A., Tripathi, R., Yadav, U. S., & Shastri, R. K. (2022). Gig economy: A paradigm shift towards digital HRM practices. *Journal of Positive School Psychology, 6*(2), 5609–5617.

Trujillo-Gallego, M., Sarache, W., & de Sousa Jabbour, A. B. L. (2022). Digital technologies and green human resource management: Capabilities for GSCM adoption and enhanced performance. *International Journal of Production Economics, 249*, 108531.

Vardarlier, P. (2020). Digital transformation of human resource management: digital applications and strategic tools in HRM. In *Digital business strategies in blockchain ecosystems: Transformational design and future of global business* (pp. 239–264).

Vasilev, V., Stefanova, D., & Popescu, C. (2023). Human Capital management and digitalization–from good practices and traditions to sustainable development. *In Digitalization, Sustainable development, and industry 5.0* (pp. 41–65). Emerald Publishing Limited.

Wang, L., Zhou, Y., & Zheng, G. (2022). Linking digital HRM practices with HRM effectiveness: The moderate role of HRM capability maturity from the adaptive structuration perspective. *Sustainability, 14*(2), 1003.

Wikhamn, W. (2019). Innovation, sustainable HRM and customer satisfaction. *International Journal of Hospitality Management, 76*, 102–110.

Yusuf, M., Satia, H. M. R., Bernardianto, R. B., Nurhasanah, N., Irwani, I., Kurniasih, D., & Setyoko, P. I. (2023). Investigating the effect of digital HRM and digital innovation and the SMEs performance in Indonesia. *International Journal of Professional Business Review, 8*(6), e02751–e02751.

Zhang, J., & Chen, Z. (2024). Exploring human resource management digital transformation in the digital age. *Journal of the Knowledge Economy, 15*(1), 1482–1498.

Zhou, J., & Cen, W. (2023). Design and application research of a digital human resource management platform based on ChatGPT. *Journal of Theory and Practice of Social Science, 3*(7), 49–57.

Zubair, D. S. S., & Khan, M. (2019). Sustainable development: The role of green HRM. *International Journal of Research in Human Resource Management, 1*(2), 1–6.

Chapter 8
Barriers to Circular Business Models in Mexico and Colombia Economies

Graciela Carrillo González, Jorge Issac Lechuga-Cardozo, and Adriana Marcela Cáceres-Martelo

Abstract The environmental crisis and the effects of climate change demand a transition towards sustainability that includes strategies such as the circular economy and concrete actions that involve governments and society in the implementation of business models with a circular approach, which implies transforming production schemes, technologies and consumption patterns. The objective of this article is to identify how these barriers to the development of NQFs are manifested in Mexico and Colombia and to locate the actions that are already present in these countries, but which are still insufficient to make way for a sustainable transition. The methodology used was qualitative based on the systematic review of the literature on sustainable transition and circular economy, as well as the analysis of official documents of the two countries studied. The conclusions indicate that both Mexico and Colombia have advanced in the construction of a legal framework and have promoted some programs from their governments to encourage the circular economy, but there are no solid financing schemes and there are no decisive and systematic policies that promote the creation of circular business models as concrete actions of the transition. The contribution of this work is focused on evidencing the weakness of the actions promoted in the countries of the South to contribute to sustainable transitions, which makes it difficult to comply with international commitments and achieve the SDG targets.

Keywords Circular business models · Circular economy · Sustainable transitions · SDGs · Sustainable business models

---

G. C. González
Department of Economic Production, Universidad Autonóma Metropolitana Xochimilco, Mexico City, Mexico
e-mail: gcarri@correo.xoc.uam.mx

J. I. Lechuga-Cardozo (✉)
School of Business, Tecnológico de Monterrey, Monterrey, Mexico
e-mail: jorgeisaac.lc@gmail.com

A. M. Cáceres-Martelo
PRIME Business School, Sergio Arboleda University, Bogotá, Colombia
e-mail: adriana.caceres@usa.edu.co

R. Singh and V. Kumar (eds.), *Sustainable Innovations and Digital Circular Economy*,
https://doi.org/10.1007/978-981-96-1064-8_8

8.1 Introduction

Business models with a circular economy approach are also called circular business models. CBMs are considered one of the important ways to advance towards a sustainable transition (Aagaard et al., 2021) that today has as a reference the 2030 horizon with the Sustainable Development Goals (SDGs). However, despite the international context and the position of international organizations and governments have made commitments in this regard, there are barriers that hinder the possibility of these models to consolidate and be representative in the economies of different countries, especially in emerging and developing countries.

The published studies point out that the main barriers are to be found in the governments themselves, particularly in the case of the countries of the Global South and emerging countries, because there is no legislation, programs to promote innovation and specific financing to drive a process of change; the second difficulty is in the market, in two ways: the first is that the technological offer and the existence of suppliers and concrete recycling markets operate in an unsystematic way, and the second is linked to consumer decisions, which in general do not yet show a real and strong preference for recycled products.

This chapter is not only interested in identifying how these barriers to the development of NQFs are manifested in two Latin American countries, Mexico and Colombia, but also in locating and describing the weak impulses that are already present but are still insufficient to move towards a sustainable transition. For this purpose, a qualitative methodology is used to identify public documents and academic works that report on the development and obstacles of NQFs in each of the countries. The results point out that although both countries are part of the same geographical region, there are important gaps in the characteristics of their economies and in the progress towards the construction of NQFs, which are evident in the barriers observed in each case.

8.1.1 Sustainable Transitions and SDGs

The concept of transition is linked to different disciplines in both the biological and social sciences. The use of this concept became generalized in the scientific and academic sphere during the 90's of the last century with the purpose of studying and explaining how a large-scale social and technical change could be achieved, marking a path towards sustainability. Based on this orientation, sustainable transitions are considered as those actions focused on the study of social regimes that, under the current operating logic, are perceived as unsustainable in the long term due to the dominant production and consumption patterns, which makes it necessary to design management strategies to achieve a transcendent change (Kemp, 1994a).

Sustainable transitions are profound alterations to a status quo, which modify institutional structures, infrastructure, and social practices in which a series of ingrained customs are inserted (Grin, 2016), emphasizing the causes of the depletion or deterioration of natural resources and the impacts this generates on society. The changes required by a transition are experienced in different dimensions such as economic, social, technological, organizational, political, and cultural so that as it progresses it involves a greater number of actors who gradually modify their perceptions, preferences and decisions. These transformations involve long periods of time (Markard et al., 2012).

In the literature on sustainable transitions, the analysis of the processes that lead to a fundamental change in socio-technical systems has been deepened, so that socio-technical transitions imply important changes in technological, organizational and institutional terms for both production and consumption, leading to the development of new products and services, as well as the emergence of a new type of organizations involving a wide range of actors (Singh et al., 2022). This process regularly takes place over periods of more than 25 years, at which point a different culture already begins to be observed in society that undergoes social changes of broad dimensions with large-scale disruptive variations (Loorbach et al., 2017) to provide solutions to new challenges.

The concept of transition also refers to the role that technology can play in overcoming challenges and achieving sustainability objectives, but meeting these objectives depends on public and private support for the development and diffusion of environmentally beneficial technologies (Kemp, 1994b). The contribution of the public and private sectors is key to drive innovations, investments, learning and technological developments, which support transformation processes in different fields; including new technologies, products and infrastructures, as well as the creation of new rules, norms and social interactions (Loorbach et al., 2017); in this transformation, some elements of the system usually remain constant at first until they eventually adjust to the assumptions, practices and norms established in the society in which they are inserted.

The process of change in transitions is not linear; transformations are slow and need to be continuously reinforced. The underlying mechanism for this variation is coevolution, as different subsystems modify each other, leading to irreversible patterns (Loorbach & Rotmans, 2006). Empirical research focused on sustainable transitions has also contributed to the development of tools to guide decision makers on how to support transformation processes. The trends set by the United Nations (UN) and the impetus it gave to the construction of the 2030 Agenda and the 17 SDGs have become the guide for transitions.

The close link between sustainable transitions and innovation raises the possibility of approaching the achievement of the SDGs; however, there is a clear disconnect between the requirements for achieving the goals proposed in the 2030 Agenda and the socioeconomic, technological and cultural dynamics prevailing in institutions and in today's society. Convergence of political decisions and social governance schemes is required for changes to be directed towards the SDG targets and for the necessary resources to be allocated for the implementation of support for the various actions required. The assessment made by the Economic Commission for Latin America and the Caribbean (ECLAC) in 2024 on the fulfillment of the SDGs in the region points out that there has been progress in the countries of the Latin American region, but it is not enough to achieve the goals by 2030.

The context of multiple crises in which the SDGs were proposed in 2015 is no better nine years later, ECLAC has already presented seven reports on the progress of the SDGs and still present conditions of slow economic growth, effects of climate change, scarcity of vital resources, forced migration, political and armed conflicts that have affected the welfare of the population and have limited the achievement of the SDGs, "ECLAC estimates that only 22% of the goals for which information is available exhibit a behavior that allows foreseeing their fulfillment in 2030. Forty-six percent of them show a correct trend, but insufficient to be achieved within the established timeframe, and the remaining 32% show a downward trend that makes their fulfillment in 2030 very unlikely" (UN-ECLAC, 2024).

Consistent progress towards the SDGs requires transition processes that lead to structural change. Sustainable transitions imply that in the long term there will be disruptive innovations that simultaneously transform various systems and dimensions such as technological, organizational, cultural, political, etc., which require public intervention. The systems that are considered priorities for the transition are: food systems, energy, digitalization, education, employment and social protection, and climate transition and biodiversity protection. The strengthening of technical and operational capacities, the design of policies, the financing of innovation and the promotion of programs with actions towards sustainability are key for the transition and in this sense the progress of the countries is differentiated, adhering to a greater or lesser extent to the application of strategies and tools that have been designed to generate change in production and consumption models, such as the circular economy (Idem).

8.1.2 Circular Economy and Business Models

The circular economy (CE) is conceived as an important strategy to achieve a change in current production and consumption patterns towards a sustainable model, as proposed in SDG 12; this implies influencing a whole lifestyle that leads to a sustainable transition and considers that while it is true that the increase in the consumption of natural resources is a response to population growth and has achieved an improvement in the standard of living of the population, it is also true that it has generated a great inequality in the availability and use of these resources, as well as in the distribution of wealth, causing a very serious impact on nature that today is expressed in climate change and the loss of natural resources, as well as in the distribution of wealth.

Production and consumption patterns in today's society are based on a traditional linear economy model based on hydrocarbon consumption. In this model, raw materials are extracted and then transformed into products that are used until they are finally discarded as waste, generating emissions into the atmosphere and water and soil contamination with the already known consequences. The linear economy is a completely unsustainable model in a finite world of limited resources and is largely responsible for climate change, biodiversity loss and resource depletion. CE is based on three principles: recover, reuse and recycle. This involves redesigning products to extend their life cycle, reducing the production of waste for a second use and regenerating nature by reducing the use of virgin raw materials.

The Ellen MacArthur Foundation has been a major promoter of recycling, defining the circular economy as "a restorative or regenerative industrial system by intention and design. It replaces the concept of "expiration" with that of 'restoration', shifts towards the use of renewable energy, eliminates the use of toxic chemicals that harm reuse and return to the biosphere" (Ellen MacArthur Foundation, 2014, p. 3). This CE initiative starts from the principle of material scarcity and the drive towards a dematerialization scheme of the economy (Bunker, 1996), which "aims to reduce both the input of virgin materials and the production of waste, closing the ecological and economic cycles from the maximum possible use of materials and energy" (Carrillo & Torres, 2019, p. 8).

CE increases the productivity and competitiveness of companies and countries. The transition to a circular production model not only contributes to environmental conservation, but also has economic and social benefits. Companies play an important role in the transition to a CE if they place the extension of product life at the center of their strategy and perceive the opportunity to reduce costs and risks for the business. CE facilitates the creation of new markets, drives innovation, attracts talent and generates new expectations among consumers, all of which require long periods of time.

In light of this strategy and in a context where innovation merges physical, environmental, biological and digital aspects, organizations opt for business models that offer them advantages considering legal, economic, environmental and social constraints (Tapia, 2018). A business model describes what an organization offers to its customers, how it creates, produces, sells and delivers value through a proposition. Lacy and Rutqvist (2015) estimate that circular-focused business models have an economic potential of 4.5 trillion dollars by 2030, for the International Labor Organization (ILO, 2018) that potential could represent about 6 million new jobs globally by 2030. Circular business models (CBMs) help companies in operationalizing CE (Nußholz, 2017); it represents a set of strategic decisions designed to preserve the embedded environmental and economic value of a product or service in the system (Centobelli et al., 2020; Nußholz, 2017).

By rethinking a business model and giving it a circular orientation, the purchase of virgin materials is combined with the recycling of waste, product optimization leading to dematerialization and on-demand production processes, which becomes a fundamental condition for taking advantage of CE opportunities (Khan et al., 2020). These business models with a circular approach that are considered novel arise from changes in the environment and allow reconfiguring the capabilities of organizations to adapt to change (Teece, 2010). Recent literature argues that business models are fundamental for a successful sustainable transition, but adaptation of existing policies, rules and regulations is required in order to compete with linear models (Singh et al., 2023).

A sustainable transition that involves changing current production and consumption patterns demands initiatives aimed at building business models that are inserted in a circular economy strategy, which implies a systemic change that involves the actions of three key actors: government, business and society (Guldmann & Huulgaard, 2020). However, although there is increasing interest in implementing circular economy actions and generating circular business models, progress is slow because there are barriers and difficulties that limit them. This circular approach has increased mainly in European Union countries, partly due to the active role played by the public sector to promote them (Christensen, 2021). Countries such as Holland, France and Denmark have already built a tradition in circular models, since the seventies of the twentieth century, as is the case of industrial ecology in Kalundborg, Denmark (Jacobsen, 2006).

8.2 Materials and Methods

This research is based on a qualitative approach based on a systematic review of the literature that provides an account of a current discussion on sustainable transition processes aimed at meeting the SDG targets. We focus on identifying the literature that gives a preponderant role to the circular economy and circular business models as the way to advance in this transition towards sustainability. Subsequently, a selective review of official documents is made to identify, in two Latin American countries, how institutions have advanced in the promotion of CE and CBMs, documents related to: legislation, support policies, financing programs and, in general, public offerings for this purpose. It also reviews opinion polls that give some signs of consumer acceptance of this type of models. Finally, criteria for comparison between countries are established in order to locate the incentives that have arisen in the promotion of circular business models.

8.3 Results

In Latin American countries, the reality is somewhat different since, unlike in the European Union, there are no policies to promote CE as a priority, let alone sufficient incentives for companies to implement circular business models; what can be observed are initiatives in this sense in multinational companies established in the countries of the South, whose motivation and incentives are in their countries of origin and in global markets (Singh et al., 2023).

8.3.1 Mexico: Barriers to CBMs

In the case of Mexico, its participation in the international environmental agenda, since 1972 at the Stockholm Conference "Environment and Human Development", has been the basis for the gradual construction of an institutional framework that has been the support of its environmental policy with the Ministry of the Environment and Natural Resources and decentralized agencies such as the National Water Commission, the National Institute of Ecology and the Federal Attorney General's Office for Environmental Protection. The publication in 1988 of the General Law of Ecological Balance and Environmental Protection (Ley General de Equilibrio Ecológico y Protección del Ambiente (LGEPPA)), with several subsequent reforms and five regulations.

A. **Legal Framework and Public Policy**

The evolution and relevance of the environmental issue in the international agenda has been a driving force for government actions and policies to adapt to new trends that respond to the same problem that today is expressed in greater dimensions. With the twenty-first century and the impulse to new trends that guide the policies of the countries, such as the green economy, the structure and legal framework for the environmental sector in Mexico is adapted and other proposals arise such as: the General Law for the Prevention and Integral Management of Waste—LGPGIR—which was enacted in 2003 (DOF, 2003); the Action Plan for a Circular Economy for Mexico City in 2019; the National Vision towards a Sustainable Management. Zero Waste in 2019 (SEMARNAT); and the General Law of Circular Economy approved in 2021 by the Senate of the Republic.

Additionally, other laws that promote sustainability oriented to the circular economy are promoted, such as: the General Law on Climate Change, which incorporates mitigation goals and the concept of eco-design of products; the Energy Transition Law that promotes circularity of materials from re-manufacturing, recycling of parts, evaluation of energy consumption and use of fossil fuels; the National Water Law linked to the Paris agreements where models of conservation and restoration of water sources are adopted; all this legislation is aligned towards the CE approach in several ways that focus on the incorporation of innovations for the recovery of materials and energy in order to minimize the impacts derived from the permanent extraction required for the production of goods and services and the reduction of waste and residues that result in a high social and economic cost.

In terms of programs, since 2012 there has been a National Climate Change Mitigation and Adaptation Strategy that governs national policy with mitigation and adaptation actions for the year 2050. Among the strategic axes to mitigate climate change is the transition to clean energy and responsible consumption, the transition to sustainable cities, the adoption of innovations in production and consumption and integrated waste management, all linked to the fulfillment of several of the objectives of sustainable development and the idea of ensuring the circularity of materials and energy. There is also the "National Circular Economy Strategy", which is part of a set of commitments that Mexico acquired as part of its international commitments.

Despite all these regulations, there is a wide-ranging problem in Mexico regarding the final disposal of waste, since, due to the lack of sufficient budget and adequate infrastructure, poor administration by local governments and the geographical dispersion of communities, open dumps still predominate and existing sanitary landfills operate inadequately. In this context, studies and tools have been developed since 2020 that contribute to the identification of solutions, such as the "Basic Diagnosis for Integrated Waste Management" and the "National Roadmap for the Transition and Adoption of CD in Mexico", which states that the country must make progress in revaluing material and energy flows, as well as reusing resources to improve productivity and national competitiveness.

At the level of the federal entities, 24 of them have an Integrated Waste Management Law, but only three have made explicit progress in incorporating the circular economy into their legal framework. The State of Quintana Roo approves in 2019 the Law for the Prevention, Integral Management and Circular Economy of Waste; in 2021 the Law for the Prevention, Integral Management and Circular Economy of Waste of the State of Baja California is approved; and that same year the Law for the Prevention, Integral Management and Circular Economy of Waste of the State of Querétaro was enacted. Two of them address the issue of waste separation and recovery to promote a by-product market, and one of them promotes the encouragement of public, private or mixed investment for the recovery and energy use of waste.

Another important program was presented in Mexico City in 2019, "The Zero Waste Action Plan, towards a Circular Economy" that promotes reducing the amount of waste received in disposal, separating and valuing it, in addition to dignifying the work of cleaning employees. Likewise, the city implements the "Mexico City Climate Action Program 2021–2030" with a strategic axis of Zero Waste. There are also at the municipal level some initiatives aimed at promoting the principles of CE in some municipalities in the states of Jalisco, Querétaro, Tamaulipas, and Puebla.

B. **Supply of Circular Products and Services**

In an approximation of the government with the private initiative, in 2021 the National Agreement for the New Plastics Economy in Mexico was signed with 85 companies in the plastics industry (Cespedes, 2020), the advances in just two years were important, more 70% of its packaging is already reusable or recyclable; 67 percent of the participants already have a management plan; About 50% of PET is collected and 40 percent for all plastic containers and packaging; 75% of companies include recycled material in their packaging and almost 15% of the material used for the production of containers and packaging is post-consumer recycled material. Another successful initiative is the "Querétaro Circular Economy System" initiated by the Querétaro automotive cluster and with the support of the Secretariat of Sustainable Development of the State in Querétaro, the topics they address are: circular energy sources; waste recovery; extension of useful life of products and materials; and the conservation of ecosystems. They started with 42 companies and reached 114 in the first year, with the goal of reaching 500 companies and 10 institutions that work on the proposal of CE initiatives. Likewise, the Mexican Network of the Global Compact carried out a study, "Mexican companies for the 2030 Agenda in the Decade of Action", where it is noted that 68% of companies affirm that they apply a circular economy model in their operations, from energy efficiency, eco-design of their products, waste management, programs for remove products with excessive consumption of energy and natural resources.

- Several companies already use recyclable packaging such as Kleenex, Tajín and Nespresso, and in the case of PETSTAR SAPI de CV. That manufactures bottles only uses recycled PET (SEMARNAT, 2020, p. 219), some other CE initiatives that are applied from large to small companies are:
  - Grupo Bimbo dedicated to the food industry, applies eco-design in the packaging of its products, reuses containers, reduces waste in plants and promotes a post-consumer recycling program
  - Heineken, a brewing company, its Mexico plant has participated since 2016 in the Ellen MacArthur Foundation's Circular Economy with the use of renewable energy, the reduction in water consumption and the local production of barley.
  - Jumex beverage manufacturing company has incorporated the use of recyclable pallets, among other actions (Córdova et al., 2021)
  - Go Trendier and Troquer are smaller companies dedicated to the buying and selling of used clothing (Carrillo & Pomar, 2021, p. 9)

Studies carried out already identify that the laws that introduce the principles of the CE and the General Law of Circular Economy itself are not linked to the National Strategy for Mitigation and Adaptation to Climate Change, the National Strategy for Energy Transition and with the SDGs, so that government and business efforts and initiatives are still in the design stage of a National Strategy to truly transition to the Circular Economy. Despite the emergence of new business models that look towards the perspective of the circular economy and the use of new technologies to compete in the markets, only 7% of the materials used in Mexico have a second use; out of 120 thousand tons of waste generated per day, about 17% are not collected and at least 13% of them are plastic (Global Circularity Report, 2023).

On the other hand, the predominance of the C to C modality is increasingly observed on social networks with a large number of second-hand product offers (Rodríguez, 2021). The new business models that are born with a vision of a circular economy are also taking advantage of the possibilities of new communication technologies and innovating in concepts such as traceability, distribution, dissemination and promotion of what is today called product-service. The demands of new consumers in Mexico force companies to rethink their value proposition and begin to consider sustainability (Barragán et al., 2020).

8.3.2 Colombia: Barriers to CBMs

In Colombia, the circular economy model has begun to gain ground, driven by government initiatives and growing market interest. This section explores the state of the circular economy in Colombia, analyzing laws, policies, financing programs, the supply of circular products and services, as well as consumer perceptions.

A. **Legal Framework and Public Policy**

Colombia has established a legal framework that supports the circular economy. Law 1715 of 2014 promotes the use of non-conventional energy sources and energy efficiency, which aligns with the sustainability principles of the circular economy (Congress of the Republic of Colombia, 2014). Likewise, Law 1252 of 2008 establishes extended producer responsibility, obliging companies to manage the life cycle of their products, which is crucial for the circular economy (Congreso de la República de Colombia, 2008).

The National Development Plan 2018–2022 includes the circular economy as a strategic axis, promoting sustainability in the use of resources. According to the Ministry of Environment and Sustainable Development (2019), this strategy seeks to integrate circular practices in different economic sectors, promoting innovation and sustainability. In addition, the Ministerio de Ambiente y Desarrollo Sostenible has launched the National Circular Economy Strategy, which seeks to promote the transition towards this model in the country.

The Colombian government has implemented financing programs to support the transition to circular business models, one of them is the National Financing Fund for Business Development (FONADE), which provides resources to projects that promote sustainability and innovation (FONADE, 2020). In addition, alliances have been established with international entities to facilitate investments in circular economy, which is crucial for the development of this model in the country.

At the local level, specific policies have been adopted to promote circular practices, such as the Integrated Solid Waste Management Plan, which includes strategies to increase recycling and reuse of materials. These local policies are fundamental to the success of the circular economy, as they allow initiatives to be adapted to the specific needs of each region.

Regarding Financing Programs, the Colombian government provides financing programs to support the transition to circular business models. The Fondo Nacional de Financiamiento para el Desarrollo Empresarial (FONADE), which seeks to provide resources to projects that promote sustainability and innovation (FONADE, 2020). In addition, alliances have been established with international entities to facilitate investments in circular economy, which is crucial for the development of this model in the country.

B. **Supply of Circular Products and Services**

At the market level, there has been a growth in the supply of products and services that adopt circular economy principles. Companies in sectors such as fashion, construction and technology are implementing models that include the reuse of materials and recycling, some brands offer garment repair and recycling services, while in the construction sector recycled materials are being used in new projects (González, 2023). Some cases of innovation in Colombia are:

- Ecovalle, dedicated to the production of furniture from recycled materials. This approach not only reduces waste, but also creates high-quality products that meet consumer expectations (Ecovalle, 2023).
- The H&M Group has implemented clothing recycling and repair programs, encouraging longer use of its products.

Colombian consumers' perception of circular business models is evolving. According to a study by the Cámara de Comercio de Bogotá (2022), 65% of consumers value sustainability and are willing to support brands that implement responsible practices. However, there is still a general lack of awareness of the benefits of the circular economy, which may limit its mass adoption. Awareness and education campaigns are necessary to increase consumer acceptance and commitment to these models (Cámara de Comercio de Bogotá, 2022).

Despite the progress made, the barriers to the development of circular business models (CBMs) in Colombia are diverse and complex. One of the main ones is the lack of access to financing, especially for small and medium-sized enterprises (SMEs). According to a study by the Cámara de Comercio de Bogotá (2022), only 44% of SMEs apply for loans from the financial system because they still find them costly. This limitation in the capacity to invest in technologies and circular processes can be overcome by offering more lines of credit and specific subsidies, as well as support programs and financial advice.

Another important barrier is insufficient infrastructure for waste collection, sorting and recycling. The Departamento Nacional de Planeación—DNP (2018) has identified deficiencies in waste management in several regions of the country, making it difficult to close material cycles and create circular supply chains. To address this barrier, investments in recycling infrastructure, selective collection systems and extended producer responsibility programs are required.

Resistance to change on the part of many companies, clinging to linear business models, also represents an obstacle. According to a study by Universidad Icesi (2023), this resistance is partly due to a lack of knowledge about the benefits of the circular economy. To overcome this barrier, it is necessary to conduct awareness and training campaigns aimed at entrepreneurs, emphasizing the economic, environmental and social benefits of CBMs.

The transition to the circular economy requires collaboration between different actors in the value chain. However, in Colombia there are still barriers to cooperation between companies, unions and government entities. To foster collaboration, information exchange platforms, circular value networks and public–private partnerships can be created.

Finally, although Colombia has a legal framework favorable to the circular economy, such as Law 1715 of 2014 and Law 1252 of 2008, there are still regulatory gaps and contradictions that hinder its implementation. To address this barrier, a review and harmonization of the legal framework is required, as well as the creation of incentives and effective compliance mechanisms.

However, there are also significant opportunities. The circular economy can generate substantial savings in material costs, estimated at up to US$11.7 million for Colombian companies (Icesi, 2023). In addition, the transition towards circular models can open new employment opportunities in sectors related to recycling and sustainability (Ministerio de Ambiente y Desarrollo Sostenible, 2019).

In conclusion, Colombia is making progress in the implementation of circular business models, but faces significant challenges in consumer perception and knowledge. Despite a favorable legal framework and government policies in support of the circular economy, many consumers still view circular products as more expensive or less accessible, which hinders their widespread adoption (Rincón, 2023).

These include:

- Lack of Infrastructure: Infrastructure for the collection and recycling of materials is still insufficient in many regions of the country (DNP, 2018).
- Resistance to Change: Many companies are still clinging to linear business models, which hinders the transition to circularity (Icesi, 2023).
- Education and Awareness: Lack of knowledge about the circular economy among consumers and companies is a significant barrier (González, 2023).

This requires the active participation of the government, the private sector and civil society, focusing on educating and raising awareness among consumers and companies about the economic and environmental benefits of these business models.

8.4 Discussion

The circular economy is a concept that has positioned itself in the last decade within the international agenda, its antecedents dating back to the seventies and eighties of the last century, which showed possible strategies at the local and regional level that pointed towards a policy design of greater magnitude. From different instances such as the European Union, the United Nations and the Ellen MacArthur Foundation, the circular economy is now being promoted as a key strategy for the construction of a route to sustainability that links innovation at different levels involving products, services, processes, new technologies and business models. This proposal for a circular approach has been, to a certain extent, adopted by different countries with the signing of commitments for the fulfillment of the SDGs.

The governments of Mexico and Colombia assumed that the promotion of the circular economy and circular economy models is a path to follow to bring together the different sectors in public and private initiatives that are inserted in the market and thus move towards a production model that reduces the consumption of materials and energy from the use of waste.

A first level of comparison corresponds to the legal framework that has been developed in both countries; in the case of Mexico, it was during the eighties of the twentieth century that a legal framework oriented to the attention to environmental issues was built but it is until the beginning of the current century that a specific legislation related to waste management is established and between 2019 and 2020 that specific laws arise at federal and state level that refer to the circular economy, in parallel at least three more laws oriented to innovation and CE have been published very recently. In the case of Colombia in 2008, there was already a law on extended producer responsibility that already showed principles of circular economy; this was strengthened with Law 1715 of 2014 referring to energy efficiency. In both countries, there is a legal weakness in not incorporating specific ecological regulations and standards in their legislation that give strength to the law and not considering circular business models as an important strategy, however, there is a greater trajectory and experience in the case of Mexico.

The second level is in the strategies and programs developed by governments to promote CE. In Mexico, a national climate change mitigation and adaptation strategy was created in 2012 with a vision for 2050, in which energy efficiency, responsible consumption and integrated waste management are considered as strategic axes. Likewise, the National CE Strategy derived from external commitments was published, and in 2012 the roadmap for the valorization of material and energy flows was defined, as well as other direct CE initiatives in at least 4 state governments. In Colombia, the topic of CE was taken since 2018 to a National Plan, in addition to the national CE strategy presented in 2020. In this aspect of public policy and specific programs, there is also an absence of specific strategies to promote circular business models, although it is worth recognizing the initiatives of both countries to place CE within their priorities.

The fourth level of comparative analysis is located in the actions of the companies and the consumer's perception to make purchasing decisions. In Mexico, many companies of different sizes issue environmentally friendly messages, and in particular, several report recycling materials, especially large companies that have innovations and new technologies, and on the other hand, the new generations opt for C to C trade in social networks; this modality opens greater possibilities for marketing and the use of new technologies in business models; the information from the surveys is that the new generations are more receptive to business models with a focus on sustainability (Carrillo & Pomar, 2021).

8.5 Implications

The circular economy has become a fundamental approach to address the environmental and economic problems facing the world today. In Latin America, government actions to promote the circular economy and circular business models have been oriented towards updating laws, creating development programs with little budget and some very incipient initiatives to promote and set guidelines, from the same public bodies, to encourage small and medium-sized national companies.

Evidently, these initiatives have been insufficient because there is a predominant disconnection between science and technology policies and environmental policies and a divergence between these and the development of data science, information systems and advances in artificial intelligence that are installed as imported technologies in some economic sectors of these countries. Nevertheless, there are already some public policy initiatives in the region to promote CE and NCMs, which derive not only from the current demands of consumer segments, but also from actions stemming from international agreements and government policies and programs.

The policies that have been promoted in several Latin American countries are: waste management and recycling policies; extended producer responsibility policies; policies on product eco-design; ban on single-use plastics; product life cycle extension policies; national circular economy strategies; circular economy roadmaps; and fiscal policies to encourage the circular economy (subsidies and taxes).

In the cases of Mexico and Colombia, it is clear that a legal structure has been built over several years to promote sustainable transition, evolving according to the strategies that are defined as the most appropriate, as is the case of the circular economy and circular business models. Likewise, the last decade has seen an environmental policy expressed in plans, programs and roadmaps aimed at promoting CE. However, the weak point is the low financial capacity and the low budget allocated to this area.

8.6 Conclusions

As explained above, the transition process towards a more sustainable production and consumption model is necessary, and therefore, the design of alternative strategies such as circular economy and circular economy models are now being promoted by international organizations, marking a route to follow for countries that have made commitments in the signing of agreements to achieve the SDGs. However, there is a wide gap between the initiatives and possibilities offered by northern countries, particularly the European Union, and the reality faced by Latin American countries where sustainability loses importance in the face of poverty, inequality, violence and in some cases even lack of governance.

Mexico and Colombia are two countries that have always stood out for being aligned with international commitments related to environmental care and people's quality of life. Both have promoted everything from the construction of a legal framework to support environmental protection to specific programs and projects financed with national and international funds. However, the success of a strategy based on the promotion of the circular economy and NCMs requires not only political will and regulation via a legal framework, but also financial resources, which are often scarce in these countries, and a change in consumer culture, which can only be achieved in the long term. Thus, the actions undertaken so far are still insufficient to take a firm step towards bringing these countries closer to the sustainable transition processes occurring in other contexts.

The efforts identified to promote CBMs are still limited; therefore, it is necessary to generate support networks with the participation of the public, private and social sectors, both nationally and internationally, and to develop a broad strategy to generate awareness of the need for change among consumers. Therefore, future goals are to focus circular economy studies on the definitions and characterization of concrete alternatives that contribute to the fulfillment of sustainability goals and to develop work that will guide the design of policies that encourage this approach.

A comprehensive summary of the findings is given in Fig. 8.1, which highlights the most important takeaways from every barrier category across the various sectors and CBM methodologies. Based on these observations, the figure also offers a number of suggestions to industry and legislators on how to get over any potential obstacles to CE under each heading. It is crucial to remember that these suggestions come from our study and aren't stated clearly in the cases. Table 8.1 can help show barriers and recommendations in circular business models.

Table 8.1 Barriers and recommendations in circular business models

	Regulatory	Supply	Market	Organizational	Technological
Barriers	The emphasis is on regulatory hurdles in industries without regulations specifically pertaining to CE. Insufficient industry expertise on present design and waste requirements in the furniture industry. Procurement bidding does not encourage service CBM approaches in product-heavy industries	Expensive secondary materials create a barrier to CBM approaches that depend on circulating materials since they are more expensive than virgin materials. The electronics industry places greater emphasis on waste CBM approaches than the furniture and clothing sectors do, primarily because of supplier limitations of non-virgin materials and components	In industries where CBM approaches are viewed as novel, the corporate market is not prepared for CE. In contrast to other industries, end users in the electronics sector are less likely to accept used goods, which has an impact on both waste and service CMB approaches	The furniture industry is characterized by the misuse of internal organizational barriers, complicated organizational systems, and a deficiency of assistance from upper management. External organizational constraints were highlighted in the apparel and electronics industries	In general, technological constraints are not given much attention in any field greater focus on the technological obstacles in industries lacking recycling infrastructure. The lack of tracing solutions in industries without established tracing systems is being emphasized more than usual
Recommendations	Policy Take into account the wide range of CBM approaches when creating new legislation. Incorporate resource efficiency and circularity requirements into the procurement process to advance services. CBM methods Industry Help the government create realistic regulations by providing industry expertise	Policy Implement market mechanisms that reduce secondary materials' relative cost to that of virgin materials. Encourage the flow of information Industry Include take-back programs in CBM strategies to guarantee a consistent supply of secondary materials from final customers	Policy Encourage the use of used electronics by implementing microeconomic policies Industry To hasten their adoption, raise investor and corporate buyer knowledge of the ecosystem's advantages through CBM approaches	Policy Determine which policy levers are targeted at particular organizational external impediments Industry In the furniture industry: Boost support for circular economy principles among upper management	Policy Provide a recycling infrastructure that is both efficient and effective for the apparel industry Industry Invest in the co-creation of tracing solutions with supply chain partners for the furniture and apparel industries

Source Own work (2024)

References

Aagaard, A., Lüdeke-Freund, F., & Wells, P. (2021). Introduction to business models for sustainability transitions. In: *Business model for sustainability transitions: How organizations contribute in societal transformation.* Palgrave MacMillan. https://doi.org/10.10007/978-3-030-77580-1_1

Barrag.n, C., Sandoval, O., & Sosa, G. (2020). Modelo de negocios para la innovación sustentable: una aproximación teórica. *International Journal of Good Conscience, 15*(1), 206–221.

Bunker, S. (1996). Materias primas y economía global: olvidos y distorsiones de la ecología industrial. *Revista Ecología Política, 12*, 81–89. https://www.jstor.org/stable/20742898

Cámara de Comercio de Bogotá. (2022). Percepción del consumidor sobre sostenibilidad y economía circular en Bogotá. https://www.ccb.org.co

Carrillo, G., & Pomar, S. (2021). La economía circular en los nuevos modelos de negocio. *Entreciencias Diálogos En La Sociedad Del Conocimiento, 9*(23), 1–16. https://doi.org/10.22201/enesl.20078064e.2021.23.79933

Carrillo, G., & Torres, L. (2019) Introducción en Biorrefienrías y Economía Circular. UAM-X.

Centobelli, P., Cerchione, R., Chiaroni, D., & Del Vecchio, P. (2020). Designing business models in circular economy: A systematic literature review and research agenda. *Business Strategy and the Environmental, 29*(4), 1734–1749. https://doi.org/10.1002/bse.2466

Cespedes, (2020). Resumen del 1er Informe del Acuerdo nacional para la Nueva Economía del Plástico en México. Diciembre 2020. Senado de la República. https://cespedes.org.mx/wp-content/uploads/2021/02/Resumen-del-1er-Acuerdo-Nacional-de-Plasticos-final.pdf

Congreso de la República de Colombia. (2008). Ley 1252 de 2008. https://www.secretariasenado.gov.co

Congreso de la República de Colombia. (2014). Ley 1715 de 2014. https://www.secretariasenado.gov.co

Córdova, M., Salgado, L., & Bravo, B. (2021). Economía circular y su situación en México. *Revista Indiciales, 1*(1), 25–37. https://doi.org/10.52906/ind.v1i1.7

Christensen, T. (2021). Towards a circular economy in cities: Exploring local modes of governance in the transition towards a circular economy in construction and textile recycling. *Journal of Cleaner Production, 305*(127058). https://doi.org/10.1016/j.jclepro.2021.127058

Departamento Nacional de Planeación-DNP. (2018). Diagnóstico de la gestión de residuos en Colombia. https://www.dnp.gov.co

DOF (2003) Ley General para la prevención y Gestión Integral de los Residuos. 8/octubre/2003. Consultado en: https://dof.gob.mx/nota_detalle.php?codigo=688657&fecha=08/10/2003#gsc.tab=0

Ellen MacArthur Foundation (2014). Hacia una Economía Circular. Resumen ejecutivo. Consultado en https://www.argentina.gob.ar/sites/default/files/hacia-una-economia-circular-resumen-ejecutivo-ellen-mac-arthur-foundation.pdf

Fondo Financiero de Proyectos de Desarrollo-FONADE. (2020). Programas de financiamiento para la sostenibilidad. https://www.fonade.gov.co

González, M. (2023). Innovaciones en la moda circular en Colombia. *Revista de Sostenibilidad, 12*(3), 45–58.

Global Circularity Report. (2023). The circularity Gap Report. We live in the overshoot era https://www.circularity-gap.world/2023

Grin, J. (2016). *Handbook on sustainability transition and sustainable peace* (Vol. 10). https://doi.org/10.1007/978-3-319-43884-9.

Grupo Bimbo. (s/f). ¿Qué hacemos en sostenibilidad? Sitio web. https://grupobimbo.com/es/sustentabilidad

Guldmann , E., & Huulgaard, R. D. (2020). Circular business model innovation for sustainable development. In *Innovation for sustainable business transformation towards a better world (Chapter 5).* Ed Palgrave Macmillan. https://doi.org/10.1007/978-3-319-97385-2_5

Heineken. (s/f). Reporte de Sostenibilidad de HEINEKEN: logros en agua, CO2 y más. https://heinekenmexico.com/noticia/reporte-de-sostenibilidad-de-heineken-logros-en-agua-co2-y-mas

Icesi. (2023). Economía circular: oportunidades de implementación de modelos de negocio. https://repository.icesi.edu.co

ILO (2018). Panorama laboral 2018. América Latina y Caribe.OIT. Perú. file:///Users/gracielacarrillo/Downloads/wcms_654969.pdf

Jacobsen, N. (2006). Industrial symbiosis in Kalundborg, Denmark: A quantitative assessment of economic and environmental aspects. *Journal of Industrial Ecology, 10*(1–2), 239–255. https://doi.org/10.1162/108819806775545411

Khan, O., Daddi, T., & Iraldo, F. (2020). The role of dynamic capabilieties in circular economy implementation and performance of companies. *Corporate Social Responsibility and Environmental Management, 27*(6), 3018–3033.

Kemp, R. (1994a). Technology and the transition to environmental sustainability. *Futures, 26*(10), 1023–1046. https://doi.org/10.1016/0016-3287(94)90071-X.

Kemp, R. (1994b). Technology and the transition to environmental sustainability: The problem of technological regime shifts. *Futures, 26*(10), 1023–1046. https://doi.org/10.1016/0016-3287(94)90071-X

Lacy, P., & Rutqvist, J. (2015). Waste to wealth: The circular economy advantage. *Environmental Economics*. https://doi.org/10.1057/9781137530707

Loorbach, D., Frantzeskaki, N., & Avelino, F. (2017). Sustainability transitions research: Transforming science and practice for societal change. *Annual Review of Environment and Resources, 42*(1), 599–626. https://doi.org/10.1146/annurev-environ-102014-021340

Loorbach, D., & Rotmans, J. (2006). Managing transitions for sustainable development. *In understanding industrial transformation* (pp. 187–206). https://doi.org/10.1007/1-4020-4418-6_10

Markard, J., Raven, R., & Truffer, B. (2012). Sustainability transitions: An emerging field of research and its prospects. *Research Policy, 41*(6), 955–967. https://doi.org/10.1016/j.respol.2012.02.013

Ministerio de Ambiente y Desarrollo Sostenible. (2019). Estrategia Nacional de Economía Circular. https://www.minambiente.gov.co

Nußholz, J. (2017). Circular business models: Defining a concept and framing an emerging research field. *Sustainability, 9*(10), 1810. https://doi.org/10.3390/su9101810

UN-ECLAC. (2024). América Latina y el Caribe ante el desafío de acelerar el paso hacia el cumplimiento de la Agenda 2030. Transiciones hacia la sostenibilidad. UN, Santiago de Chile.

Rodríguez, E. (2021). Los residuos de polietileno-tereftalato (PET) en la Ciudad de México, 2010–2018: Análisis y recomendaciones desde la Economía Circular [Tesis de Maestría. Universidad Nacional Autónoma de México]. http://132.248.9.195/ptd2021/junio/0813431/Index.html

Secretaría de Medio Ambiente y Recursos Naturales-SEMARNAT. (2020). Diagnóstico Básico para la Gestión Integral de los Residuos. https://www.gob.mx/cms/uploads/attachment/file/554385/DBGIR-15-mayo-2020.pdf

Singh, R., Khan, S., & Dsilva, J. (2023). A framework for assessment of critical factor for circular economy practice implementation. *Journal of Modelling in Management, 18*(5), 1476–1497. https://doi.org/10.1108/JM2-06-2021-0145

Singh, R., Khan, S., & Centobelli, P. (2022). Investigating the interplay between social performance and organisational factors supporting circular economy practices. *Sustainability, 14*, 16781. https://doi.org/10.3390/su142416781

Tapia, G. (2018). Modelos de negocios sustentables. http://sadaf.com.ar/index.php?act=doDescargarPublicacion&id=333

Teece, D. (2010). Business models, business strategy and innovation. *Long Range Planning, 43*(1), 172–194.

Chapter 9
Digitalisation and Its Impact on the Organisation Performance and Sustainability: A Systematic Literature Review

S. N. Rosheni Fiona Dias, Narayanage Jayantha Dewasiri, and K. L. Wasantha Perera

Abstract Literature reviews are a substantial prerequisite to assessing existing knowledge, identifying gaps, and identifying feasible future developments. However, conducting structured reviews of the studies on digitalisation and its impact on organisations in terms of performance and sustainability has been scarce. Over the past few years, digitalisation has impacted numerous aspects of the economy, including its intense inference on sustainability in different domains. This study aims to present a systematic literature review (SLR) following PRISMA (Preferred Reporting Items for Systematic Reviews and Meta-Analyses) guidelines to conglomerate the existing literature to examine the diverse impact of digitalisation on organisational performance and sustainability. This chapter systematically reviews 38 journal articles published in the Scopus database from 2014 to 2024. (1) Digitalisation's impact on firms' performance metrics, (2) digitalisation's impact on firm performance and reputation, (3) green information systems and technology on organisation performance, and (4) technological innovations and corporate sustainability are the key themes

S. N. R. F. Dias (✉)
Department of Banking and Finance, Faculty of Business Studies and Finance, Wayamba University of Sri Lanka, Kuliyapitiya, Sri Lanka
e-mail: fiona@wyb.ac.lk

N. J. Dewasiri
Department of Accountancy and Finance, Faculty of Management Studies, Sabaragamuwa University of Sri Lanka, Belihuloya, Sri Lanka

University Center for Research and Development (UCRD), Chandigarh University, Mohali, Punjab, India

N. J. Dewasiri
e-mail: jayantha@mgt.sab.ac.lk

K. L. W. Perera
Department of Finance, Faculty of Management Studies and Commerce, University of Sri Jayewardenepura, Nugegoda, Sri Lanka
e-mail: wasantha.perera@sjp.ac.lk

R. Singh and V. Kumar (eds.), *Sustainable Innovations and Digital Circular Economy*,
https://doi.org/10.1007/978-981-96-1064-8_9

identified under knowledge clusters. The study findings revealed that digital technologies have become a key enabler in promoting organisational performance metrics and sustainability. Thus, this compound of digitalisation and sustainability emphasises the importance of an integrated approach and a policy framework to utilise digital innovations in promoting organisational performance and sustainability. This review contributed to the existing literature by offering several contemporary issues and the scope of future research in digitalisation and sustainability in the economy, which may help policymakers, practitioners, and researchers investigate the solutions.

Keywords Digitalisation · Organisation performance · Organisational sustainability · Systematic literature review · PRISMA

9.1 Introduction

Digitalisation has gained considerable attention across global economies in recent years as it is an essential influencer of individuals, business organisations, and the overall economy (Bhuiyan, 2024). During the last decade, the digital transformation of business operations has drastically increased, and the use of digital technologies has accelerated organisational performance and sustainability. Technological advancements like artificial intelligence, blockchain, Internet of Things, social media, cloud computing, big data analytics, information systems, etc. have become driving forces of small to large-scale business organisations, enabling them to alter their business processes and outcomes.

The studies have highlighted that sustainability has obtained growing global attention after the launching of 17 Sustainable Development Goals (SDGs) by the United Nations in 2015 to be achieved by 2030 (Siddik et al., 2023; Rosamartina et al., 2022; Johl & Toha, 2021). This has enabled organisations to achieve corporate sustainability objectives to enhance environmental, social, and economic performance (Rosamartina et al., 2022; Oware & Mallikarjunappa, 2021). The crucial role of digital adoption in strengthening the organisation's performance and sustainability has been obtaining worldwide attention from researchers and business firms for its significant role in increasing the efficiency and productivity of the internal practices of businesses through its innovative digital platforms (Oware & Mallikarjunappa, 2021; Zhang et al., 2020). Among the digital innovations, financial technology (FinTech) is a blooming research area that enhances access to financial services such as credit, savings, insurance, financial market investments, digital payments, and settlement of business transactions using mobile technology, blockchain, artificial intelligence, etc. Use of social media platforms such as Linkedin, Facebook, YouTube, and Twitter are digital platforms that support organisations to broaden their network of clients, suppliers, and investors at a mass level with a minimum cost that allows them to increase their sales growth, service quality, and organisation reputation at a competitive edge (Bruce et al., 2023). Big data analytics is another digital innovation especially adopted by large firms that must manage extensive amounts of data from various sources (Chatterjee et al., 2022). The use of information systems for different organisation functions, such as supply chain management, customer relationship management, knowledge management, etc., is also a key focus in literature for achieving the green objectives of the organisation (Yang et al., 2017, 2019; Zhang et al., 2020). Cloud Computing is another emerging digital advancement during the last decade that has been adopted by large-scale to small and medium enterprises (Olayah, 2019). It is the on-demand availability of computer system resources over the internet, such as servers, data storage, networking, databases, and software, where organisations do not want to make considerable investments in software and hardware and could increase their productivity, communication, collaboration, expandability, and finally drive towards enhancing organisation performances and sustainability (Dušková, 2021; Olayah, 2019).

Rationale: It has been empirically found that digitalisation plays a unique role in uplifting organisations' decision-making, efficiency, and productivity, enhancing their performance in finance, operations, and innovations, and driving the organisation towards enhancing its sustainability performances (Siddik et al., 2023). Consequently, systematically evaluating the impact of digitalisation on the organisation's performance and sustainability is very significant for the stakeholders who navigate this influence of digitalisation to accelerate the organisation's performance to promote sustainability, yet insufficiently studied in the literature (Gusarova et al., 2023). Several systematic literature reviews have separately examined the impact of digitalisation on these critical dimensions (Vann Yaroson et al., 2024; Bhuiyan, 2024; Memon & Ooi, 2023). Thus, the current literature lacks a comprehensive literature review that observes the influence of digital technologies in these key domains, namely, organisation performance and organisational sustainability, and what competitive advantages and challenges have been created during this digital transformation. Thus, conducting a comprehensive systematic review of the impact of digitalisation on organisational performance and sustainability is imperative. The study's objectives are to investigate (1) current knowledge on digitalisation and its impact on organisational performance and sustainability and (2) lacking areas of empirical research on digitalisation and its impact on organisational performance and sustainability.

The remaining review sections are organised as follows: Sect. 9.2 discusses the study's methodology, including how the studies were selected and how the review was carried out systematically. Section 9.3 discusses the study findings under four subsections: study selection, study characteristics, key findings, and bias assessment. The rest of the sections include the findings, discussion, practicality and research implications and conclusion of the study.

9.2 Methods and Methodology

9.2.1 Study Selection Process and Methods

The SLR was conducted to explore the impact of digitalisation on the organisation's performance and sustainability, cohering with the PRISMA guidelines to ensure a transparent report and methodological rigor of the SLR as stated by Karunarathna et al. (2024), Dewasiri et al. (2024), Rathnasiri et al. (2024), Jayawardena and Dewasiri (2023), Dewasiri et al. (2023), and Dewasiri et al. (2022). The article selection procedure was followed, as shown in Fig. 9.1, which is the PRISMA flow diagram as it is recommended for SLR (Liberati et al., 2009).

The "identification", "screening", and "include" were the significant steps followed in the article selection procedure. Under the identification stage, the search terms, eligibility criteria, and data extraction methods were decided. The considerable search terms of the study were "digitalisation", "organisation performance", and "organisation sustainability". When developing the search criteria for the first key term "digitalisation", terms such as "digitalisation", "digital" and "technology". The search criteria for "organisation performance" were "organisational performance", "business performance", "firm performance", "company performance", and "financial performance". The search criteria for the third term, "organisation sustainability",

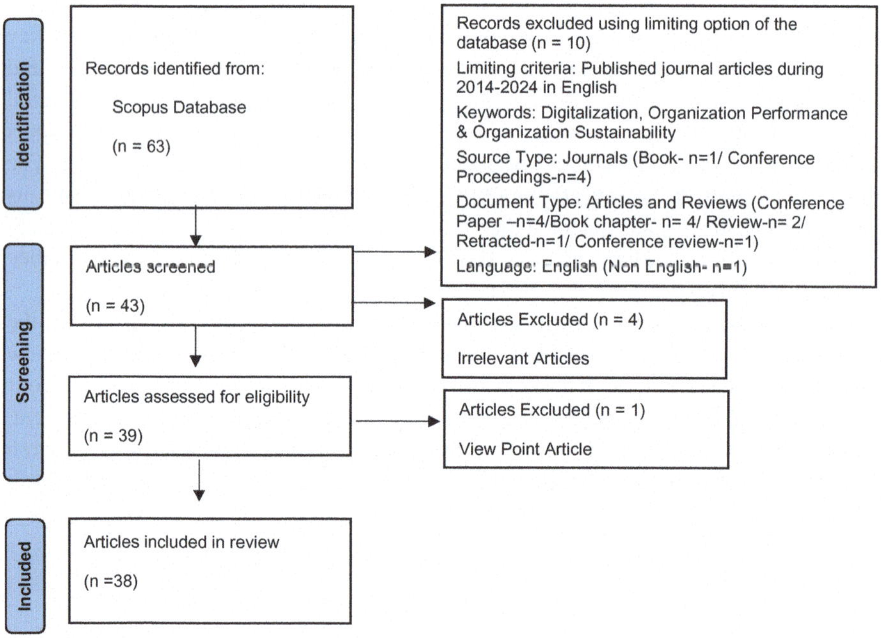

Fig. 9.1 Article selection process. *Source* Authors' own

were "organisation sustainability", "corporate sustainability", and "firm sustainability". Each search criteria of all key search terms has been typed in the Scopus database using the "OR" operative between each term.

Per the PRISMA guideline, the next step was screening the identified articles based on several inclusion criteria. If any paper did not meet the inclusion criteria, it was removed from the study (Priyashantha et al. 2021a, b, c, 2022). Accordingly, the inclusion criteria of the present SLR, all "empirical studies" published in "English" in the "Journal" articles from the year "2014 to 2024" were screened. Only the empirical journal articles were identified because they ensured uniformity in the methodological quality to extract findings and internal soundness (Okoli & Schabram, 2015), which is recommended for use in SLR (Priyashantha et al. 2021a, b, c, 2022).

The articles were automatically screened in the Scopus database using study type, language, report, and publication date. Accordingly, the "empirical studies" published in "English" "Journals" from the year "2014 to 2024" were retrieved for the study. Consequently, non-English publications outside the specified period, such as book chapters, conference proceedings, book reviews, or any other type other than empirical journal articles, were automatically excluded.

The automatically identified articles were then used for the manual screening for the selection as the next screening stage. The abstracts and the full version of the automatically screened articles were reviewed and assessed for their methodological quality (Meline, 2006). Other articles were excluded from the empirical studies which have not followed a quantitative methodology.

9.2.2 *Article Risk of Bias Assessment*

The researcher's biases in selecting and analysing the articles can compromise the review's quality (Priyashantha et al., 2022). The current study has employed a robust reviewing process to mitigate these biases, with two more researchers assessing the articles to ensure systematic and objective selection.

9.2.3 *Method of Analysis*

The SLR has employed a bibliometric analysis technique to quantitatively analyse the impact and trend in the empirical literature. The two specialised tools used in the bibliometric analysis were Biblioshiny and VOSviewer. Biblioshiny, a web-based interface for the bibliometric R package, was performed to identify the critical areas evoked in the literature. The retrieved bibliometric data of the empirical studies in Scopus on digitalisation and its impact on organisation performance and sustainability were imported to Biblioshiny. It employed various analyses to obtain descriptive statistics to identify the citation patterns, publication trends, author productivity, etc.

VOSviewer software visualised the bibliometric networks, including co-authorship networks, citation networks, and keyword co-occurrence maps. This also assisted in identifying key research trends and clusters in the given literature landscape. Consequently, both tools assisted in performing a comprehensive bibliometric analysis, which has allowed us to understand the literature landscape on digitalisation and its impact on organisation performance and sustainability.

9.3 Results and Findings

9.3.1 Study Selection

The PRISMA article selection flow diagram was used to select the articles for the review. As the initial step, articles for the review were screened using the search keywords explained in Sect. 2.1, and then articles were excluded using the Scopus database's automatic screening option. The inclusion criteria used at this stage are presented in Table 9.1.

At the first time screening of articles, Scopus generated 63 articles, and 8 out of them were excluded under the first inclusion criteria of the review. Then, 1 Portuguese article was excluded under the second inclusion criteria. Under the basis of the third inclusion criteria, which is source type, conference proceedings (n = 4) and books (n = 1) were excluded from the review, and conference papers (n = 4), book chapters (n = 4), and retracted (n = 1). Conference reviews (n = 1) were excluded under the fourth inclusion criteria.

Consequently, 43 articles were retrieved for manual screening. The authors independently assessed the abstracts of the articles to check suitability based on the study inclusion criteria. Four articles were excluded as irrelevant to the review, and based on the fifth inclusion criteria; another article was excluded (viewpoint articles n = 1). Finally, 38 articles were screened for the bibliographic analysis. The PRISMA article selection flow diagram is presented in Fig. 9.1.

Table 9.1 Article inclusion criteria

S/N	Inclusion Criteria	Focus on
1	Publication year range	2014–2024
2	Language	English
3	Source type	Journals
4	Type of study	Articles and reviews
5	Methodological quality	The articles that follow a quantitative methodology

Source Authors' Conception, 2024

9.3.2 Study Characteristics

Using the Biblioshiny in R software, the authors generated the essential information for the article set selected for review. As depicted in Table 9.2, from 2014 to 2024, 38 documents were published in 33 Journals. Among these 38 documents, 37 are research articles with a single review article, which signifies the prime focus of researchers on original research rather than comprehensive literature reviews and meta-analyses. The average publication age of 3.53 years indicates that the studies are recent and reflect the recent developments and ameliorations in the field of study. The significant influence on the field is that it can be re-ensured using the high average citation per document, which is 40.74. This signals that these studies have been frequently cited and referred to in later studies, ensuring their relevance and significance to the ongoing discourse in the academic community. The collaboration of 110 authorships representing 29 countries in the selected publications emphasises the diverse international collaboration network of numerous perspectives and expertise of the field. The 153 author keywords highlight that a wide array of subfields and research interests have been covered in the selected publications, and the 3178 references in 38 documents reflect the rigorous commitment to the existing literature.

Figure 9.2 explains the annual article publication trend in the study from 2014 to 2024. The annual publication rate was relatively low during the initial years, showing a notable gap in 2015 and after an upward trend in research outputs. After 2018, overall, it manifests an upward trend. In 2022, it remarked the highest number of publications, nine articles. A publication slowdown can be observed in 2023 and early 2024.

Figure 9.3 manifests the average citation behaviour during the given period. The years 2014, 2020, and 2018 were recorded as the highest average citations per article per year, which reflects greater attention on digitalisation's impact on organisational behaviour and sustainability from scholars. The average citation levels have been

Table 9.2 Main information about data

Description	Results
Timespan	2014:2024
Documents	38
Article	37
Review	1
Average year from publication	3.53
Average citations per doc	40.74
References	3178
Author's keywords	153
Authors	110
Countries	29

Source Review Data, 2024

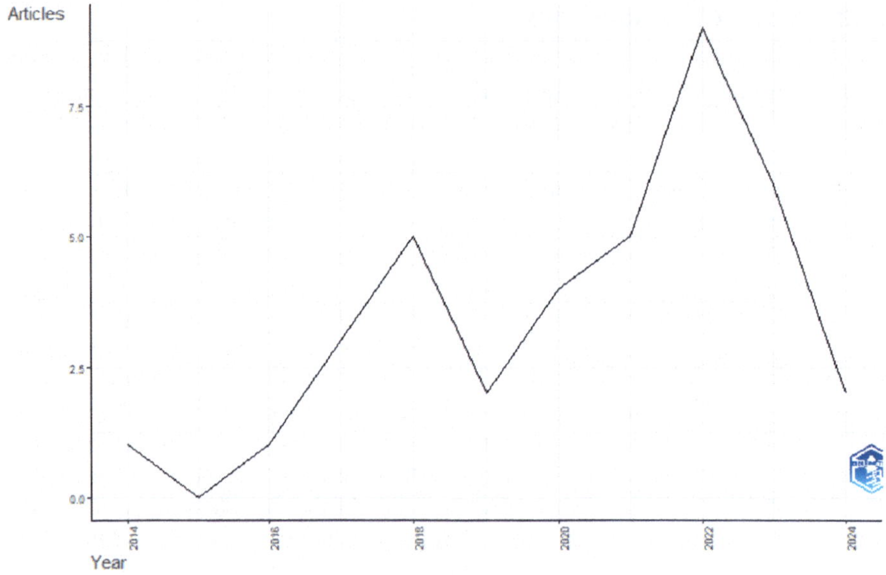

Fig. 9.2 Annual article publication. *Source* Review Data, 2024

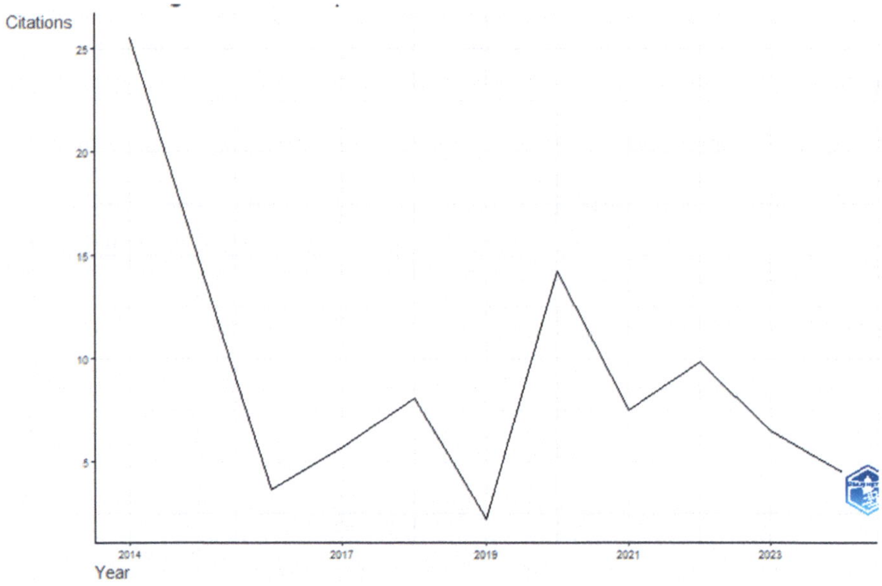

Fig. 9.3 Average citations received. *Source* Review Data, 2024

Fig. 9.4 Most relevant sources of the article publications. *Source* Review Data, 2024

sensibly low in recent years, 2023 and 2024; this is a typical, expected pattern due to the limited time to be cited.

The most relevant sources of article publications are demonstrated in Fig. 9.4. The Journal of Cleaner Production has produced four articles, and sustainability (Switzerland) has produced three articles, the most frequently published sustainability-focused sources. In contrast, all other journals have published only one article from each. Many single-article journals highlighted the broad dissemination of research across disciplines such as finance, economics, technology, business, environmental management, etc.

Figure 9.5 exhibits the representation of country-wise publications and the country-wise citation interdependency. The size of the figure's nodes represents the number of publications by the particular country. Accordingly, China has the most significant node, which manifests the highest number of publications, and the United States has the second-highest number. Table 9.3 summarises the ranking of the countries based on their number of publications and citations received. Accordingly, it confirms that China, the United States, and India are the key contributors to the journal articles. Vietnam, the United Kingdom, South Korea, Malaysia, and Italy equally hold the following significant positions: each country contributing three articles. From the rankings of citations received, the United States leads, followed by China and Switzerland. Even though Switzerland holds a single article, the high number of citations underscores the influence of the research. Vietnam, Italy, Germany, the United Kingdom, and Malaysia also record a high number of citations compared to

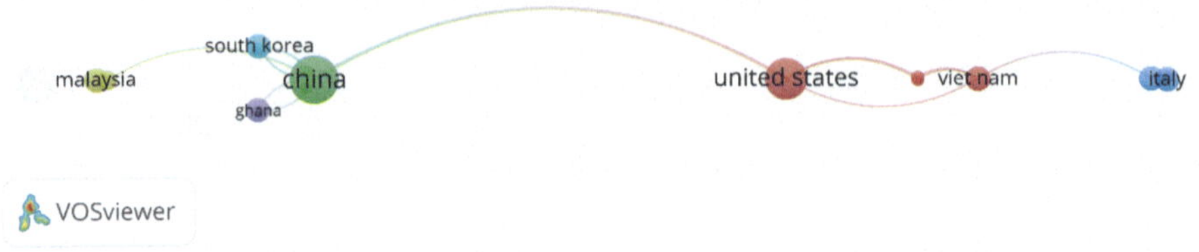

Fig. 9.5 Country-wise article publication. *Source* Review Data, 2024

fewer publications. This represents a notable gap in the number of publications and citations received.

9.3.3 Key Findings of the Study

The key objectives of the study were achieved using 'keyword co-occurrence network visualisation' and 'keyword density visualisation'. The first objective of investigating the current knowledge was assessed using the keyword co-occurrences network. The second objective of identifying the areas lacking in empirical research on particular research area was achieved using keyword density visualisation.

9.3.3.1 The Current Empirical Knowledge of Digitalisation's Impact on Organisational Performance and Sustainability

As presented in Table 9.4, the minimum keyword occurrences functionality of VOSviewer software has produced a given threshold number of keyword occurrences starting from single-time occurrences of a keyword to 9-time occurrences of a keyword. Out of the 271 keywords, the researchers selected the 42-threshold keyword level of a minimum of two occurrences of a keyword, as the higher levels of minimum keyword occurrences produce fewer threshold keyword occurrences. The 42 keywords and their dependency are presented in Fig. 9.6 of the keyword co-occurrence network visualisation. Table 9.5 depicts the frequencies of the occurrences of each keyword.

In the network visualisation of Fig. 9.6, the size of the circle and the label explain the weight of each item. Accordingly, corporate sustainability holds the highest weight, followed by firm performance, innovation, and financial performance. This can be further confirmed by the keyword occurrence frequency in Table 9.5. This demonstrates that these keywords have primarily occurred in publications during the given timeframe, where corporate sustainability is the most widespread keyword in the given research landscape. Thus, existing research focuses more on sustainability and performance indicators.

Table 9.3 Country-wise article publications and citations received

Rank	Country	Journal articles	Rank	Country	Citations
1	China	9	1	United States	510
2	United States	7	2	China	432
3	India	4	3	Switzerland	281
4	Vietnam	3	4	Vietnam	184
5	United Kingdom	3	5	Italy	173
6	South Korea	3	6	United Kingdom	73
7	Malaysia	3	7	Germany	73
8	Italy	3	8	Malaysia	72
9	Czech Republic	3	9	India	66
10	Bangladesh	2	10	South Korea	66
11	Taiwan	1	11	Bangladesh	59
12	Syrian Arab Republic	1	12	Taiwan	52
13	Switzerland	1	13	Macao	52
14	Spain	1	14	Syrian Arab Republic	34
15	South Africa	1	15	Spain	33
16	Saudi Arabia	1	16	Portugal	29
17	Portugal	1	17	Pakistan	22
18	Pakistan	1	18	Czech Republic	19
19	Nigeria	1	19	Nigeria	17
20	Macao	1	20	Ghana	9
21	Latvia	1	21	Finland	9
22	Jordan	1	22	Cyprus	9
23	Indonesia	1	23	Canada	9
24	Hungary	1	24	South Africa	6
25	Ghana	1	25	Indonesia	6
26	Germany	1	26	Hungary	6
27	Finland	1	27	Jordan	3
28	Cyprus	1	28	Saudi Arabia	1
29	Canada	1	29	Latvia	1

Source Review Data, 2024

The lines between the items in the figure represent the link between each, and the thickness of these lines indicates the solidity of the linkages. Thus, corporate sustainability has thicker linked lines with innovations, environmental technology, and green information systems, indicating that they are highly related. Further, technology adoption also manifests a relationship with firm performance and sustainability.

The colour of the items represents the clusters they belong to. Figure 9.6 shows four colour themes: red, green, blue, and yellow. These clusters have keywords, as

Table 9.4 Occurrences of keywords

Minimum number of occurrences of a keyword	Threshold keyword level
1	271
2	42
3	17
4	9
5	7
6	6
7	5
8	4
9	3

Source Review Data, 2024

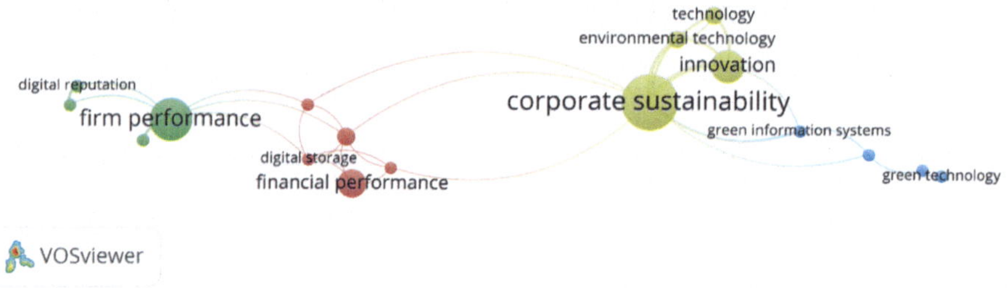

Fig. 9.6 The keyword co-occurrence network visualisation. *Source* Review Data, 2024

presented in Table 9.6. The keywords clustered into one indicate a high chance that these keywords can be found in the same topic. Thus, as in Table 9.6, four themes can be identified by grouping the keywords under 5 clusters which are "Digitalization impact on firms' performance metrics", "Green technology and information management on organisation performance", "Digitalization's impact on firm performance and reputation", "Technological innovations and corporate sustainability", and "Development in green innovations and environmental performance" respectively.

Cluster 1 digital storage, sustainability performance, finance performance, industry performance, technology adoption (Digitalization impact on firms' performance metrics).

Technological innovations such as artificial intelligence, mobile technology and blockchain, big data, cloud computing, social media, industrial internet, cyber security, analytics, etc., have set organisations at a competitive edge in enhancing the organisation performance in every aspect and driving it to the sustainable performance (Kim & Lee, 2017; Siddik et al., 2023; Velinov & Konovalova, 2022). The theory of Resource View (RBV) has been discussed in many studies as it is a managerial framework to control an organisation's strategic resources. The researchers have

Table 9.5 Keyword with a minimum of two occurrences

Keywords	Occurrences
Corporate sustainability	15
Firm performance	9
Innovation	6
Finance performance	5
Sustainability performance	3
Environmental technology	3
Technology	3
Digital storage	2
Green information systems	2
Sustainable products	2
Information management	2
Technology adoption	2
Green technology	2
Digital reputation	2
Industry performance	2
Organisational performance	2
Non-financial performance	2
Firm sustainability	2

Source Review Data, 2024

Table 9.6 Clustering of keywords

Cluster	Cluster color and No. of items	Keywords	Theme
1	Red (5 items)	Digital storage, sustainability performance, finance performance, industry performance, technology adoption	Digitalisation impact on firms' performance metrics
2	Green (4 items)	Digital reputation, firm performance, non-financial performance, sustainable products	Digitalisation's impact on firm performance and reputation
3	Blue (4 items)	Green information systems, green technology, information management, organisational performance	Green information systems and technology on organisation performance
4	Yellow (4 items)	Corporate sustainability, environmental technology, innovation, technology	Technological innovations and corporate sustainability

Source Review Data, 2024

emphasised that visible resources like financial, technological, and physical and non-visible resources like information, knowledge, and reputation enable organisations to gain outstanding financial and sustainable performance (Khan et al., 2022). Sustainability performance refers to achieving a firm's sustainability in terms of its organisational, financial, technological, economic, social, environmental, etc. (Hanelt et al., 2017). Corporate sustainability performance measures how well the organisations integrate the environmental, social, and governance (ESG) perspectives along with the economic aspects of the business operations to enhance the organisation's performance and the community (Miao et al., 2023; Siddik et al., 2023; Wang, 2018). The financial performance of an organisation measures the comprehensive standing of the firm's overall standing in terms of assets, equity, liabilities, income, expenses, and profitability, and the key indicators are return on assets, return on capital, return on sales, and return on equity (Santhi et al., 2024). Accordingly, the critical focus of the existing research landscape was to observe the interconnection of corporate sustainability with different performance metrics of a firm, such as firm performance and financial performance, and to test how the integration of technological advancement is capable of uplifting these performance metrics (Hanelt et al., 2017; Oware & Mallikarjunappa, 2021; Santhi et al., 2024). Many studies have observed these relationships between digitalisation integration and performance metrics in different industry-wide performances, such as performance in the manufacturing industry, automobile industry, wine industry, healthcare industry, mining industry, and information and telecommunication industry (Jung et al., 2018; Mandal & Bagchi, 2016).

Cluster 2 Digital reputation, firm performance, non-financial performance, sustainable products (Digitalization's impact on firm performance and digital reputation)

The key focus of the third cluster is the studies on digitalisation and its influence on enhancing the non-financial performance metrics, which are environmental, social and corporate governance (ESG) performance or corporate sustainability, and how the influence of digital processes enhances the organisations' reputation. Organisations' non-financial performance can be derived through the scores generated from the reported information on ESG pillars, and ESG disputes are also considered (Schena et al., 2022). Closely resembling that digitalisation of organisation processes has opened up the wings to work out of the borders, increasing communication and simplifying information management through different technological adoptions such as cloud computing, social media, information systems, big data, cyber security, analytics, etc. (Rosamartina et al., 2022). This has driven the organisations toward sustainability goals and created a digital reputation for the firms (Schena et al., 2022; Rosamartina et al., 2022). The conventional definition of reputation is the overall impression that an organisation has gained from its stakeholders. It holds the firm at a competitive edge in retaining employees, attracting clients, increasing prices, etc. (Schena et al., 2022). Consequently, the digital reputation is the impression of the stakeholders towards the company's characterisation developed in digital platforms such as social media, big data visualisation, analytics, etc. (Schena et al.,

2022; Rosamartina et al., 2022). The current research studies propose that creating a digital reputation would enhance the company's performance and drive towards more corporate sustainability.

Cluster 3 green information systems, green technology, information management, organisational performance (Green information systems and technology on organisation performance)

The cluster emphasises green information systems, technology, information management, and organisational performance. Maintaining environmentally friendly information flow using digital platforms can be characterised as green information systems (Yang et al., 2019). The Technology-Organisation-Environment (TOE) framework is a theoretical framework developed in information systems that supports how adopting new technologies, organisational context, and environmental factors influence green innovations and lead to corporate sustainability—the efficient allocation of resources for various operational activities. Designing, production, and logistics can be done through information systems to enhance the organisation's performance and achieve the environmental sustainability goal (Yang et al., 2017, 2019). Information is vital for the overall organisation's performance as it plays a crucial role in decision-making and knowledge generation (Mandal & Bagchi, 2016). Thus, organisations invest in different information systems to capture, store, process, retrieve, and communicate information to the decision-making bodies effectively and efficiently (Mandal & Bagchi, 2016). Some of the significant information systems used for different business tasks were enterprise recourse planning (ERP) systems, customer relationship management (CRM), supply chain management (SCM), enterprise application integration, web services, etc. and support the ecological goals (Mandal & Bagchi, 2016; Yang et al., 2017; Yang et al., 2017). Green innovations like green technology green information systems allow firms to perform their operations in a way that reduces environmental pollution, enhance resource efficiency and, reduce waste, etc., and promote organisational performance and sustainability (Le, 2022; Mate et al., 2022; Zhang et al., 2020).

Cluster 4 Corporate Sustainability, environmental technology, Innovation, technology (Technological innovations and corporate sustainability)

The fourth cluster indicates that existing literature has researched technological innovations and their impact on corporate sustainability. Empirical literature showcases that technological innovations have a more decisive influence over the performance metrics of small and medium enterprises (SMEs) as technological innovations like social media (Twitter et al., etc.) have become a competitive marketing tool to reach their customer and investors and then to grow overall performance and foster sustainability of the firms (Bruce et al., 2023; Johl & Toha, 2021). Studies have been carried out on cloud computing, which is another digital innovation used by SMEs to large-scale corporates that provide ample storage, networking, servers, intelligence, analytics, and software services to enhance the firm performance and sustainability (Dusková, 2021; Olayah, 2019). FinTech is another technological innovation examined by studies that can improve firms' performance and sustainability (Siddik et al.,

2023). Organisations are provided financial services such as capital raising, insurance, digital payment and settlements, and credit through FinTech innovations such as mobile technology, artificial intelligence, and blockchain (Meiling et al., 2021). Further, the Ecological Modernisation (EMT) Theory supports this argument that the environmental problems that arise with economic development can be managed through technological advancements like FinTech. Large-scale firms like Microsoft, Uber, Netflix, and Facebook have also adopted big data analytics as they handle large amounts of Data from different sources (Chatterjee et al., 2022). The empirical literature has observed how big data analytics supports collecting, analysing, and interpreting data systematically to make rational decisions and increase business value. Studies have found that this has positively influenced the performance metrics of the firms and then positively impacted the firm's sustainability performance (Chatterjee et al., 2022).

9.3.3.2 Areas Where Empirical Research is Insufficient on Digitalisation and Its Impact on Organisation Performance and Sustainability

The keyword co-occurrence density visualisation map created by VOSviewer, as depicted in Fig. 9.7, supports the study's second objective: to identify the critical areas of research lacking in digitalisation and its impact on organisation performance and sustainability. According to the VOSviewer manual, the keyword nodes are highlighted by three colour backgrounds based on their density of research. These colour backgrounds indicate the density of the particular research area. The red indicates that sufficient studies have been conducted in the relevant research area, and the green indicates the lower research density (Chen et al., 2016). As in Fig. 9.7, the keywords corporate sustainability, firm performance, innovation and financial performance keyword nodes are in red, which says that the empirical literature has covered sufficient knowledge on particular research areas.

Here, the keyword innovation mainly indicated the technological innovations in the literature, which emphasises that the key focus of many studies was to look at the sustainability performance of the firms with the integration of technological innovations into the business processes. Looking at Table 9.5, it can be further confirmed that corporate sustainability is the most commonly used keyword in research as it has the highest number of keyword occurrences, which is 15, followed by firm performance, innovations, and financial performance, respectively. The remaining keywords appear from a yellow to green background, indicating that fewer studies have been conducted. Accordingly, studies on green technology, green information systems, and digital reputation do not create enough knowledge in a particular field. The keywords environment technology also works as another synonym for green technology, and the other terms technology and digital storage are also similar keywords used by researchers on digitalisation. Accordingly, the research on green technology adoption, like green information systems that enhance environmental sustainability performance, needs further investigation for the body of knowledge in

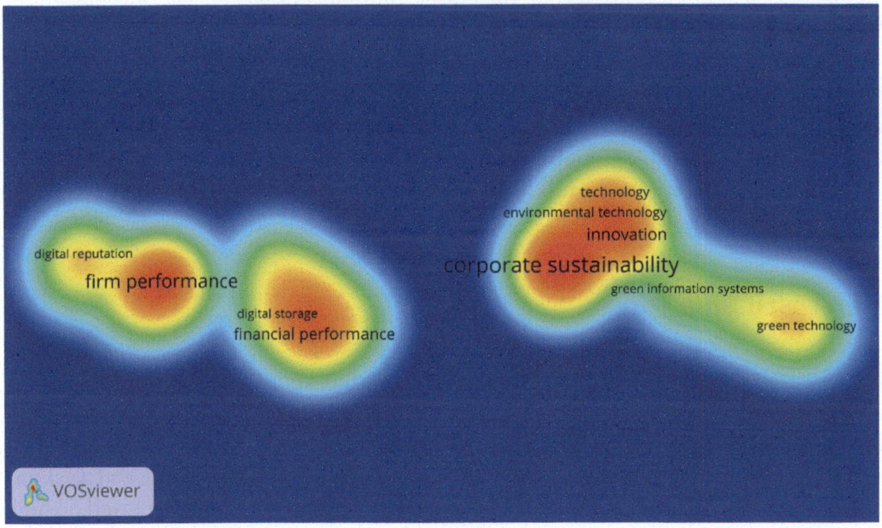

Fig. 9.7 The keyword co-occurrence density visualisation. *Source* Review Data, 2024

the literature. Further, research on how digitalisation can enhance the digital reputation of firms is another crucial emerging research background that needs further investigation. Figure 9.8 of overlay visualisation also confirms the interconnection between digital reputations, digital storage and the firms' performance metrics. These keywords are in yellow, and based on the colour bar indication at the bottom, they have emerged very recently, during the year 2022. Still, very few studies have been conducted and need further investigation.

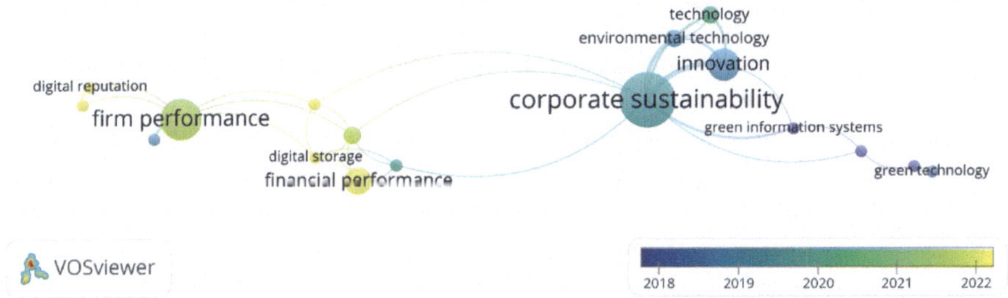

Fig. 9.8 The keyword co-occurrence overlay visualisation. *Source* Review Data, 2024

9.3.4 Reporting Bias Assessment

Per the PRISMA guidelines, a biased assessment is needed due to missing pertinent results (Priyashantha et al., 2022). A systematic assessment was not followed for this task. However, PRISMA guidelines and systematic and objective tools were used to avoid bias in reporting the results.

9.4 Discussion

The research articles used in the study were reported under the two primary objectives of "investigating the current knowledge on digitalisation's impact on organisation performance and sustainability" and "identifying the lacking areas of empirical research on digitalisation and its impact on organisation performance and sustainability". In general, critical findings under both these objectives have reported on how integrating digital innovations into the business processes has enhanced the organisation's performance metrics with a particular emphasis on enhancing the organisation's sustainability performances. The findings disclose several vital themes: the impact of digitalisation on the critical performance metrics of organisations, the link between digital reputations, firm performance, and corporate sustainability, the role of green information systems, and the influence of digital innovations on corporate sustainability.

The review findings underscore critical theories such as EMT theory, RBV theory, and TOE theoretical framework, which lay the foundation for research on digitalisation's impact on organisational performance and sustainability. The EMT theory suggests that resource efficiency can be enhanced through digital adoption. In contrast, RBV theory suggests that integrating tangible resources, including technology and intangible resources, supports overcoming environmental and sustainability challenges (Khan et al., 2022). The TOE framework suggests that green innovations are positively influenced by the availability of technology, organisation readiness and environmental fitness (Singh et al., 2023; Zhang et al., 2020). The study findings have emphasised that digitalisation has positively affected the performance metrics of the organisations, including financial, operational and overall business performance and corporate sustainability in terms of ESG along with the economic aspects (Kim & Lee, 2017; Singh et al., 2022; Velinov & Konovalova, 2022). The findings reveal that many of the studies have focused on achieving corporate sustainability and have observed how ESG performances have been influenced by digital integrations and other organisational aspects (Miao et al., 2023; Wang, 2018). More significantly, many studies have tested the direct impact of digitalisation on corporate sustainability, organisation performance, and another aspect of the organisation (Bruce et al., 2023; Oware & Mallikarjunappa, 2020; Wang, 2018; Kim & Lee,

2017) where few of the studies have tested digitalisation as a moderator or a mediator between organisation performance metrics and corporate sustainability (Miao et al., 2023; Schena et al., 2022; Yang et al., 2019; Yang et al., 2017).

The general outcomes of the study revealed that FinTech, cloud computing, social media platforms, Information Systems, and big data analytics are the key technological advancements that the studies have researched (Siddik et al., 2023; Bruce et al., 2023; Chatterjee et al., 2022; Meiling et al., 2021; Zhang et al., 2020; Yang et al., 2019; Olayah, 2019; Yang et al., 2017). Further, digital reputation and green technology are other blooming areas of research, yet fewer studies have been carried out in this area (Zhang et al., 2020). The studies on digital reputation and organisation performance and sustainability have been researched very recently and conducted very little. The studies on how different digital innovations like cloud computing and big data analytics have not been researched at a satisfactory volume, and the impact of digital innovations like the Internet of Things (IoT), artificial intelligence, and blockchain adoption on firm performances and sustainability has not been researched during the last decade and are the identified gaps in the review.

One limitation of the review is that the authors reached only 38 articles using only one database. Further, only empirical research articles and reviews were considered in the study. This exclusion has reduced a considerable amount of literature in the field of study.

9.4.1 Practicality and Research Implications

The study's practical implications signify that digital innovations like cloud computing, FinTech, big data analytics, information systems and social media have positively supported the improvement of the organisation's performance metrics and sustainability and digital reputation. Thus, small, medium and large-scale organisations and policymakers can adopt these digital innovations to enhance the firm's financial, operational and overall performance and sustainability regarding ESG and economic operations.

Future research implications from the study are that a few studies are carried out on digitalisation aspects. Significantly, the studies on digital reputation are significantly less, and the studies on separate digital technologies such as big data analytics and cloud computing are minimal, which would not be enough to establish the knowledge required in the research landscape. The impact of digital technologies like artificial intelligence, Internet of Things and blockchain have not been tested empirically in the selected research articles.

The present study comprises only 38 research articles, implying the limitedness of the existing research on the particular subject area. Thus, it signifies the need for more research on digitalisation's impact on organisational performance and sustainability. Studies can investigate which digital advancements impact organisational performance and sustainability so that organisations can invest in the right digital transformation to positively impact firm performance and sustainability.

The studies have highlighted that firms' sustainability should be achieved economically, environmentally, and socially. They have mainly discussed integrating digital technologies into different green concepts, such as green human resource management, green information management, and other green technological innovations to promote environmental sustainability (Yang et al., 2019; Zhang et al., 2020). However, there is less recognition of how digital integration can promote social sustainability. Thus, future studies should investigate this prime area of study.

Regarding the interconnection between organisation performance and digitalisation research, the empirical studies have mainly studied how financial performance, such as a firm's profitability, has been influenced by digital integration (Johl & Toha, 2021; Schena et al., 2022; Oware & Mallikarjunappa, 2020). The influence of digitalisation on firms' operational performance, such as how it affects improving the production processes, customer services, supply chain management, and market-based performance, such as influence on market share and return to shareholders, has not been observed in the identified studies. Thus, future studies can open up their investigation on digitalisation in these untouched areas.

One of the studies also discussed responsible technological innovation, focusing primarily on the importance of regulating technological innovations to avoid unmanageable and revolutionary technological innovations that can negatively influence an organisation's performance and sustainability (Memon & Ooi, 2023). This is a very emerging and vital area for future researchers to pay attention to as it is similarly essential to investigate the negative aspects of this digital transformation. Thus, all these implications are possible avenues for future studies on digitalisation and its influence on organisational performance and sustainability.

9.5 Conclusion

This study has conducted an SLR to achieve two primary objectives: identifying (1) the current body of knowledge and (2) the areas of research lacking on digitalisation's impact on organisational performance and sustainability. The study adopted the PRISMA guidelines for the article selection procedure and reporting the results. Based on the study inclusion criteria, 38 journal articles published during the 2014 to 2024 period were selected for the review. A bibliometric analysis was performed using the bibliophile and VOSviewer software to analyse the selected articles in terms of the annual article publication, average article citation, country-wise publications, publication sources, keyword map, keyword density visualisation, etc.

The study's first objective, to investigate the current knowledge of the respective research area, was achieved based on keyword clustering using VOSviewer. Accordingly, four major themes were identified, namely: (1) digitalisation's impact on firms' performance metrics, (2) digitalisation's impact on firm performance and reputation, (3) green information systems and technology on organisation performance, and (4) Technological innovations and corporate sustainability. Consequently, achieving organisational sustainability through digitalisation is the key focus of all these themes. The findings revealed that during the last decade, studies have primarily focused on increasing the performance metrics of the firms, including the sustainability performance in terms of ESG and economics, by integrating digital advancements into the business processes. Another critical area that the studies have discussed is achieving green objectives and the digital reputation of the organisation by adopting different technological advancements. The selected studies have primarily studied the adoption of various digital advancements by the organisation, the impact of digitalisation on the critical aspects of the organisation, and the mediating or moderating role of digitalisation in the relationship between business aspects and corporate sustainability. According to the study findings, the influence of digital innovations like cloud computing, big data analytics, FineTech, Social Media, and information systems on organisation performance and sustainability have been separately studied.

The study's second objective is to identify areas lacking for future research. The findings demonstrate that the research studies on digitalisation's impact on organisational performance and sustainability are weak and need more investigation to establish the specific body of knowledge. Thus, all the research areas identified under objective one and some other areas associated with digitalisation and organisation performance metrics were identified for future researchable areas. Accordingly, only a single article was found on cloud computing, social media influence, big data analytics and its influence on the organisation's performance and sustainability. Further, areas like artificial intelligence, blockchain and the Internet of Things have not been researched in the selected articles for the study. Based on the keyword density visualisation, keywords like green technology, green information systems, and digital reputation were less researched and needed further investigation. Further, many studies have investigated how digital innovations can promote green objectives and drive organisations towards environmental sustainability. However, none of the studies have researched social sustainability, a key aspect of corporate sustainability. Thus, future researchers can study the identified areas to fill the knowledge gap in the particular research panorama. Further, looking at the country-wise research publications, none of the research studies have been carried out in the Sri Lankan context, which reveals a knowledge gap that needs to be addressed.

The review findings manifest that digitalisation is an emerging element that strengthens organisational performance and sustainability. To achieve an organisation's sustainability goals, it needs to inform, adopt, and upgrade new technologies while regulating them to gain a competitive advantage over them. Thus, this review would become a substructure for practitioners and researchers to underpin the current body of knowledge to uplift the organisation's performance and sustainability and further explore the interplay among them.

References

Bhuiyan, M. R. I. (2024). Examining the digital transformation and digital entrepreneurship: A PRISMA based systematic review. *Pakistan Journal of Life and Social Sciences, 22*(1), 1136–1150. https://doi.org/10.57239/PJLSS-2024-22.1.0077

Bruce, E., Keelson, S., Amoah, J., & Bankuoru Egala, S. (2023). Social media integration: An opportunity for SMEs sustainability. *Cogent Business & Management, 10*(1), 2173859. https://doi.org/10.1080/23311975.2023.2173859

Chatterjee, S., Chaudhuri, R., Vrontis, D., & Thrassou, A. (2022). Impacts of big data analytics adoption on firm sustainability performance. *Qualitative Research in Financial Markets, 15*(4), 589–607. https://doi.org/10.1108/QRFM-01-2022-0005

Chen, X., Chen, J., Wu, D., Xie, Y., & Li, J. (2016). Mapping the research trends by Co-word analysis based on keywords from funded project. *Procedia Computer Science, 91*, 547–555. https://doi.org/10.1016/j.procs.2016.07.140

Dewasiri, N. J., Baker, H. K., Banda, Y. K. W., & Rathnasiri, M. S. H. (2022). The dividend decision model: A possible solution for the dividend puzzle. In S. Rana, Sakshi, & J. Singh (Eds.), *Exploring the latest trends in management literature* (Review of Management Literature, Vol. 1, pp. 249–267). Emerald Publishing Limited, Leeds. https://doi.org/10.1108/S2754-586520220000001013

Dewasiri, N. J., Karunarathne, K. S. S. N., Menon, S., Jayarathne, P. G. S. A., & Rathnasiri, M. S. H. (2023). Fusion of artificial intelligence and blockchain in the banking industry: Current application, adoption, and future challenges. In A. Saini, & V. Garg (Eds.), *Transformation for sustainable business and management practices: Exploring the spectrum of industry 5.0* (pp. 293–307). Emerald Publishing Limited, Bingley. https://doi.org/10.1108/978-1-80262-277-520231021

Dewasiri, N. J., Dharmarathna, D. G., & Choudhary, M. (2024). Leveraging artificial intelligence for enhanced risk management in banking: A systematic literature review. In R. Singh, S. Khan, A. Kumar, & V. Kumar (Eds.), *Artificial intelligence enabled management: An emerging economy perspective* (pp. 197–213). De Gruyter, Berlin. https://doi.org/10.1515/9783111172408

Dušková, M. (2021). Key enabling technologies and measuring of the company performance in relation to sustainable development: Evaluation model design. *International Journal of Innovation and Sustainable Development, 15*(1), 1–34. https://doi.org/10.1504/IJISD.2021.111551

Gusarova, S. A., Gusarov, I. V., & Smeretchinskii, M. S. (2023). Digital transformation of the South Asia region: Challenges and prospects. In *E3S web of conferences* (Vol. 403, p. 06016). EDP Sciences. https://doi.org/10.1051/e3sconf/202340306016

Hanelt, A., Busse, S., & Kolbe, L. M. (2017). Driving business transformation toward sustainability: Exploring the impact of supporting IS on the performance contribution of eco-innovations. *Information Systems Journal, 27*(4), 463–502. https://doi.org/10.1111/isj.12130

Johl, S. K., & Toha, M. A. (2021). The nexus between proactive eco-innovation and firm financial performance: A circular economy perspective. *Sustainability, 13*(11), 6253. https://doi.org/10.3390/su13116253

Jung, S., Nam, C., Yang, D. H., & Kim, S. (2018). Does corporate sustainability performance increase corporate financial performance? Focusing on the information and communication technology industry in Korea. *Sustainable Development, 26*(3), 243–254. https://doi.org/10.1002/sd.1698

Jayawardena, N. S., & Dewasiri, N. J. (2023). Food acquisition and consumption issues of South Asian countries: A systematic literature review and future research agenda. *FIIB Business Review*. https://doi.org/10.1177/23197145231194113

Karunarathna, K. S. S. N., Dewasiri, N. J., Singh, R., & Rathnasiri, M. S. H. (2024). What does artificial intelligence–powered ChatGPT bring to academia? A review. In R. Singh, S. Khan, A. Kumar, & V. Kumar (eds.), *Artificial intelligence enabled management: An emerging economy perspective* (pp. 3–13). De Gruyter, Berlin. https://doi.org/10.1515/9783111172408

Khan, S., Ali, S. S., & Singh, R. (2022). Determinants of remanufacturing adoption for circular economy: A causal relationship evaluation framework. *Applied System Innovation, 5*(4), e62. https://doi.org/10.3390/asi5040062

Kim, H., & Lee, H. (2017). Architecture innovation in the DRAM industry: How it affects firms' sustainable competence. *Solid State Technology, 60*(2), 12–16.

Liberati, A., Altman, D. G., Tetzlaff, J., Mulrow, C., Gøtzsche, P. C., Ioannidis, J. P., ... & Moher, D. (2009). The PRISMA statement for reporting systematic reviews and meta-analyses of studies that evaluate health care interventions: Explanation and elaboration. *Annals of Internal Medicine, 151*(4), W–65. https://doi.org/10.1371/journal.pmed.1000100

Le, T. T. (2022). How do corporate social responsibility and green innovation transform corporate green strategy into sustainable firm performance? *Journal of Cleaner Production, 362*, 132228. https://doi.org/10.1016/j.jclepro.2022.132228

Mate, D., Oláh, J., Erdei, E., Estiyanti, N. M., Bacs, Z., & Kovacs, S. (2022). The impacts of sustainable industrial revolution (IR) on the profitability of Hungarian food companies. *Frontiers in Sustainable Food Systems, 6*, 1034010. https://doi.org/10.3389/fsufs.2022.1034010

Mandal, P., & Bagchi, K. (2016). Strategic role of information, knowledge and technology in manufacturing industry performance. *Industrial Management & Data Systems, 116*(6), 1259–1278. https://doi.org/10.1108/IMDS-07-2015-0297

Meiling, L., Yahya, F., Waqas, M., Shaohua, Z., Ali, S. A., & Hania, A. (2021). Boosting sustainability in healthcare sector through Fintech: Analysing the moderating role of financial and ICT development. *Inquiry: The Journal of Health Care Organization, Provision, and Financing, 58.* https://doi.org/10.1177/00469580211028174

Meline, T. (2006). Selecting studies for systemic review: Inclusion and exclusion criteria. *Contemporary Issues in Communication Science and Disorders, 33*(Spring), 21–27. https://doi.org/10.1044/cicsd_33_S_21

Memon, K. R., & Ooi, S. K. (2023). Responsible innovation and resource-based theory: Advancing an antecedent-outcome model for large manufacturing firms through structured literature review. *Asian Journal of Business Ethics, 12*(2), 441–467. https://doi.org/10.1007/s13520-023-00181-6

Miao, Y., Shi, Y., & Jing, H. (2023). Effect of servitisation on performance in manufacturing firms: A mediating effect model of digitalisation moderated by ESG performance. *Heliyon, 9*(10). https://doi.org/10.1016/j.heliyon.2023.e20831

Okoli, C., & Schabram, K. (2015). A guide to conducting a systematic literature review of information systems research. https://doi.org/10.2139/ssrn.1954824

Olayah, F. (2019). A conceptual modeling for achieving a performance and sustainability on SMEs using cloud computing technology. *International Journal of Advanced Trends in Computer Science and Engineering, 8*(1.4), 236–242. https://doi.org/10.30534/ijatcse/2019/3681.42019

Oware, K. M., & Mallikarjunappa, T. (2020). Disability employment and financial performance: The effect of technological innovation of listed firms in India. *Social Responsibility Journal, 17*(3), 384–398. https://doi.org/10.1108/SRJ-09-2019-0299

Oware, K. M., & Mallikarjunappa, T. (2021). Disability employment and financial performance: The effect of technological innovation of listed firms in India. *Social Responsibility Journal, 17*(3), 384–398. https://doi.org/10.1108/SRJ-09-2019-0299

Priyashantha, K. G., De Alwis, A. C., & Welmilla, I. (2021a). Gender stereotypes change outcomes: A systematic literature review. *Journal of Humanities and Applied Social Sciences, ahead-of-print*(ahead-of-print). https://doi.org/10.1108/JHASS-07-2021-0131

Priyashantha, K. G., De Alwis, A. C., & Welmilla, I. (2021b). The facets of gender stereotype change: A systematic literature review. In *Proceedings of 12th international conference on business and information, faculty of commerce and management studies*, University of Kelaniya. p. 70. https://doi.org/10.2139/ssrn.4117570

Priyashantha, K. G., De Alwis, A. C., & Welmilla, I. (2021c). Gender stereotypes change outcomes: A systematic literature review. *Journal of Humanities and Applied Social Sciences* (ahead-ofprint). https://doi.org/10.1108/JHASS-07-2021-0131

Priyashantha, K. G., De Alwis, A. C., & Welmilla, I. (2022). Disruptive human resource management technologies: A systematic literature review. *European Journal of Management and Business Economics, 33*(1), 116–136. https://doi.org/10.1108/EJMBE-01-2022-0018

Rathnasiri, M. S. H., Dewasiri, N. J., & Kumar, A. A. (2024). Policy Framework and implementation strategies for Sri Lanka's transition to a net-zero economy. In R. Singh, S. Khan, A. Kumar, S. Luthra, & H. Chokshi (Eds.) *Net zero economy, corporate social responsibility and sustainable value creation* (pp. 43–60). CSR, Sustainability, Ethics & Governance. Springer, Cham. https://doi.org/10.1007/978-3-031-55779-8_3

Rosamartina, S., Giustina, S., & Angeloantonio, R. (2022). Digital reputation and firm performance: The moderating role of firm orientation towards sustainable development goals (SDGs). *Journal of Business Research, 152*, 315–325. https://doi.org/10.1016/j.jbusres.2022.07.025

Santhi, P., Sasirekha, P., & Anija, J. (2024). A causal linkage between corporate sustainability performance and financial performance of select IT & ITeS companies in India. *Indian Journal of Finance, 18*(2), 43–59. https://doi.org/10.17010/ijf/2024/v18i2/173520

Singh, R., Khan, S., & Centobelli, P. (2022). Investigating the Interplay between social performance and organisational factors supporting circular economy practices. *Sustainability, 14*(24), e16781. https://doi.org/10.3390/su142416781

Singh, R., Khan, S., & Dsilva, J. (2023). A framework for assessment of critical factor for circular economy practice implementation. *Journal of Modelling in Management, 18*(5), 1476–1497. https://doi.org/10.1108/JM2-06-2021-0145

Schena, R., Russo, A., & Pinkse, J. (2022). Employee domain and non-financial performance: The moderating effect of digital reputation. *Meditari Accountancy Research, 30*(4), 893–913. https://doi.org/10.1108/MEDAR-02-2021-1205

Siddik, A. B., Rahman, M. N., & Yong, L. (2023). Do fintech adoption and financial literacy improve corporate sustainability performance? The mediating role of access to finance. *Journal of Cleaner Production, 421*, 137658. https://doi.org/10.1016/j.jclepro.2023.137658

Vann Yaroson, E., Chowdhury, S., Mangla, S. K., Dey, P., Chan, F. T., & Roux, M. (2024). A systematic literature review exploring and linking circular economy and sustainable development goals in the past three decades (1991–2022). *International Journal of Production Research, 62*(4), 1399–1433. https://doi.org/10.1080/00207543.2023.2270586

Velinov, E., & Konovalova, N. (2022). Top management team diversity and firm performance in the digital era. *Polish Journal of Management Studies, 26*. https://doi.org/10.17512/pjms.2022.26.2.24

Wang, D. D. (2018). Unravelling the effects of the environmental technology portfolio on corporate sustainable development. *Corporate Social Responsibility and Environmental Management, 25*(4), 457–472. https://doi.org/10.1002/csr.1472

Yang, Z., Sun, J., Zhang, Y., & Wang, Y. (2019). Perceived fit between green IS and green SCM: Does it matter? *Information & Management, 56*(7), 103154. https://doi.org/10.1016/j.im.2019.02.009

Yang, Z., Sun, J., Zhang, Y., & Wang, Y. (2017). Green, green, green: A triad model of technology, culture, and innovation for corporate sustainability. *Sustainability, 9*(8), 1369. https://doi.org/10.3390/su9081369

Zhang, Y., Sun, J., Yang, Z., & Wang, Y. (2020). Critical success factors of green innovation: Technology, organisation and environment readiness. *Journal of Cleaner Production, 264*, 121701. https://doi.org/10.1016/j.jclepro.2020.121701

Chapter 10
Product Circularity in the Digital Circular Economy

Balgopal Singh

Abstract Realising sustainability through adopting a circular economy is a strategic intent for manufacturing companies. Historically, adopting a circular economy means being regenerative and restorative. The realisation of a circular economy rests in successfully eliminating waste generation through the circular use of planatory resources. The emergence of digital technology contributed to innovation in the traditional circular economy model. Digital technology has facilitated business firms to develop innovative practices that enable product design with circular features. Business models like 'digital servitisation' and 'digital product passport' systems contributed significantly to achieving sustainability goals. These business practices contributed to environmental sustainability and offered a competitive advantage for the firm that successfully built its competency around looping its value delivery system.

Keywords Digital circular economy · Product circularity · Digital servitisation · Digital product passport · ReSolve framework

10.1 Introduction

Circular product design enables a smooth transition from the dominant linear model of economic growth characterised as 'take-make-use-dispose' to a circular model of economic development through 'reduce-reuse-recycle' (Aguiar & Jugend, 2022; Ghisellini et al., 2016; Khan et al., 2022). The circular product design facilitates circular economy by preserving the product use either by lengthening the product use lifecycle of the product or by reusing or recycling the product indefinitely (Alcayaga & Hansen, 2024; Bocken et al., 2016; Singh et al., 2022). Further, with the emergence of digitalisation (Industry 4.0), the circular economy concept as a regenerative and restorative industrial system leveraged digital technologies in its operational process (Govindan, 2023; George et al., 2015; Rajput & Singh, 2019).

B. Singh (✉)
Faculty of Management Studies, Banasthali Vidyapith, Jaipur, Rajasthan, India
e-mail: balgopalsingh@banasthali.in

R. Singh and V. Kumar (eds.), *Sustainable Innovations and Digital Circular Economy*,
https://doi.org/10.1007/978-981-96-1064-8_10

Adopting digital technologies in a circular economy operational process facilitated monitoring (real-time) and control, resulting in increased resource efficiency and productivity; additionally, optimisation capabilities of digital technology further improved the production process, contributing to sustainable resource consumption (Kristoffersen et al., 2020; Singh et al., 2023). Similarly, digital technologies are vital in monitoring and optimising product circularity, enabling a digital circular economy (Diaz & Baumgartner, 2024; Jugend et al., 2024).

Product circularity in the circular economy is a logical consequence of creating a 'closed cycle of flows' with two possibilities: i.e. recycling or reusing the products at the end of life (Diaz & Baumgartner, 2024). The 'circular approach' to product design is central to enabling product circularity. The 'circular approach' concept is borrowed from the work of McDonough and Braungart's (2002) famous book—'*Cradle to Cradle: Remaking the Way We Make* Things'—which emphasises looping technical and biological attributes of products facilitating reuse or recycling of the materials products built on. Attaining product circularity in a digital circular economy necessitates a new conceptualisation of product design equipped with digital technology (Dagilienė et al., 2024; Han et al., 2023).

Digital technologies such as IOT, block chain, artificial intelligence, big data, and data analytics are the recent technological advancement that provides possibilities for collecting, processing, and disseminating data in near-real time and at scale, providing enormous potential for decision-making to eliminate inefficiencies in the industrial ecosystem (Han et al., 2023; Langley et al., 2023). The extant literature provides sufficient evidence of the sustainable impact of digital transformation in dealing with product circularity; notably, it facilitated an accurate and verifiable set of information about a product's environmental impact during its entire lifecycle (Alcayaga & Hansen, 2024). Nevertheless, this information highlights design engineers and product managers for product lifecycle management and formulating life extension strategies (Toth-Peter et al., 2023; Rusch et al., 2023).

Product circularity is in contrast to the prevalent throw-away mentality in society. The product design for circularity stresses design for 'extended use, reuse and recovery' (Jugend et al., 2024 den Hollander et al., 2017). The extended use and reuse strategies product design emphasise quality characteristics, maintenance, repairability, remanufactured or refurbished. However, despite these efforts, the future of 'product design for circularity' depends on consumers' aspirations, i.e. consumer decision on product 'obtainment, use and riddance' (Selvefors et al., 2019). Thus, to truly ripe the fruit of product circularity, it necessitates fostering circular consumption behaviour; this requires conscious effort to break commonly held beliefs in society at large, i.e. '*throw away—what is not working*' (Mestre & Cooper, 2017; Toth-Peter et al., 2023). Coping with this user-related challenge demands a smart solution.

Implementing a circular economy transforms a linear business model into a circular one. Likewise, the digital circular economy demands business models with digital capabilities (Pagoropoulos et al., 2017; Rusch et al., 2023). The digital transformation of business models necessitates instituting data management systems, data analytics capabilities, and leadership and strategic capabilities to participate in the digital circular economy. The circular business model built on a high level of digital maturity innovates in value creation and delivery using smart processes (Alcayaga & Hansen, 2024). The product proposition of such digital circular business models embraces digitally enabled product circularity (Wang et al., 2022).

The main focus of this chapter is to present a conceptual framework for 'product circularity' in a 'digital circular economy'.

10.2 Circular Economy/Digital Circular Economy

The circular economy concept was primarily introduced by Pearce and Turner (1990) as a prerequisite to achieving sustainability. Ellen Macarthur Foundation (2013) introduced circular economy as 'an industrial economy that is restorative or regenerative by intention and design' (p. 14). According to the literature, circular economy emerged as three 'actions' for realising sustainability goals. The three actions are distinctively known as '3R's Principles', i.e. Reduce, Reuse, and Recycle. The 3R's Principles were concerned with industrial initiatives towards sustainable waste management. These include using resources through multiple cycles, product life extension, and prevention of landfills (Ghisellini et al., 2016; Geissdoerfer et al., 2017; Singh et al., 2022).

The contemporary understanding of the circular economy, according to Ellen Macarthur Foundation (2013), is attributed to biomimicry (Benyus, 2002); blue economy (Pauli, 2010); cradle-to-cradle (McDonough & Braungart, 2002); regenerative design, (Lyle, 1996); industrial ecology (Graedel & Allenby, 2003); and law of ecology (Commoner, 1971); performance economy (Stahel, 2010).

A first comprehensive definition of circular economy was given by Webster (2015) through Ellen McArthur Foundation publication that 'a circular economy is one that is restorative and regenerative by design, and which aim to keep products, components and materials at their highest utility and value, at all times' (p. 16). This definition was summarised into three main principles: regenerating natural systems, keeping products and materials in use, and designing out waste and pollution (Ellen MacArthur Foundation, 2015). Bocken et al. (2016) complemented this definition by adding circular economy as a 'design and business model strategy that slows, closes, and narrows resource loops' (p. 309). Observing the literature on contributions towards sustainability in his review, Geissdoerfer et al. (2017) put forth a comprehensive definition of circular economy as a 'regenerative system in which resource input and waste, emission, and energy leakage are minimised by slowing, closing, and narrowing material and energy loops. This can be achieved through long-lasting design, maintenance, repair, reuse, remanufacturing, refurbishing and recycling' (p. 759).

With the emergence of digital technology, the R-framework gained traction. Digital technology-based tools and processes like the Internet of Things (IoT), artificial intelligence (AI), blockchain, big data, and business analytics (BA) contributed a wide range of support to traditional 3R's circular strategy. In the extant literature, adequate evidence is available to connect digital technologies with the circular economy. Recently conducted systematic literature reviews by Farace and Tarabella (2024), Seyyedi et al. (2024), Toth-Peter et al. (2023), Rejeb et al. (2022), and Chauhan et al. (2022), show that the use of digitalisation is progressively growing and its connection with the CE is becoming more prominent.

Digital technology also nurtured various circular economy-oriented innovations under the 'Digital Circular Economy' domain. Digital circular economy embraced with digital prowess efficiently applied digital technologies in enabling product circularity. Product circularity is designing new offerings with circular attributes (Aguiar & Jugend, 2022; Dagilienė et al., 2024).

10.3 Circular Product Design and Business Model

With the increase in the sustainability concern among policymakers and companies, 'sustainable design' emerges as the most reliable means to deal with depleting resources and environmental concerns arising due to landfill of industrial and household waste (Bocken et al., 2016; den Hollander et al., 2017; Diaz & Baumgartner, 2024). 'Sustainability' is concerned with meeting the needs of the present without compromising the ability of future generations to meet their own needs, i.e. development that mainly maintains the indefinite perpetuation of natural resources and ecology (Diaz et al., 2022; Geissdoerfer et al., 2017). Sustainable design in the form of 'eco-design' is concerned with the 'creation of new, added value, eco-efficient products or services that can stimulate the economic competitiveness of industries, while contributing more sustainable forms of consumption and lifestyle scenarios' (Mestre & Cooper, 2017, p. 2). The eco-design strategy initially includes using materials of low impact and in lesser quantity, optimisation of production and distribution, reduction of impact during use, optimisation of initial lifetime and end-of-life system, and developing new concept (Han et al., 2023; Mestre & Cooper, 2017).

The guiding principle of eco-design functions was 'waste hierarchy', i.e. first efforts are on preventing waste, then reuse, recycle, recover and finally, disposal (Jugend et al., 2024; den Hollander et al., 2017). Thus, the idea of eco-design products is conceived with the prevention of waste as the primary goal, incorporating a product development strategy for product lifecycle longevity (Aguiar & Jugend, 2022). The lifecycle design strategy for sustainable products integrates a product lifecycle approach that focuses on the longevity of the product that prevents the product from becoming waste; however, failing to do this further strives for subsequent stages goal to be achieved (Bocken et al., 2016). Nevertheless, despite all efforts, products inevitably become obsolete after a specific time. The emergence of the circular economy concept has drawn the attention of product designers towards verifying the validity of this waste hierarchy guiding principles in a new phase of sustainability thinking; this further motivated them to design a product development strategy meeting the circularity principles (Wang et al., 2022; den Hollander et al., 2017).

The new circular economy paradigm demands new product concepts and designs for enabling circularity. In the extant literature, a range of strategies for circular product design is conceptualised (Aguiar & Jugend, 2022; Bocken et al., 2016; Jugend et al., 2024)—the following section first addresses product design for businesses craving to move in the circular economy model. Subsequently, the product design for the digital circular economy is presented.

10.3.1 Product Design for a Circular Economy Model

The search for extant literature brings together various frameworks conceptualised by the researcher for circular product design. The noticeable and early framework developed on product design for the circular economy was promulgated by Bocken et al. (2016), den Hollander et al. (2017) and Mestre and Cooper (2017).

Bocken et al. (2016) framework was based on the terminology of '*slowing, closing, and narrowing*' resource loops. *Slowing resource loops* and *closing resource loops* were the fundamental strategies towards resource sustainability while narrowing resource loops aimed to use fewer resources per product. The slowing resource loops design strategy was designing products with long-life and extended product life. The design strategy for slowing the resource loop was supported by a design for reliability and physical durability, as well as a design for consumer attachment and trust, i.e. emotional durability. Design for product life extension was enabled through design for ease of repair and maintenance, design for upgradability and adaptability, design for standardisation and compatibility and design for dis- and reassembly. This design strategy aims to increase the utilisation period of products by customers, resulting in a slowdown of resource usage (Bocken et al., 2016).

The closed resource loop design sought to recycle the products after post-use, resulting in a closed loop or circular flow between post-use and production. The design strategy for a closed resource loop includes design for the technological cycle, i.e. developing products with those materials that can be continuously and safely recycled into new materials or products; here, waste resources are recycled into ingredients having properties the same as the original. Another strategy for closing the resource loop is designing for biological cycle or biodegradability; this design favours the use of biodegradable materials for developing products as they are safe and healthy for consumption and create food for natural systems across their lifecycle. Another design strategy for closed-loop is 'design for disassembly and reassembly', i.e. ensuring that product and parts can be separated and reassembled easily. This strategy also contributes to other design strategies for closed loops (Bocken et al., 2016).

den Hollander et al. (2017) focused on 'product integrity', i.e. the extent to which the product remains identical to its original, and suggested a 'Product integrity framework' for circular designing. This framework was intended to disassociate from

Bocken et al. (2016) 'slowing, closing and narrowing' framework; however, there is significant conceptual overlapping. den Hollander et al. (2017) circular product design framework also emphasises extending product lifetime by 'preventing and reversing obsolescence'. Product obsolescence can be due to aesthetic, social, technological and economic reasons. Preventing and reversing product obsolescence demands extending product lifetime or increasing product use cycle. Product lifetime is the timespan a product is functional, starting from the instant a product is put into use after production and ending when it becomes obsolete beyond recovery at the product level. The product use cycle is the time interval between a product offered for use after manufacture or recovery and when a product becomes obsolete.

den Hollander et al. (2017) design for product integrity posits circular product design through preventing and reversing obsolescence; this framework advocates developing a strategy for resisting obsolescence, postponing obsolescence and reversing obsolescence of products for circularity. Resisting obsolescence includes product design for prolonged use, postponing obsolescence for extended use and reversing obsolescence for product recovery. Prolonged product use is ensured by design for physical and emotional durability. Extended use is ensured by design for maintenance, upgrading and recovery design for recontextualising, repair, refurbishment, and remanufacturing.

'Mother Nature herself believe in circularity, copying Mother Nature's way for sustainability inspired 'biomimicry' concept of circular economy (Benyus, 2002). This circular design approach is based on imitating the mother nature model, which is 'design for a biological cycle' and 'design for a technical cycle'. The 'design for a biological cycle' resembles the mother nature model of circularity, which believes in returning to the biosphere at the end of one's life cycle to provide nutrients to other biological life. Similarly, 'design for biological cycle' prefers the use of biodegradable materials that can serve as biological nutrients for micro-organisms and also safe and healthy residuals for natural ecosystems and bio-organism lifecycle, where 'design for technical cycle' supports designing products with materials (technical nutrients that can be safely and continuously recycled into new materials or products (Benyus, 2002).

10.3.2 Product Circularity for a Digital Circular Economy Model

With the digital transformation of the companies, the traditional linear flow of products as cradle-to-grave is looped as cradle-to-cradle. The use of digital technology facilitates data flow, data transformation, and resource optimisation. These capabilities accelerate circularity by slowing, closing and narrowing the resource loops and enabling the extended product use cycle. The innovative business model with greater capability for product circularity is supported by digital technology. Implementing digital technology with a strong sustainability perspective results in reducing limited planetary resource consumption. The digital transition of the circular economy further embraces socio-techno innovations, improving product design, process design and delivery of products and services to customers. *The product-service systems, Servitisation*, and *Digital passport systems* are digitally enabled business model innovations for product circularity.

A. **Product-Service System**

The product-service system stands for a combination of 'product(s) and service(s) system' with user-friendly functionality (Baines et al., 2007, p. 3). The contemplation for recommending this joint set of packages, i.e. product(s) and service(s) together to users, is supported because the consumer is only concerned with obtaining 'value' on consumption. During consumption, whether they are the owner of the physical product or not is immaterial to them. Thus, the same 'value' can be offered through the provision of service. This service mode of value delivery by the manufacturer (owner) enables the realisation of circular economy principles by enhancing the product use cycle (Baines et al., 2007; Wu et al., 2021; Wu & Pi, 2023).

The extant literature explains three types of product-service system models that are on a continuum from pure product to pure service, viz. product-oriented, use-oriented, and *performance-oriented* product-service systems. The '*product-oriented*' product-service system business model delivers ownership of tangible products to users with additional services of repair, maintenance, consumables, and suggestions for optimum product use. The contribution of this product-service system to circularity is bestowing extended use of products; nevertheless, repair reverses the product obsolescence. However, the unpredictable behaviour of the owner can jeopardise the realisation of circularity. Next, the 'use-oriented' product-service system business model offers access to physical products but not ownership here; the firm delivers the physical products for use only but keeps the regular schedule for maintenance, upgrading, and repair by getting hold of the product back from the customer, besides this at the end of life, remanufacture, recycle for reuse, or properly disposed of product can be ensured. Therefore, this model of product circularity first resists obsolescence, enhancing the product use cycle and, at the end of life, guaranteeing biodegradability of unused or unrecycled material or residuals.

Meanwhile, the 'performance-oriented' product-service system offers an agreed-upon result without specifying tangible products. This system can materialise circularity by closing loops and enhancing the product use cycle. The delivery of performance based on quality characteristics (i.e. functionality, reliability, relevance, timeliness, suitability, completeness, and consistency), on the one hand, ensures product use longevity and on the other consumer emotional durability (Chuang & Chen, 2022; Le-Dain et al., 2023; Li et al., 2020; Pirola et al., 2020; Sai et al., 2023; Sgambaro et al., 2024).

The product-service system for circular economy requires digital technology with production capability to meet circular economy principles. The application of digital technology in product-service systems enables firms to collect data remotely from multiple sources regarding product condition, consumption, and location. Data collection aims to accurately determine maintenance and repair requirements, meet customer needs, and provide online services based on product lifecycle and product use cycle data. The data collected is also used for preventive maintenance and repair requirements; the same is also communicated to the users and the firm's service staff.

The Internet of Things (IoT), Big data analytics and Cloud computing have received the most attention (Upadhyay et al., 2021). These digital technologies operate on massive amounts of consumer and resource usage data. These digital technologies provide information to segment and target consumers for offering products and services. Predictive data analysis powered by these digital technologies supports firms in predicting consumer behaviour and resource optimisation, i.e. analysing the probability of product failure or the probable reason for failure. These digital technologies also assist in IT-driven value co-creation that endlessly endeavours to fulfil consumer needs in an effective circular way (Diaz & Baumgartner, 2024; Bertoni and Bertoni, 2022; Negash et al., 2023).

B. **Digital Servitisation**

The economic prosperity that modern society enjoys is at the cost of various forms of environmental problems such as greenhouse gas emissions, climate crisis and diminishing natural capital. To address environmental and ecological related issues due to overconsumption and over-production, an innovative business model, i.e.' digital servitisation' powered with digitalisation for environmental sustainability, was proposed. Servitisation is an amalgamation of products, services, a network of 'players', and supporting infrastructure meant to fulfil customer needs, ensuring circularity that impacts environmental sustainability less than traditional linear business models. Adopting digital servitisation facilitates firms in their industry to gain a competitive advantage by developing distinct competency in providing more value through add-on services than tangible items and stand-alone products. This business model realises environmental sustainability through applying 'servitisation' through digital technology (Chen et al., 2023; Minaya et al., 2024).

The rapid development of digital technology facilitates significant changes in business models, product and service design, value delivery processes, and innovation that shape the character of commercial operations in industrial ecosystems for circularity. Digital servitisation is an innovative process of reducing environmental pollution and wastage of raw materials through adopting business models that shift business definition from product-centric to service-centric. Digital servitisation is a more sustainable and circular value delivery system that steadily pursues a use-oriented and performance-oriented product-service system. Digital servitisation is applied mainly by product-based companies intended to offer their product to consumers for use only as a service. The services bestowed to the consumer are incorporated into the product, which can be simple repair or maintenance contracts, operational services or more complex performance-based offerings such as rental or pay-per-use. Companies pursuing circularity goals can achieve dematerialisation, extended product use lifecycle, customer retention, and engagement through digital servitisation. However, servitisation requires the firm to develop and manage the network of people, products, organisation and systems (Lee et al., 2024; Li et al., 2024).

The improvement in network externalities due to digital technology acceptance in society for connecting people and companies encouraged 'servitisation'. Digital technologies facilitate firms' innovation in designing a more comprehensive marketing mix or a customer-centric 'bundle' of goods and services, adding more value to the core products. Also, digital technology facilitates a looped value delivery process connecting consumers, manufacturers, computer systems, products, and services. This value delivery process provides a high potential for manufacturing firms to enable product integrity (González Chávez et al., 2024; Ruffoni & Reichert, 2024).

Manufacturers who participate in the digital circular economy by adopting digital servitisation gain competitive differentiation for their offerings due to enabling their business process to innovate in product delivery and digitally enabled service delivery. Adopting digital servitisation thus aids organisations in gaining competitiveness through enhancing benefits to the customer by providing them more value with add-on service than compared to stand-alone tangible items provided by competitors (Kolagar et al., 2022; Smania et al., 2024).

The companies interested in developing digital servitisation solutions exploit the promising potential of several digital technologies such as artificial intelligence, blockchain, big data, cloud computing, and Internet of Things to create and deliver service value. The rapid growth in companies adopting digital technologies for digital servitisation ensures new potential for achieving circularity (Paiola et al., 2021).

Manufacturing companies responding to opportunities for realising circular economy through digital servitisation are gradually embarking on digital technologies such as artificial intelligence (AI), big data, cloud computing, and the Internet of Things (IoT). Adopting these technologies transforms a firm's capabilities for effective design, efficient production and delivery. Firms involved in digital servitisation steadily capture, create, and deliver service value to the customer, evolving due to a broad range of network externalities from within the industry and other industrial firms as well as with their associate ecosystems (Sklyar et al., 2019; Benedettini, 2024).

Digital technologies for successful servitisation must facilitate the bundle of product-service offerings to consumers digitally and must be repositioned with the benefits of digitally offering existing services for obtaining consumers' emotional durability. Digital servitisation built on digital business models generates data from connected devices about products, components, and materials, as well as product use and location through digital devices, tools, and processes; this facilitates product use cycle optimisation and engineers new ways for value co-creation. Firms using digital servitisation gain an advantage over the competition and improve their triple bottom lines, i.e. *people, profit, and planet (*Kohtamäki et al., 2019).

Digital technologies offer an immense opportunity for manufacturing enterprises wishing to shift to service-based strategies. Digital servitisation needs businesses to emphasise 'product, service, and software', which must operate in coherence. Thus, the traditional concept of producing stand-alone products in a factory and transferring ownership to the consumer for consumption is pointless in this concept of circularity (Paiola et al., 2021; Sklyar et al., 2019).

The success of digital servitisation from the conservation of ecology perspective depends upon the extent of integrating the product's core value designed with network functionality for servitisation to satisfy individual customer needs. The configuration of products with digital technologies for servitisation is exceptionally complex; managing front-end and back-end technologies, platforms for offering servitisation, and product-service software imposes significant challenges at different levels of value creation and delivery, this can jeopardise the realisation of sustainability goals through circularity if not appropriately managed (Kolagar et al., 2022; Minaya et al., 2024).

C. **Digital Passport System**

The principle of sustainability is the heart and soul of the circular economy. Realising sustainability in the true sense is not just mitigating the ill effect of industrialisation but bringing about constructive changes in using depleting natural resources to protect people and the planet. To this end, digital technologies offer tremendous opportunity to design *'data-driven tracing'* of the value chain activity for enabling transparency in achieving circularity. In the *inbound-outbound logistic phase*, the data-driven traceability accounts for supply chain contribution in mitigating prospective carbon footprint through sourcing responsibly, further in *manufacturing process* tracing production process in near-real time and building database result in identifying and eradicating potential waste, i.e. inefficiencies from the process. Next, in *marketing*, for consumers, traceability assists in informed purchasing decisions based on information about the product's origin and possible environmental impact; moreover, at the *after-sales-service* stage, traceability facilitates usage, maintenance and repair description, while at the *end-of-life stage,* traceability facilitates recycling, refurbishing, remanufacturing and dematerialising ensuring harmful materials are safely disposed and valuable materials are recovered. To this end, *Digital Product Passport* is a repository that enables recording and monitoring the value-adding activity throughout the value chain (Langley et al., 2023; Walden et al., 2021).

A digital product passport is the digital traceability of a product's information from its creation to its disposal. This concept embodies a product's digital information throughout its lifecycle, including origin, material, usage, and carbon footprint. The information about the beginning-of-life, i.e. manufacturing details of products, are supplied to marketing channels, policymakers and end users; also, digital product passports enable aggregating crucial data about the complete product's lifecycle, i.e. middle-of-life and end-of-life information in a database. The role of this database is primarily informative and instructive but also serves corrective and predictive purposes. Informative and instructive information ensures extended product use, mainly concerned with product use-and-misuse and maintenance schedule. Corrective and predictive sources of information are concerned with applying data analytics to instruct about product needs for repairability, predict product failure, and support product use optimisation; thus, digital product passport through digital traceability of product lifecycle enables sustainability and circularity (Berger et al., 2022; Hulea et al., 2024; Jensen et al., 2023).

The contribution of digital product passport concept is that it resists, postpones and reverses obsolescence by defending product integrity. The realisation of a sustainable product lifecycle through applying a digital product passport is apparent in product longevity, extended use and recovery, mitigating the detrimental impact of economic growth on the environment. Important to achieving sustainability through product and process circularity is enabling product reuse and repurpose, recycling of base materials, refurbishing and remanufacturing, and creating demand for newly recycled material; here, for all these purposes, the digital product passport could support circularity (Adisorn et al., 2021; Jensen et al., 2024).

The digital passport system's primary function is to store data from various product life cycle stages. The decision to deploy a digital product passport depends on the intended purpose, i.e. information to the manufacturer, intermediaries, consumers, recyclers, value chain partners or policymakers. The standard database data classes in digital product passports are material, environmental, manufacturing, value network, maintenance, circularity, and end-user data. *Material data* pertains to information about raw materials or parts used in manufacturing the product. This information is helpful in the recycling process. *Environmental data* is related to the carbon footprint of a product in its entire lifecycle. *Manufacturing data* pertains to the product manufacturing process, i.e. real-time processing data generated through various sensors that monitor the manufacturing process. *Value network data* stores all information about the product's value network. *Maintenance data* is related to the nature and schedule of maintenance service used by the consumer. *Circularity data* store information about circularity products, i.e. reusing, recycling, or repurposing. In contrast, *end-user data* is produced by the product's end-user by owning or availing service (Voulgaridis et al., 2024; Panza et al., 2023; Zhang & Seuring, 2024).

The digital product passport, as the name signifies, leverages the power of information and communication technology tools like Artificial Intelligence, the Internet of things, Big Data, blockchain, data analytics and other emerging digital technologies. Information and communication technology enables circularity focusing on continuous collection and processing of data from various stages of a firm's value chain. The data collected at the beginning-of-life are primarily concerned with product design and manufacturing specification. The manufacturer provides this data while conceiving an item's digital product passport. The middle-of-life data is the user's use or maintenance-related data. Meanwhile, end-of-life data is about activities that occur when the product is discarded due to obsolescence. The data collected for the entire lifecycle of the product by using digital technologies are processed using innovative digital technologies and further shared to facilitate monitoring of circularity achieved by individual products, also simultaneously assessing the potential influence of the product on the environment (Rumetshofer et al., 2024; Psarommatis and May, 2024).

10.4 Monitoring Product Circularity

The Ellen MacArthur Foundation (2015) proposed a ReSOLVE framework that aids firms in making business decisions for sustainability (Sell et al., 2023; Tedesco et al., 2022). The firms embarking on product circularity primarily design product specifications for slowing, narrowing or closing the loops (Benyus, 2002; Mestre & Cooper, 2017) further apply digital technology for sustainable product lifecycle management focusing on the 'concept of designing for product integrity' suggested by den Hollander et al. (2017).

The important question related to the efforts by the firms for circularity is—*Do digitally enabled innovative approaches for product circularity contribute to sustainability?* The ReSOLVE framework provides practical and specific metrics, i.e. regenerate, share, optimise, loop, virtualise, and exchange, to assess the viability of initiatives for circular product designing. Here, 'Regenerate' embodies a shift towards renewability in material and energy. '*Share*' epitomises asset sharing. '*Optimise*' is using technology for performance improvement. '*Loop*' implies enabling biological and technological cycles. '*Virtualise*' signifies replacing physical goods with virtual ones. '*Exchange*' implies transferring obsolete non-renewal goods to renewable ones. Thus, the ReSOLVE framework functions as a guiding framework for sustainability and a tool to analyse the sustainability initiatives practices by the firms (Jabbour et al., 2019; Payne & Kwofie, 2024; Ta et al., 2024).

10.5 Conclusion

The circular economy concept is a systemic vision for achieving sustainability. Primary schools of thought like Blue Economy, Biomimicry, Cradle to Cradle, Industrial Ecology, and Natural Capitalism all advocate the necessity for preserving the depleting natural resources. The main feature of the circular economy is being regenerative through slowing, closing and narrowing the loop. The success of the circular economy rests in eliminating waste generation through the circular use of resources. The emergence of digital technology contributed to an innovative circular economy model, i.e. a digital circular economy. Artificial intelligence, big data, blockchain, data analytics, and cloud computing powered firms to develop innovative business practices like 'servitisation' and the advent of the 'digital product passport' system. Digital servitisation offers a competitive advantage to the business firm by providing stand-alone product use as a service, which ensures circularity at the end of life. A digital product passport is a sustainable product lifecycle management system embarked on by the power of a database built through data collection at every stage of the product lifecycle that enables the sharing information with users to facilitate circularity in offerings. Thus, the chapter concludes by providing a conceptual framework for enabling product circularity in the digital circular economy, as shown in Fig. 10.1.

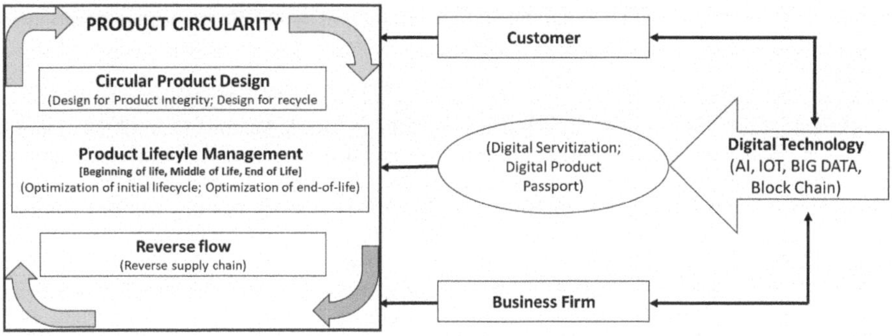

Fig. 10.1 Product circularity in digital circular economy

References

Adisorn, T., Tholen, L., & Götz, T. (2021). Towards a digital product passport fit for contributing to a circular economy. *Energies, 14*(8). https://doi.org/10.3390/en14082289

Aguiar, M. F., & Jugend, D. (2022). Circular product design maturity matrix: A guideline to evaluate new product development in light of the circular economy transition. *Journal of Cleaner Production, 365*, 132732. https://doi.org/10.1016/j.jclepro.2022.132732

Alcayaga, A., & Hansen, E. G. (2024). Smart circular economy as a service business model: An activity system framework and research agenda. *R&D Management*. https://doi.org/10.1111/radm.12707

Baines, T. S., Lightfoot, H. W., Evans, S., Neely, A., Greenough, R., Peppard, J., Roy, R., Shehab, E., Braganza, A., Tiwari, A., Alcock, J. R., Angus, J. P., Bastl, M., Cousens, A., Irving, P., Johnson, M., Kingston, J., Lockett, H., Martinez, V., Wilson, H., et al. (2007). State-of-the-art in product-service systems. *Proceedings of the Institution of Mechanical Engineers, Part B: Journal of Engineering Manufacture, 221*(10), 1543–1552. https://doi.org/10.1243/09544054JEM858

Benedettini, O. (2024). Strategic emphasis, outsourcing intensity, and financial performance in digital servitisation. *Industrial Marketing Management, 122*, 1–12. https://doi.org/10.1016/j.indmarman.2024.08.001

Benyus, J. M. (2002). *Biomimicry: Innovation inspired by nature*. Mariner Books.

Berger, K., Schöggl, J. P., & Baumgartner, R. J. (2022). Digital battery passports to enable circular and sustainable value chains: Conceptualisation and use cases. *Journal of Cleaner Production, 353*. https://doi.org/10.1016/j.jclepro.2022.131492

Bertoni, M., & Bertoni, A. (2022). Designing solutions with the product-service systems digital twin: What is now and what is next? In *Computers in industry* (Vol. 138). Elsevier B.V. https://doi.org/10.1016/j.compind.2022.103629

Bocken, N. M. P., de Pauw, I., Bakker, C., & van der Grinten, B. (2016). Product design and business model strategies for a circular economy. *Journal of Industrial and Production Engineering, 33*(5), 308–320. https://doi.org/10.1080/21681015.2016.1172124

Chauhan, C., Parida, V., & Dhir, A. (2022). Linking circular economy and digitalisation technologies: A systematic literature review of past achievements and future promises. *Technological Forecasting and Social Change, 177*, 121508. https://doi.org/10.1016/j.techfore.2022.121508

Chen, Y., Wang, Z., & Ortiz, J. (2023). A sustainable digital ecosystem: Digital servitization transformation and digital infrastructure support. *Sustainability (Switzerland), 15*(2). https://doi.org/10.3390/su15021530

Chuang, Y. C., & Chen, Y. M. (2022). Digital servitisation of symbiotic service composition in product-service systems. *Computers in Industry, 138*. https://doi.org/10.1016/j.compind.2022.103630

Commoner, B. (1971). *The closing circle: Nature, man, and technology*. Random House Inc.

Dagilienė, L., Čeičytė-Pranskūnė, J., Telešienė, A., Valušytė, R., & Varžinskas, V. (2024). Developing a circular design framework: Co-creation and validation of a circular product and service design tool. *Journal of Industrial Ecology, 28*(4), 783–799. https://doi.org/10.1111/jiec.13494

den Hollander, M. C., Bakker, C. A., & Hultink, E. J. (2017). Product design in a circular economy: Development of a typology of key concepts and terms. *Journal of Industrial Ecology, 21*(3), 517–525. https://doi.org/10.1111/jiec.12610

Diaz, A., & Baumgartner, R. J. (2024). A managerial approach to product planning for a circular economy: Strategy implementation and evaluation support. *Journal of Cleaner Production, 442*, 140829. https://doi.org/10.1016/j.jclepro.2024.140829

Diaz, A., Reyes, T., & Baumgartner, R. J. (2022). Implementing circular economy strategies during product development. *Resources, Conservation and Recycling, 184*, 106344. https://doi.org/10.1016/j.resconrec.2022.106344

Ellen MacArthur Foundation. (2013). *Toward the circular economy: Economic and business rationale for an accelerated transition* (Vol. 1). https://www.ellenmacarthurfoundation.org/towards-the-circular-economy-vol-1-an-economic-and-business-rationale-for-an

Ellen MacArthur Foundation. (2015). *Growth within: A circular economy vision for a competitive Europe*. https://www.mckinsey.com/capabilities/sustainability/our-insights/growth-within-a-circular-economy-vision-for-a-competitive-europe

Farace, B., & Tarabella, A. (2024). Exploring the role of digitalisation as a driver for the adoption of circular economy principles in agrifood SMEs—an interpretive case study. *British Food Journal, 126*(1), 409–427. https://doi.org/10.1108/BFJ-12-2022-1103

Geissdoerfer, M., Savaget, P., Bocken, N. M. P., & Hultink, E. J. (2017). The circular economy—a new sustainability paradigm? *Journal of Cleaner Production, 143*, 757–768. https://doi.org/10.1016/j.jclepro.2016.12.048

George, D. A. R., Lin, B. C., & Chen, Y. (2015). A circular economy model of economic growth. *Environmental Modelling and Software, 73*, 60–63. https://doi.org/10.1016/j.envsoft.2015.06.014

Ghisellini, P., Cialani, C., & Ulgiati, S. (2016). A review on circular economy: The expected transition to a balanced interplay of environmental and economic systems. *Journal of Cleaner Production, 114*, 11–32. https://doi.org/10.1016/j.jclepro.2015.09.007

González Chávez, C. A., Brynolf, S., Despeisse, M., Johansson, B., Rönnbäck, A. Ö., Rösler, J., & Stahre, J. (2024). Advancing sustainability through digital servitisation: An exploratory study in the maritime shipping industry. *Journal of Cleaner Production, 436*, https://doi.org/10.1016/j.jclepro.2023.140401

Govindan, K. (2023). How digitalisation transforms the traditional circular economy to a smart circular economy for achieving SDGs and net zero. *Transportation Research Part E: Logistics and Transportation Review, 177*, 103147. https://doi.org/10.1016/j.tre.2023.103147

Graedel, T. E., & Allenby, B. R. (2003). *Industrial ecology and sustainable engineering* (2nd ed.). Prentice Hall.

Han, Y., Shevchenko, T., Yannou, B., Ranjbari, M., Shams Esfandabadi, Z., Saidani, M., Bouillass, G., Bliumska-Danko, K., & Li, G. (2023). Exploring how digital technologies enable a circular economy of products. *Sustainability, 15*(3), 2067. https://doi.org/10.3390/su15032067

Hulea, M., Miron, R., & Muresan, V. (2024). Digital product passport implementation based on multi-blockchain approach with decentralized identifier provider. *Applied Sciences (Switzerland), 14*(11). https://doi.org/10.3390/app14114874

Jabbour, C. J. C., de Sousa Jabbour, A. B. L., Sarkis, J., & Filho, M. G. (2019). Unlocking the circular economy through new business models based on large-scale data: An integrative framework and research agenda. *Technological Forecasting and Social Change, 144*, 546–552. https://doi.org/10.1016/j.techfore.2017.09.010

Jensen, S. F., Kristensen, J. H., Adamsen, S., Christensen, A., & Waehrens, B. V. (2023). Digital product passports for a circular economy: Data needs for product life cycle decision-making. *Sustainable Production and Consumption, 37*, 242–255. https://doi.org/10.1016/j.spc.2023.02.021

Jensen, S. F., Kristensen, J. H., Christensen, A., & Waehrens, B. V. (2024). An ecosystem orchestration framework for the design of digital product passports in a circular economy. *Business Strategy and the Environment*. https://doi.org/10.1002/bse.3868

Jugend, D., dos Santos, H. H., Garrido, S., Siqueira, R. M., & Mesa, J. A. (2024). Circular product design challenges: An exploratory study on critical barriers. *Business Strategy and the Environment, 33*(5), 4825–4842. https://doi.org/10.1002/bse.3703

Khan, S., Ali, S. S., & Singh, R. (2022). Determinants of remanufacturing adoption for circular economy: A causal relationship evaluation framework. *Applied System Innovation, 5*(4), 62. https://doi.org/10.3390/asi5040062

Kohtamäki, M., Parida, V., Oghazi, P., Gebauer, H., & Baines, T. (2019). Digital servitisation business models in ecosystems: A theory of the firm. *Journal of Business Research, 104*, 380–392. https://doi.org/10.1016/j.jbusres.2019.06.027

Kolagar, M., Parida, V., & Sjödin, D. (2022). Ecosystem transformation for digital servitisation: A systematic review, integrative framework, and future research agenda. *Journal of Business Research, 146*, 176–200. https://doi.org/10.1016/j.jbusres.2022.03.067

Kristoffersen, E., Blomsma, F., Mikalef, P., & Li, J. (2020). The smart circular economy: A digital-enabled circular strategies framework for manufacturing companies. *Journal of Business Research, 120*, 241–261. https://doi.org/10.1016/j.jbusres.2020.07.044

Langley, D. J., Rosca, E., Angelopoulos, M., Kamminga, O., & Hooijer, C. (2023). Orchestrating a smart circular economy: Guiding principles for digital product passports. *Journal of Business Research, 169*, 114259. https://doi.org/10.1016/j.jbusres.2023.114259

Le-Dain, M. A., Benhayoun, L., Matthews, J., & Liard, M. (2023). Barriers and opportunities of digital servitisation for SMEs: The effect of smart Product-Service System business models. *Service Business, 17*(1), 359–393. https://doi.org/10.1007/s11628-023-00520-4

Lee, M. J., Kim, Y., & Roh, T. (2024). Exploring the role of digital servitisation for green innovation: Absorptive capacity, transformative capacity, and environmental strategy. *Technological Forecasting and Social Change, 207*. https://doi.org/10.1016/j.techfore.2024.123614

Li, A. Q., Rich, N., Found, P., Kumar, M., & Brown, S. (2020). Exploring product–service systems in the digital era: A socio-technical systems perspective. *TQM Journal, 32*(4), 897–913. https://doi.org/10.1108/TQM-11-2019-0272

Li, H., Zhao, J., Cao, Y., Su, L., Zhao, Z., & Zhang, Y. (2024). Servitisation and product service system: A literature review on value creation. *Technological Forecasting and Social Change, 208*. https://doi.org/10.1016/j.techfore.2024.123724

Lyle, J. T. (1996). *Regenerative design for sustainable development*. John Wiley & Sons.

McDonough, W., & Braungart, M. (2002). *Cradle to cradle: Remaking the way we make things* (1st ed.). North Point Press.

Mestre, A., & Cooper, T. (2017). Circular product design. a multiple loops life cycle design approach for the circular economy. *The Design Journal, 20*(sup1), S1620–S1635. https://doi.org/10.1080/14606925.2017.1352686

Minaya, P. E., Avella, L., & Trespalacios, J. A. (2024). Synthesising three decades of digital servitisation: A systematic literature review and conceptual framework proposal. *Service Business, 18*(2), 193–222. https://doi.org/10.1007/s11628-024-00559-x

Negash, Y. T., & Sarmiento, L. S. C. (2023). Smart product-service systems in the healthcare industry: Intelligent connected products and stakeholder communication drive digital health service adoption. *Heliyon, 9*(2). https://doi.org/10.1016/j.heliyon.2023.e13137

Pagoropoulos, A., Maier, A., & McAloone, T. C. (2017). Assessing transformational change from institutionalising digital capabilities on implementation and development of product-service systems: Learnings from the maritime industry. *Journal of Cleaner Production, 166*, 369–380. https://doi.org/10.1016/j.jclepro.2017.08.019

Paiola, M., Schiavone, F., Grandinetti, R., & Chen, J. (2021). Digital servitisation and sustainability through networking: Some evidences from IoT-based business models. *Journal of Business Research, 132*, 507–516. https://doi.org/10.1016/j.jbusres.2021.04.047

Panza, L., Bruno, G., & Lombardi, F. (2023). Integrating absolute sustainability and social sustainability in the digital product passport to promote Industry 5.0. *Sustainability (Switzerland), 15*(16). https://doi.org/10.3390/su151612552

Pauli, G. (2010). *The blue economy: 10 years, 100 innovations, 100 million jobs*. Paradigm Publications.

Payne, A., & Kwofie, E. M. (2024). Unleashing circular economy potential in agriculture: Integrating social impact assessment with the ReSOLVE framework as a tool for sustainable development. *Sustainable Development*. https://doi.org/10.1002/sd.2952

Pearce, D. W., & Turner, R. K. (1990) *Economics of natural resources and the environment*. Harvester Wheatsheaf, Hemel Hempstead.

Pirola, F., Boucher, X., Wiesner, S., & Pezzotta, G. (2020). Digital technologies in product-service systems: A literature review and a research agenda. In *Computers in industry* (Vol. 123). Elsevier B.V. https://doi.org/10.1016/j.compind.2020.103301

Psarommatis, F., & May, G. (2024). Digital product passport: A pathway to circularity and sustainability in modern manufacturing. *Sustainability (Switzerland), 16*(1). https://doi.org/10.3390/su16010396

Rajput, S., & Singh, S. P. (2019). Connecting circular economy and industry 4.0. *International Journal of Information Management, 49*, 98–113. https://doi.org/10.1016/j.ijinfomgt.2019.03.002

Rejeb, A., Suhaiza, Z., Rejeb, K., Seuring, S., & Treiblmaier, H. (2022). The internet of things and the circular economy: A systematic literature review and research agenda. *Journal of Cleaner Production, 350*, 131439. https://doi.org/10.1016/j.jclepro.2022.131439

Ruffoni, E. P., & Reichert, F. M. (2024). Exploring configurations of digital servitisation capabilities for value creation: An fsQCA approach. *Industrial Marketing Management, 122*, 131–144. https://doi.org/10.1016/j.indmarman.2024.08.010

Rumetshofer, T., Straka, K., & Fischer, J. (2024). How the digital product passport can lead the plastics industry towards a circular economy—a case study from bottle caps to Frisbees. *Polymers, 16*(10). https://doi.org/10.3390/polym16101420

Rusch, M., Schöggl, J., & Baumgartner, R. J. (2023). Application of digital technologies for sustainable product management in a circular economy: A review. *Business Strategy and the Environment, 32*(3), 1159–1174. https://doi.org/10.1002/bse.3099

Sai, E., Koide, R., & Murakami, S. (2023). Assessing the environmental impacts of product service systems in the digital-devices market: An application of attributional and consequential life cycle assessment. *Sustainable Production and Consumption, 38*, 331–340. https://doi.org/10.1016/j.spc.2023.04.021

Sell, L. G., Tedesco, M., Simioni, F. J., Sehnem, S., Soares, J. F., & Coelho Junior, L. M. (2023). Circular economy practices using the ReSOLVE framework: An assessment by sector and scale in the Brazilian planted tree industry. *Journal of Cleaner Production, 433*, 139780. https://doi.org/10.1016/j.jclepro.2023.139780

Selvefors, A., Rexfelt, O., Renström, S., & Strömberg, H. (2019). Use to use—a user perspective on product circularity. *Journal of Cleaner Production, 223*, 1014–1028. https://doi.org/10.1016/j.jclepro.2019.03.117

Sgambaro, L., Chiaroni, D., & Urbinati, A. (2024). The design and servitisation of products according to the circular economy principles: An ecosystem perspective in the building industry. *Journal of Cleaner Production, 454*, 142322. https://doi.org/10.1016/j.jclepro.2024.142322

Singh, R., Khan, S., & Centobelli, P. (2022). Investigating the Interplay between social performance and organisational factors supporting circular economy practices. *Sustainability, 14*(24), 16781. https://doi.org/10.3390/su142416781

Singh, R., Khan, S., & Dsilva, J. (2023). A framework for assessment of critical factor for circular economy practice implementation. *Journal of Modelling in Management, 18*(5), 1476–1497. https://doi.org/10.1108/JM2-06-2021-0145

Sklyar, A., Kowalkowski, C., Tronvoll, B., & Sörhammar, D. (2019). Organising for digital servitisation: A service ecosystem perspective. *Journal of Business Research, 104*, 450–460. https://doi.org/10.1016/j.jbusres.2019.02.012

Smania, G. S., Ayala, N. F., Coreynen, W., & Mendes, G. H. S. (2024). Data-related tensions in digital servitisation ecosystems: A systematic literature review. *Industrial Marketing Management, 123*, 31–48. https://doi.org/10.1016/j.indmarman.2024.09.002

Stahel, W. R. (2010). *The performance economy*. Palgrave Macmillan.

Ta, T. Y., Nguyen, K. L. P., Nguyen, T. H. T., Le, D. T., Nguyen, M. T., Nguyen, Q. L., Nguyen, H.-Q., & Bui, T.-K.L. (2024). Integrated life cycle assessment with the ReSOLVE framework for environmental impacts mitigation in mushroom growing: The case in Lam Dong province, Vietnam. *Environmental Development, 51*, 101024. https://doi.org/10.1016/j.envdev.2024.101024

Tedesco, M., Simioni, F. J., Sehnem, S., Soares, J. F., & Coelho Junior, L. M. (2022). Assessment of the circular economy in the Brazilian planted tree sector using the ReSOLVE framework. *Sustainable Production and Consumption, 31*, 397–406. https://doi.org/10.1016/j.spc.2022.03.005

Toth-Peter, A., Torres de Oliveira, R., Mathews, S., Barner, L., & Figueira, S. (2023). Industry 4.0 as an enabler in transitioning to circular business models: A systematic literature review. *Journal of Cleaner Production, 393*, 136284. https://doi.org/10.1016/j.jclepro.2023.136284

Upadhyay, A., Mukhuty, S., Kumar, V., & Kazancoglu, Y. (2021). Blockchain technology and the circular economy: Implications for sustainability and social responsibility. *Journal of Cleaner Production, 293*, 126130. https://doi.org/10.1016/j.jclepro.2021.126130

Voulgaridis, K., Lagkas, T., Angelopoulos, C. M., Boulogeorgos, A. A. A., Argyriou, V., & Sarigiannidis, P. (2024). Digital product passports as enablers of digital circular economy: A framework based on technological perspective. In *Telecommunication systems* (Vol. 85(4), pp. 699–715). Springer. https://doi.org/10.1007/s11235-024-01104-x

Walden, J., Steinbrecher, A., & Marinkovic, M. (2021). Digital product passports as enabler of the circular economy. *Chemie-Ingenieur-Technik, 93*(11), 1717–1727. John Wiley and Sons Inc. https://doi.org/10.1002/cite.202100121

Wang, J. X., Burke, H., & Zhang, A. (2022). Overcoming barriers to circular product design. *International Journal of Production Economics, 243*, 108346. https://doi.org/10.1016/j.ijpe.2021.108346

Webster, K. (2015). *The circular economy: A wealth of flows*. Ellen MacArthur Foundation.

Wu, C., Chen, T., Li, Z., & Liu, W. (2021). A function-oriented optimising approach for smart product service systems at the conceptual design stage: A perspective from the digital twin framework. *Journal of Cleaner Production, 297*. https://doi.org/10.1016/j.jclepro.2021.126597

Wu, D., & Pi, Y. (2023). Digital technologies and product-service systems: A synergistic approach for manufacturing firms under a circular economy. *Journal of Digital Economy, 2*, 37–49. https://doi.org/10.1016/j.jdec.2023.04.001

Zhang, A., & Seuring, S. (2024). Digital product passport for sustainable and circular supply chain management: A structured review of use cases. *International Journal of Logistics Research and Applications*. https://doi.org/10.1080/13675567.2024.2374256

Seyyedi, S. R., Kowsari, E., Gheibi, M., Chinnappan, A., & Ramakrishna, S. (2024). A comprehensive review integration of digitalisation and circular economy in waste management by adopting artificial intelligence approaches: Towards a simulation model. *Journal of Cleaner Production, 460*, 142584. https://doi.org/10.1016/j.jclepro.2024.142584

Chapter 11
Empowering Sustainable Human Capital for Digital Circular Economy in the Gulf Cooperation Council HealthCare Sector

Rubee Singh, Amit Joshi, Godwin Francis, and Jacinta Dsilva

Abstract The main purpose of the chapter is to investigate human capital empowerment within the Gulf Cooperation Council (GCC) healthcare sector, accentuating its role in nurturing a sustainable digital circular economy. The chapter identifies barriers to effective human capital development and assess the impact of such challenges on healthcare outcomes in the GCC. The chapter employs a comprehensive literature review, synthesizing prior and contemporary research to contextualize human capital empowerment in the GCC healthcare setting. It highlights the relevance of strategic leadership, continuous learning and interdisciplinary collaborations to address ongoing workforce challenges. Preliminary findings reveal significant barriers to human capital empowerment in the form of accelerated turnover rates and a severe dearth of trained local medical professionals. It underscores the imperativeness for robust, ongoing training programs with a cultural shift towards acknowledging and valuing continuous learning to enhance operational efficiency and patient care quality in the GCC healthcare sector. This chapter contributes to sustainable human capital development literature by concentrating on the unique challenges faced by the GCC healthcare sector. It integrates concepts of digital innovation and sustainability, offering a dynamic and novel perspective on how these elements can enhance workforce capabilities and resilience. The findings suggest critical pathways and vital directions for policy development aimed at reinforcing and strengthening human capital in the GCC healthcare system. Future research should investigate context-specific strategies to overcome identified barriers and assess the long-term impacts of human capital initiatives on healthcare outcomes and sustainability.

R. Singh (✉) · A. Joshi
Department of Management Studies, Kumaun University, Nainital, Uttarakhand, India
e-mail: dr.rubeerajput@gmail.com

R. Singh · G. Francis
Institute of Business Management, GLA University, Mathura, India

J. Dsilva
The British University in Dubai, Dubai, UAE

SEE Research Centre, SEE Institute, The Sustainable City, Dubai, UAE

R. Singh and V. Kumar (eds.), *Sustainable Innovations and Digital Circular Economy*,
https://doi.org/10.1007/978-981-96-1064-8_11

Keywords Sustainable human capital · Digital circular economy · Gulf Cooperation Council · Healthcare sector · Resilience

11.1 Introduction

The last century has witnessed a significant rise in automation, mass production and implementation of the state-of-the-art technology across regions and sectors. More specifically, the rapidly evolving and progressive healthcare landscape calls for a critical and urgent convergence of digital innovation and sustainability. Central to this progression is the concept of a sustainable digital circular economy, which not only intents to augment operational efficiency and mitigate environmental impact, but also endeavours to rebuild a resilient and adaptive healthcare ecosystem. That said, although this transformational potential of technology is unfathomable, its realization, acceptance and adoption are being hampered by fundamental challenges (Esmaeilzadeh, 2024), with respect to aggravated costs, lack of key resources and soaring demands for clinical excellence whilst enhancing value based and improved care quality (Emanuele & Koetter, 2007). However, perennial concerns persist regarding the propensity of employees to be amenable to these sophisticated changes, thereby diluting their capacity to respond to potential threats and opportunities (Rod, 2019) effectively and efficiently. This is where human capital empowerment serves as the heart of this transformation with the imperativeness of healthcare professional being equipped with the knowledge, skills and tools obligated to navigate and leverage the opportunities presented by digital technologies and circular economy principles, the success of which purely hinges on workforce capabilities and commitment (Yee and Rahman, 2019; Singh et al., 2021). This further entails fostering a culture of continuous learning by investing in comprehensive education and promoting inter-disciplinary approaches that bridge gaps between healthcare, technology and environmental stewardship as well as integrating sustainability into every facet of healthcare operations and decision-making which calls for a robust leadership and strategic vision to drive these systematic changes that align with both economic viability and ecological responsibility.

Human capital refers to the knowledge, skills and health that people accumulate in their lifetime (El-Saharty et al., 2020). Human capital strength is undebatable in terms of dual accrued benefits such as enhanced efficiency and productivity, both serving as vital factors to enhanced economic growth. In fact, the sustained investment in training and developing high quality human capital has changed the status of several countries from underdeveloped/developing to the emerging status (Pereira et al., 2020), especially the Gulf regions.

The Gulf Cooperation Council (GCC) established on May 25th 1981 is an alliance of six countries, including Bahrain, Kuwait, Oman, Qatar, Saudi Arabia and United Arab Emirates, whose enduring advantages are rooted in cultural commonalities, strategic positioning, independent Sharia-based judiciary system, incessant socio-economic reforms and a minimal corporate tax regime (as cited in Pillai & Al-Malkawi, 2018, p. 395). These countries have now embarked on a mission to value and acknowledge human capital development for a sustainable future by reconsidering and rejuvenating their development frameworks which were earlier confined to exploitation of non-renewable resources such as oil and gas. The GCC governments are now active participants in the World Bank's Human Capital Project (focussed on three vital areas namely knowledge, skills and health measured by the Human Capital Index (HCI)) (World Bank, 2018), a testament of their solid commitment to the sustainable transformation (El-Saharty et al., 2020). More specifically, GCC countries stage an expanding and ageing population apart from an escalating prevalence of disease from sedentary lifestyle, thus prioritizing their attention to the healthcare sector. Moreover, the over-emphasis on customer centricity, assurance of ultimate patience experience, medical tourist influx, technological revolutions and unprecedented patient requirements are all serving as critical factors for the bourgeoning GCC healthcare landscape. Surprisingly, despite the proliferation of healthcare workers in the GCC, high turnover rates and retention issues have led to an evident void with respect to trained physicians and nurses, specifically local professionals (Brownie et al., 2015). At this point, Khoja et al. (2017) posit the requirement for the formulation and implementation of innovative approaches in tackling these emerging healthcare challenges with a specific focus on building human resource capacity and evidence-based guidelines to supplement the same. This entails development of training programs, establishing a competitive curriculum and embarking on an in-depth investigation in unveiling innovative, timebound, achievable and practical strategies to develop a competitive human capital base (Tripathi et al., 2021).

Contextualized within the extant research on human capital empowerment, this chapter draws on early and contemporary literature to examine scholarly perspectives on empowering human capital for a sustainable digital circular economy in the GCC, juxtaposed with the challenges and potential impact within the healthcare sector. Existing research for GCC remain scarce despite the fact of the region being a host to a significant number of diverse expatriate population from around the world (El-Saharty et al., 2020), especially in the health sector, all of which serve as an interesting and unique ground for further research. Amidst the absence of concrete evidence on empowering human capital development and its association with sustainable digital circular economy apart from failure to explore context-specific challenges of human capital development in the digital era create significant research avenues for deriving region-specific best practices. Accordingly, the objectives of the chapter are four-fold in nature. First to accumulate theoretical perspectives centring around human capital development, second to unearth the challenges posing as a barrier to healthcare human capital empowerment in a sustainable digital circular GCC economy, third to assess the success/failure stories as potential impacts of balanced/unbalanced healthcare human capital empowerment in a digital circular economy and finally to set policy implications to strengthen efforts for human capital empowerment in the GCC healthcare sector while advancing a sustainable digital circular economy.

The chapter contributes to the human capital empowerment domain within the context of a digital circular economy in healthcare for several reasons. First, it provides clarity on the theoretical evolution of human capital practices and its variations in various regulatory environments with respect to the healthcare sector. Secondly, it highlights the urgency evident in the convergence of digital innovation and sustainability in the rapidly progressing healthcare landscape. Thirdly, the chapter highlights the relevance of empowering healthcare professionals with the knowledge, skills and tools necessary to adapt to unprecedented volatilities, all of which are pivotal for leveraging the opportunities presented by digital technologies. Katoue et al. (2022) specifically recommend the necessity for healthcare force development to ensure successful healthcare reforms. Finally, the pertinence of the concept of a sustainable digital circular economy underscores how the framework can enhance operational efficiency, mitigate environmental impacts and foster a resilient healthcare ecosystem in the GCC economic setting.

The chapter then proceeds as follows: Sect. 11.2 will provide a theoretical background and Sect. 11.3 will highlight prior literature on human capital development challenges and the relevance of the human capital aspects in the GCC context. Section 11.4 will outline policy implications with Sect. 11.9 concluding the chapter by enlisting limitations and future research prospects.

11.2 Theoretical Overview

The healthcare sector in the Gulf Cooperation Council (GCC) region is at a revolutionary phase driven by trends such as digitalization, sustainability and circular economy principles. Consequently, the preparation of human capital has been more important than ever to have a qualified and motivated workforce that can meet these new challenges. The theoretical review encompasses a comprehensive analysis of existing theories to evaluate and synthesize the underpinning frameworks that inform practice and research in the specific area.

11.2.1 Theories Related to Human Capital

Human capital theory is a core part of the economic literature, which explains how the skills we learn, education and experiences make us more productive and therefore more valuable in labour markets. Yet since its inception, the human capital concept has grown and taken up multiple dimensions which mirror wider social, spatial and economic shifts. An understanding of these dimensions is crucial for the labour force strategy forged within healthcare by an understanding about how human capital drives innovation, sustainability or digital transformation.

11.2.1.1 Traditional Human Capital Theory

The first point of view that originated from the field of economy—human capital theory by Becker (1964) gave stress to improvements in individual productivity thanks to education and practice. Prior to the wave of studies of employer perspectives on job readiness (and subsequent consideration, assessment and publication over a period exceeding 25 years), cost–benefit analysis was structured by traditional human capital theory that underlying education often enhanced with expected increases in wages due to productivity potential (Becker, 1964). Nowhere is this more glaringly true than in healthcare where the competence of one's professional knowledge and training, translates to both the overall health of a patient as well the sustainability within that system economically.

In the GCC, the conventional view of human capital is important as highly trained professionals are needed in some sectors (e.g. healthcare). The GCC healthcare arena needs significant investments in education and training to cater for the burgeoning skilled labour supply needed to cope with the rapid expansion of technologies.

11.2.1.2 Human Capital Versus Social Capital

Newer human capital theory also reflects the realization that one person's human capital can be a by-product of another person's social capital; in other words, your "ability" and resources may just be gifts from White society if you just ponder on it. Emergence Theory of Human Capital Resources (HCR) (Wolfson & Mathieu, 2018) posits that organizational boundary-spanning social interactions and collaborations must occur to foster the knowledge and shared practice necessary for managing emergent complexity within organizations. Given the team-based care models and interdisciplinary collaboration necessary in healthcare, social capital becomes a crucial factor to enhance human capital. Social capital also allows dissemination of best practices that can promote more efficient and desirable healthcare.

In addition, social capital plays a major component in human capital development, which suits the GCC context both culturally and organizationally as networking, social interaction and collaboration affect how their labour market operates. Healthcare organizations in the GCC are learning that they can only share knowledge and innovate properly through enabling environments for social capital.

11.2.1.3 Human Capital Through the Lens of Evolution

Cañibano and Potts (2019) suggest an evolutionary read of human capital theory, with a focus on the force that drives change in skills and knowledge. In this way of seeing things, human capital is anything but ephemeral; it develops and contracts in accordance with technological disruption, industry-specific clustering and long-term economic progression. The notion is more apt in the healthcare context, where medical technology evolves at a fast pace and digital health systems are increasingly being incorporated into practice, necessitating periodic upskilling among healthcare professionals. As digital transformation changes the way healthcare is delivered in the GCC, it opens an evolutionary viewpoint on human capital that stresses on uninterrupted professional education and lifelong learning.

11.2.1.4 Space and Place in Human Capital

Within human capital theory, a spatial turn has occurred among studies that analyses how the geography of society and location affects the creation and mobilization of human capital. Yang (2023), for example, discusses the importance of labour mobility and migration patterns which are unique to individual regions in his concept of Human Capital Space (HCS) and its influence on regional economic dynamics. Of course, this is something a lot more pertinent to the GCC region due to majority of the healthcare industry here extremely dependent on expatriate workers who travel from other nations. A primary use of these data is to understand the spatial dynamics of human capital in the region for workforce planning and efforts to invest in place-based strategies that hold, grow and energize homegrown talent.

Spatial dynamics in the GCC are also indicative of mobility of healthcare professionals into and out of the region. With countries like UAE, Saudi Arabia, Qatar, etc. vying for the same pool of qualified professionals, retention challenges persist thus creating an unstable healthcare workforce. Thus, human capital development policies need to examine the spatial and regional dynamics of labour mobility—a critical consideration in the context of GCC dependence on migrant healthcare workforce.

11.2.1.5 People, Planet and Profit (Human Capital and Sustainability)

The coupling of human capital with sustainability is an aspect of critical importance especially as the GCC tries to transform their economies towards diversification and sustainability. In an environmentally heavy industry as healthcare, a reasoning factor here could be access to human capital, where sustainable habits can be driven by way of innovation and sustainability practices central to the organization. The realization of a circular economy in healthcare that minimizes waste, maximizes resource use and encourages sustainability envisages this new cohort to have competencies on healthcare delivery as well as around environmental sustainability (Binsuwadan et al., 2023).

As the GCC region moves toward a more sustainable economy, it is imperative that one understands that it requires human capital development, inclusive of environmental stewardship and sustainability knowledge. There is a need for enhanced skills and knowledge for healthcare professionals to practise sustainability, reduce wastage and promote the circular economy within the health sector.

11.3 Human Capital Empowerment Challenges in the GCC HealthCare Sector

Several obstacles exist to hinder the empowerment and development of human capital in the GCC healthcare sector despite its crucial role in driving healthcare innovation and sustainability. Such challenges include an extensive reliance on expatriate workers, gender inequality, educational and training deficiencies and the ever-increasing speed of digital transformation.

11.3.1 Dependence on Foreign Labor

Over-reliance on expat workers is one of the most pressing challenges in the GCC healthcare sector. Foreign-trained healthcare workers (HCWs) have compensated for the deficiency in the native workforce supply of key staff. In countries such as Saudi Arabia, Qatar and UAE foreign HCWs step in to take on primary responsibilities due to a limited number of locally trained HCWs. Overdependence has led towards increased workforce instability, as most expats are employed on short-term admission contracts with higher turnover (Sheikh et al., 2019). This disrupts continuity of care and adds to pressure on healthcare systems seeking to recruit and train staff.

In addition, the strong reliance on expatriate labour has implications for the development of long-term human capital in the region. While expatriates provide valuable knowledge, they may not increase local capacity-building or the transfer of knowledge into regions necessary for a self-sustaining health workforce.

Countries like Saudi Arabia are beginning to tackle this challenge through initiatives such as Vision 2030—to increase the Saudi National workforce in healthcare. Nonetheless, it carries significant burdens and financial weights that need to be met by the state via investments in education, training and workforce development to decrease dependence on foreign forces (Al-Hanawi et al., 2019).

11.3.2 Gender Disparities

The lack of gender equality is yet another hindrance to human capital empowerment in the healthcare industry in GCC. Although the involvement of women in the health workforce has increased in some GCC countries, a major gender gap continues across many areas, such as leadership roles and specialities within medical fields. Though the UAE leads in female empowerment in healthcare, countries like Saudi Arabia, Oman and Bahrain still suffer from societal and cultural restrictions in women joining the workforce (Maben et al., 2010; Mehrajunnisa & Jabeen, 2019).

This is particularly important for realizing the human capital we are talking of in healthcare. Research indicates that organizations with more diverse and inclusive workforces are not only exponentially more innovative and productive, but they also provide better outcomes for the people they care for empowering women for a healthier self and other women in healthcare should be a top pick for healthcare organizations, as well as government policy making in the health sectors of GCC.

11.3.3 Educational and Training Gaps

Challenges in the GCC healthcare sector include a lack of standardized education and training to develop a qualified local talent base. The region has scarce healthcare education institutions making healthcare education to manufacture a smaller number of the skilled human resources in health (Sheikh et al., 2019). Therefore, GCC nations end up employing expatriate workers to handle other essential hospital and healthcare facility duties.

This lack of schools as well as programs not training quality healthcare personnel to comply with industry standards has negatively affected the healthcare industry. In the field of digital health, much research points to a skills gap in these hybrid capabilities (Zakaria et al., 2022), as the expansion of digital technologies is driving escalation in necessary teacher and students blending healthcare professions with technological know-how, particularly involving knowledge of the use new technology. However, only a few academic programs across the region have included these new digital competencies in their curricula, collapsing the graduating student with the inability to develop into modern healthcare workforce demands. More specifically, Qureshi (2020) emphasizes the pressing need for knowledge integration to develop a resilient healthcare human capital to successfully field technological advancement related challenges in the UAE.

The GCC countries face the interrelated challenges of insufficient educated healthcare personnel, foreign dependency for training and lack of coordination among providers at both the secondary vocational and higher education levels. Solutions could include increased investments in healthcare education and related facilities, and upgrading educational curricula guided by the needs of industry. Similarly, collaborations among academic institutions and healthcare organizations may be key to connecting education with practice.

11.4 Digital Transformation and Workforce Readiness

Compared to other sectors, digital transformation in healthcare is moving at a pace that not only promises enormous opportunities but comes with the kind of challenges in human capital development which impart from regulation-change-management-government incentives. Digital health technologies, which include telemedicine (digital patient consultations), electronic medical records and health informatics, are changing the face of medicine in ways that make healthcare delivery more efficient and care more accessible. But to ensure these technologies are successful, you need people who have the digital skills needed.

In Saudi Arabia, the implementation of digital health services was not as brisk as anticipated mainly because there were not enough competent individuals employed in digital health (Alanezi, 2021). It recognizes the urgency to ensure competency frameworks in digital health are as critical as they are for standards-based after-training programs. Adding to this complexity is the very nature of digital transformation processes, along with resistance-to-change resistance and concern about data security (Tawk & Daw, 2024). Healthcare organizations must fund initiatives designed to develop a digitally literate workforce and offer continuous training on new technologies.

To address these challenges, collaborative organizations including the GCC Taskforce on Workforce Development in Digital Healthcare have been formed to specify competencies and identify career tracks for digital health professionals (Zakaria et al., 2022). These efforts are essential to developing a resilient healthcare workforce able to adapt and respond quickly in an ever-changing healthcare environment.

Moreover, there are economic and policy hurdles as well that restrict the empowerment of human capital in the GCC healthcare sector. Availability of data is critical for the decision-making process and in many GCC countries, workforce planning at healthcare level is limited due to a lack of standardized, quality information at an acceptable cost (Almazrou et al., 2021). In addition, government revenues and fiscal policy heavily contribute to the pool of resources earmarked for workforce development in health. Fiscal policies prioritize infrastructure development rather than appropriate investment in education and training, restricting the capacity to develop a sustainable health workforce (Khan et al., 2022).

GCC governments need to invest in developing human capital through education, training and workforce planning to overcome these obstacles. Policy makers in the Middle East will also need to make certain that healthcare workforce strategies are consistent with wider economic and sustainability goals, given the regional focus on economic diversification and environmental sustainability.

11.5 GCC Healthcare Sector and the Importance of Human Capital

Human capital is key in defining the performance, resilience and digitalization of healthcare in the GCC. The GCC can enrich healthcare outcomes and the resilience of systems and create long-term sustainable healthcare by investing in human capital development.

11.5.1 Agile Organizational Performance

The improvement of organizational performances and their capabilities and adaptability in the health sector are dependent on human capital. The relationship of human resource management policies and practices to organizational resilience is well-established in the literature (Alhosani & Ahmad, 2024); however, we still have much to learn about this relationship. For example, in the United Arab Emirates (UAE), hospitals implementing high performance work practices like compensation based on capability and performance related pay are more responsive to changes in healthcare market and better patient outcomes (Al-Taweel, 2021).

Moreover, the healthcare providers that put money into human capital end up being better able to handle disasters like Covid-19 going forward. Research has displayed those places of employment containing well-trained, able workforces were better prepared to address the problems related with the pandemic including understaffing challenges, increasing patient loads and a demand for immediate incorporation of new technologies.

11.5.2 Workforce Development and Capacity Building

Capacity building and workforce development are critical to filling the gap in human resources for health in the GCC. There are best practices in the literature to get locally trained health workers through continuing medical education (CME) and professional development programs that will contribute to the overall quality of care; these programs would have to be custom-tailored based on potential worker's levels of education (Sheikh et al., 2019). For instance, investment in nursing education has been found to lower turnover rates and enhance work which results in improved patient care quality and decreased healthcare expenditures (Rondeau et al., 2009).

Capacity building initiatives together with HR can assist healthcare organizations in nurturing their niche talent areas like digital health, health informatics competence and advanced medical technologies. This is a tempting development for the GCC countries which can turn more of their expatriate healthcare workers into locals if there is regional room to enhance skills and move up their career ladder (Singh et al., 2022).

11.5.3 Digital Transformation and e-Health

Digital technologies are shaping the future of healthcare and redefining it, which also holds great implications for talent. The GCC region is witnessing an increasing adoption of electronic health records (EHRs), telemedicine and artificial intelligence (AI) tools by healthcare systems in the region due to which there is a requirement for professionals who have knowledge about digital health practices. Nevertheless, the mismatch between academic programs and industry requirements remains a challenge (Zakaria et al., 2022).

To bridge this divide, healthcare institutions and academic organizations must have a joint effort to create universal digital health competencies throughout the curriculum. Initiatives like the GCC Taskforce on Workforce Development in Digital Healthcare are essential in describing digital health competencies and developing a pipeline of jobs for existing professionals within this field (Zakaria et al., 2022).

Digital healthcare transformation also offers an opportunity to deliver better outcomes for patients while simultaneously lowering the cost of care. For example, one significant near-term opportunity is in expanding access to care via telemedicine, particularly when it comes to rural and underserved areas. This is made more profound since digital health initiatives require effective use of technology by its users, which emphasises the significance of human capital.

11.5.4 Health Implications and Economic Consequences

The development of human capital holds considerable economic ramifications, especially for the healthcare industry in the GCC. Findings have demonstrated health benefits as well as enhanced productive efficiencies and cost-effectiveness in healthcare institutions among states that establish structures for workforce development (Iorember et al., 2022). An investment in professional development and building capacity can, for example, reduce staff turnover which could be a cost saving to healthcare facilities as well as contributing to higher satisfaction of patients resulting from better quality care (Rondeau et al., 2009).

Human capital development, besides from enhancing healthcare quality, facilitates in fostering the sustainability of healthcare systems by advocating eco-friendly practices and principles of the circular economy. Therefore, using green human resources management (HRM) to incorporate environmental sustainability into HR practices can improve the performance and decrease the harmful environmental outcomes of healthcare organizations (Obeidat et al., 2022). Using energy-efficient technologies and implementing waste reduction strategies helps healthcare organizations save cost, in turn lessening their carbon footprints.

In conclusion, the path of empowerment for human capital is preliminary and an important factor in the establishment of a sustainable health system within the GCC region. By addressing expatriate dependency, gender gap and educational gap alongside investments in education, training and digital competencies, the GCC can build a resilient healthcare workforce that will enable the seamless evolution of regional healthcare systems towards sustainability and digital transformation.

11.6 Implications of Digital Circular Economy

11.6.1 Contribution to the Development of Existing Literature

This research attempts to contribute to the existing literature and ongoing discourse on empowering human capital for the healthcare sector in the GCC, with an added focal lens on digital transformation and sustainability driving a digital circular economy. Much of the Human Capital Theory (HCT) discussed earlier in this chapter addresses western and developed economies, with this chapter attempting to highlight on new insights specific to the unique socioeconomic and cultural dynamics of the Gulf Cooperation Council (GCC) countries. By investigating the role of human capital in fostering a sustainable digital circular economy, this chapter integrates critical concepts from digital innovation, sustainability, and healthcare management.

While previous theoretical studies talk extensively about the importance of healthcare workforce development and training, few have addressed the challenges faced by the healthcare sector in the GCC like high turnover rates, over-dependence on expatriate healthcare workers and a scarcity of locally trained medical professionals. As a result, this chapter attempted to enrich the discourse by linking the mentioned challenges to the opportunities ahead for developing a digitally equipped and resilient workforce within the healthcare sector. Continuous learning is discussed as a cultural necessity positing that strategic leadership and interdisciplinary collaboration are not merely supportive mechanisms, but essential drivers of systemic change. Through this lens, human capital empowerment is viewed not just as a tool for improved workforce efficiency, but as a transformative element for enhancing patient care and institutional resilience.

11.6.2 Opportunities for Development of New Theories and Frameworks

A significant new contribution of this chapter is to showcase an opportunity for the development of a Continuous Learning and Digital Innovation Framework presents continuous learning not as an adjunct but as a fundamental part of professional practice, particularly in the fast-paced, technologically driven context of GCC healthcare systems. The framework encourages training models for healthcare workers to seamlessly integrate with the new digital technological advancements of the sector. It emphasizes the need for healthcare professionals to continuously engage with digital tools such as telemedicine platforms, AI-powered diagnostics, and advanced data analytics to enhance their competencies and adapt to the evolving healthcare environment and develop the digital circular economy.

Additionally, this chapter examined human capital at the heart of efforts to create a sustainable digital circular economy within healthcare. This model highlights the interdependence between workforce development and the adoption of digital innovation, moving beyond the traditional notion of retention of talent. It encompasses creating an ecosystem where digital tools and human capital development work in sync to enhance healthcare delivery and efficiency. By aligning the principles of sustainability with human capital empowerment, this model reimagines the role of healthcare professionals as active agents in driving systemic sustainability, thereby advancing both workforce resilience and long-term healthcare outcomes across the GCC.

These frameworks advance the academic conversation on human capital development by providing a fresh, context-specific approach to overcoming the challenges unique to the GCC healthcare sector, while also offering a forward-thinking perspective on the integration of digital innovation into healthcare practices.

11.7 Practical Recommendations

The chapter presents several practical recommendations for stakeholders involved in the healthcare sector in the GCC primarily policy makers, healthcare and academic institutions. Several challenges have been highlighted and addressing them will require targeted strategies and inter-disciplinary collaborations across sectors.

11.7.1 Localizing the Workforce

The GCC countries must take cognizance of the issue of an absence of a locally skilled health workforce that has led to an over-reliance on expatriate healthcare workers. Investments are required in improving national education and training programs tailored to the healthcare sector. Initiatives such as Saudi Arabia's Vision 2030 offer a blueprint for reducing this reliance by fostering local talent. Government policies should emphasize creating clear career pathways for local healthcare workers, coupled with attractive compensation packages to encourage retention.

It is essential to increase funding for healthcare education infrastructure to develop a healthcare workforce trained within the community. This includes setting up new medical universities, nursing institutions, and specific training facilities designed to meet the healthcare requirements of the area. The curricula need to be revised to match changing industry needs, especially in new fields like digital health, telemedicine, and health informatics (Sheikh et al., 2019). Studies indicate that matching educational programs with labour market demands can ensure graduates have the necessary skills for addressing healthcare challenges in the region (Qureshi, 2020). Investment in education not only improves training quality but also helps in establishing a self-sufficient healthcare system (Al-Hanawi et al., 2019).

To draw in and keep local healthcare workers, it is important to offer financial rewards and establish definite opportunities for career advancement. These actions could consist of healthcare education scholarships, programs to forgive loans, and compensation structures based on performance (Al-Hosani & Ahmad, 2024). Studies show that offering competitive pay and chances for career growth can enhance job satisfaction and employee retention in the healthcare sector (Rondeau et al., 2009). Additionally, providing incentives to local professionals not only decreases turnover but also establishes a steadfast healthcare workforce for the future (Al-Taweel, 2021).

11.7.2 Promoting Gender Equality in Healthcare

Encouraging gender equality in healthcare is essential for the GCC region, where the involvement of women in the workforce, especially in leadership positions and specialized medical roles, is still restricted. There are ongoing obstacles in countries like Saudi Arabia, Oman, and Bahrain, where women's professional growth is often limited by cultural and social factors (Mehrajunnisa & Jabeen, 2019). Nations can use the UAE as a successful example of promoting women's empowerment in the healthcare industry. Acknowledging this disparity is crucial not only to make the most of the human resources at hand but also to improve healthcare results, as inclusive and diverse teams are usually more creative and efficient. Authorities in the GCC region need to introduce specific actions to promote gender equality, such as mentorship schemes, leadership training opportunities, and policies supporting women's involvement in healthcare across all areas. Supporting women in the healthcare field will strengthen health systems and help the region to meet the increasing need for qualified professionals in the sector.

11.7.3 Accelerating Digital Transformation for Workforce Preparedness

The implementation of AI, machine learning and robotics in medical diagnosis, planning treatment and managing patients is transforming the way healthcare functions. An example is when AI helps healthcare providers diagnose diseases more precisely and create personalized treatment plans using data-driven insights. Healthcare professionals must possess the necessary skills to effectively utilize these advanced technologies, highlighting the significance of digital literacy at every stage of healthcare services (Obeidat et al., 2022). Furthermore, automating administrative functions like billing and patient record management through digital transformation can lead to a decrease in errors and a more efficient operation within the healthcare system (Zakaria et al., 2022).

Creating digital health competency frameworks is essential for successful incorporation of digital technologies in healthcare. These models outline the essential skills and information required for healthcare professionals to operate effectively in a more digitized setting, including telemedicine, electronic health records (EHRs), and health informatics. Through incorporating these skills into academic and professional training, healthcare organizations can guarantee their employees are equipped for the demands of digital advancement (Zakaria et al., 2022). Studies suggest that nations utilizing competency-based digital health frameworks typically demonstrate increased efficiency in delivering healthcare services and improving patient outcomes (Alanezi, 2021).

To help improve the digital skills of healthcare professionals, organizations should put in place targeted Continuous Professional Development (CPD) programs that reflect the changing technology in healthcare. Consistent workshops and training sessions centred on digital health tools like telemedicine platforms, AI-driven diagnostics, and EHR management help healthcare staff stay skilled in using new technologies. These workshops need to be regularly scheduled to make sure that all employees are informed about the newest advancements. In addition, healthcare institutions should promote healthcare professionals to participate in online learning platforms and obtain certifications in digital health areas such as data management and cybersecurity. Healthcare institutions may partner with academia to provide accredited courses that can help healthcare workers to improve knowledge and practical skills. Utilizing virtual reality (VR) and other immersive technologies for simulation-based learning can provide professionals with practical experience before incorporating these tools into clinical practice.

Peer-to-peer learning and mentorship programs, in which knowledgeable staff in digital health can mentor their colleagues, is also a useful CPD approach. Institutions should encourage professionals to participate in digital health conferences and webinars by providing financial assistance for attending global events, where they can learn about the most recent industry advancements. Integrating digital skills into routine performance evaluations can also guarantee ongoing enhancement, with specific objectives defined for every employee. Furthermore, healthcare organizations must develop custom learning plans tailored to each person's skill level, guaranteeing that digital skills training is both applicable and adjustable to everyone's requirements and expertise.

Telemedicine and remote care have become crucial parts of contemporary healthcare, particularly following the COVID-19 crisis. The outbreak of the pandemic accelerated the process of digitizing healthcare systems, increasing the importance and availability of telemedicine services. Using digital platforms, healthcare providers can now provide prompt care to patients in distant or underserved locations, surpassing geographical obstacles and improving the effectiveness of healthcare distribution. This has enhanced medical service availability while also cutting down on operational expenses for healthcare providers and patients alike. The incorporation of telemedicine into current healthcare systems help to ease the strain on traditional healthcare centres and provides a flexible answer to the increasing worldwide need for urgent medical care, diagnostics and related health services (Alanezi, 2021).

Even though telemedicine is becoming more popular, it is still a new idea with many uncharted territories, such as enhancing diagnostic tools, offering specialized care through virtual consultations and using artificial intelligence in remote healthcare. Furthermore, rules and recommendations regarding telemedicine are constantly changing, as governments and organizations strive to develop structures for protecting data, patient confidentiality and the uniformity of telehealth procedures.

The complete benefits of telemedicine have not been fully achieved, as there are current obstacles like incorporating telemedicine into current healthcare systems, guaranteeing fair technology access and adjusting to the varying demands of patients in various locations.

Adequate training for healthcare providers is essential to maximize the advantages of telemedicine. Healthcare professionals must be knowledgeable in utilizing telehealth technologies, such as digital consultations, remote diagnostics, and secure management of patient data. Insufficient training can lead to miscommunication, diagnostic mistakes or violations of patient privacy, potentially weakening the success of telemedicine. In addition, training can assist healthcare providers in feeling more at ease with technology and increasing patient involvement to ensure that the transition to digital care is beneficial.

Challenges often arise in the adoption of digital technologies in healthcare due to resistance to change. Issues like worries about job stability, privacy of information, and the intricacy of unfamiliar systems can cause hesitancy among healthcare professionals to adopt digital technologies (Tawk & Daw, 2024). Healthcare organizations must promote a culture of flexibility and creativity to deal with this problem. Change management initiatives that focus on the advantages of digital tools, along with ongoing training and assistance, can facilitate the shift to a digital healthcare setting leading to developing closed loop strategies for digital circular economy.

11.7.4 Data Governance and Cyber Security risks in e-Healthcare

The growing use of digital technologies in healthcare also brings up worries about data governance and cybersecurity. Policymakers need to create precise rules about data privacy and security to safeguard sensitive patient information and encourage the ethical utilization of digital health technologies.

Establishing a clear data governance structure in healthcare will foster trust with providers and patients and guarantee the secure integration of digital technologies (Almazrou et al., 2021). Healthcare providers need to be educated in cybersecurity protocols, including ensuring the security of patient data and efficiently handling digital systems to avoid security breaches.

By putting these practical recommendations into practice, the GCC nations can attempt to strengthen their healthcare personnel, decrease their dependence on foreigners, and more closely match their healthcare systems with the objectives of sustainability and digital circularity. This will enhance both the quality of care and the overall resilience and performance of the healthcare sector in response to future challenges.

11.8 Directions and Scope for Future Research

There is a vast and evolving future for research on enhancing human capital in the healthcare sector of the GCC for a sustainable digital circular economy. One avenue for future exploration involves incorporating advanced digital health technologies like artificial intelligence (AI) and blockchain into workforce development.

It will be essential to examine how these technologies can improve healthcare operations and enhance human capital skills as the region moves towards a more digitalized healthcare system.

Furthermore, there is potential for further research on the impact of customized continuous professional development initiatives on healthcare workers within the framework of digital circular economy. The chapter evaluated how these initiatives affect workforce preparedness, sustainability efforts and diminishing reliance on foreign workers. Exploring the lasting impacts of gender diversity and inclusion strategies on empowering healthcare staff is a potential area for further research.

Future research could investigate the impact of reducing the gender disparity in leadership positions and healthcare fields on maintaining a strong workforce and enhancing patient results.

Moreover, going forward, upcoming research could investigate the economic effects of a digital circular economy in the healthcare sector, specifically looking at cost reduction, resource efficiency, and environmental consequences. This would offer a better grasp of how human capital empowerment is in line with economic growth and sustainability goals in the GCC. Similarly, the study can be replicated in other sectors to analyse the impact of digital circularity in people empowerment.

11.9 Conclusion

This chapter has investigated the crucial importance of human capital empowerment in promoting a sustainable digital circular economy in the GCC healthcare industry. The healthcare systems of the GCC are currently facing a critical moment, as issues like dependence on foreign workers, gender inequalities, gaps in education and training, and the fast adoption of digital technology impede sustainability and innovation in the long run. It is essential to invest strategically in education, workforce development, and policy reform in order to create a strong workforce that can effectively respond to contemporary needs and develop resilience.

Important discoveries emphasize that the effectiveness of digital health technologies such as telemedicine and AI relies on having a skilled workforce that is well-trained in digital skills. Moreover, promoting gender diversity and inclusivity in the healthcare sector is not only about fairness, but also a critical necessity that can improve innovation, efficiency and results leading to higher productivity.

Nevertheless, there are still various shortcomings in the healthcare industry of the GCC, particularly in aspects like insufficient local workforce training, absence of uniform education systems and disjointed policy efforts. The economic and policy obstacles further complicate the challenges, restricting investments in human capital. To overcome these restrictions, GCC governments and healthcare organizations need to improve cooperation between academic institutions and healthcare providers to match educational programs with industry requirements, especially in digital skills. Future studies could examine the role of continuous professional development programs tailored to the healthcare workforce in the context of the circular economy. Research could assess the impact of these programs on improving workforce readiness, sustainability practices and reducing dependency on expatriates.

Furthermore, policy makers should prioritize making long-term investments in human capital development to ensure that efforts are sustainable and aligned with wider economic and sustainability objectives. In summary, enhancing human resources in the healthcare industry in the GCC is crucial for better health results and for transitioning to a sustainable, technology-focused circular economy. Through strategic planning of workforce, gender inclusion and educational reforms, the GCC can develop a healthcare system that is strong and prepared for the future by overcoming current challenges. This chapter thus has attempted to set the foundation for future investigation into combining digital technologies and sustainable practices in healthcare, providing opportunities for additional innovation and growth.

References

Alanezi, F. (2021). Factors influencing decision making for implementing eHealth in light of the COVID-19 outbreak in Gulf Cooperation Council countries. *International Health, 13*(5), 457–464. https://doi.org/10.1093/inthealth/ihab003

Alhosani, F. H., & Ahmad, S. Z. (2024). Role of human resource practices, leadership, and intellectual capital in enhancing organizational performance: The mediating effect of organizational agility. *Journal of Intellectual Capital*. https://doi.org/10.1108/jic-08-2023-0183

Al-Hanawi, M. K., Khan, S. A., & Al-Borie, H. M. (2019). Healthcare human resource development in Saudi Arabia: Emerging challenges and opportunities—a critical review. *Public Health Reviews, 40*(1), 1–16. https://doi.org/10.1186/s40985-019-0112-4

Almazrou, S., AlAujan, S. S., & Al-Aqeel, S. (2021). Barriers and facilitators to conducting economic evaluation studies in Gulf Cooperation Council (GCC) countries: A survey of researchers. *Health Research Policy and Systems, 19*(1), 1–10. https://doi.org/10.1186/s12961-021-00721-1

Al-Taweel, I. R. (2021). Impact of high-performance work practices in human resource management of health dispensaries in Qassim region, Kingdom of Saudi Arabia, towards organizational resilience and productivity. *Business Process Management Journal*. https://doi.org/10.1108/BPMJ-11-2020-0498

Becker, G. S. (1964). *Human capital: A theoretical and empirical analysis, with special reference to education*. University of Chicago Press.

Binsuwadan, J., Yousif, G. M., Abdulrahim, H., & Alofaysan, H. B. (2023). The role of the circular economy in fostering sustainable economic growth in the GCC. *Sustainability, 15*(15), 1–20. https://doi.org/10.3390/su152215926

Brownie, S. M., Hunter, L. H., Aqtash, S., & Day, G. E. (2015). Establishing policy foundations and regulatory systems to enhance nursing practice in the United Arab Emirates. *Policy, Politics, & Nursing Practice, 16*(1–2), 38–50. https://doi.org/10.1177/1527154415583396

Cañibano, C., & Potts, J. (2019). Toward an evolutionary theory of human capital. *Journal of Evolutionary Economics, 29*(1), 65–89. https://doi.org/10.1007/s00191-018-0588-y

El-Saharty, S., Kheyfets, I., Herbst, C., & Ajwad, M. I. (2020). *Fostering human capital in the Gulf Cooperation Council countries*. World Bank Publications.

Emanuele, J., & Koetter, L. (2007). Workflow opportunities and challenges in healthcare. In *2007 BPM & workflow handbook* (Vol. 1, p. 157). https://www.researchgate.net/profile/Laura-Koetter/publication/252065707_Workflow_Opportunities_and_Challenges_in_Healthcare/links/552290170cf2f9c13052e464/Workflow-Opportunities-and-Challenges-in-Healthcare.pdf

Esmaeilzadeh, P. (2024). Challenges and strategies for wide-scale artificial intelligence (AI) deployment in healthcare practices: A perspective for healthcare organizations. *Artificial Intelligence in Medicine, 151*, 102861. https://doi.org/10.1016/j.artmed.2024.102861

Iorember, P. T., Iormom, B. I., Jato, T. P. J., & Abbas, J. (2022). Understanding the bearable link between ecology and health outcomes: The criticality of human capital development and energy use. *Heliyon, 8*(9), e12611. https://doi.org/10.1016/j.heliyon.2022.e12611

Katoue, M. G., Cerda, A. A., García, L. Y., & Jakovljevic, M. (2022). Healthcare system development in the middle East and North Africa region: Challenges, endeavors and prospective opportunities. *Frontiers in Public Health, 10*, 1045739. https://doi.org/10.3389/fpubh.2022.1045739

Khan, S., Ali, S. S., & Singh, R. (2022). Determinants of remanufacturing adoption for circular economy: A causal relationship evaluation framework. *Applied System Innovation, 5*(4), 62. https://doi.org/10.3390/asi5040062

Khoja, T., Rawaf, S., Qidwai, W., Rawaf, D., Nanji, K., & Hamad, A. (2017). Health care in Gulf Cooperation Council countries: A review of challenges and opportunities. *Cureus, 9*(8). https://doi.org/10.7759/cureus.1586

Maben, J., Al-Thowini, K., West, E., & Rafferty, A. M. (2010). Uneven development: Comparing the indigenous health care workforce in Saudi Arabia, Bahrain, and Oman. *International Journal of Nursing Studies, 47*(3), 392–400. https://doi.org/10.1016/j.ijnurstu.2009.07.008

Mehrajunnisa, M., & Jabeen, F. (2019). Ranking the enablers promoting female empowerment in the UAE healthcare sector. *International Journal of Gender and Entrepreneurship, 11*(1), 51–74. https://doi.org/10.1108/IJGE-12-2018-0135

Obeidat, S. M., Abdalla, S., & Al Bakri, A. A. A. (2022). Integrating green human resource management and circular economy to enhance sustainable performance: An empirical study from the Qatari service sector. *Employee Relations, 44*(2), 234–251. https://doi.org/10.1108/er-01-2022-0041

Pillai, R., & Al-Malkawi, H. A. N. (2018). On the relationship between corporate governance and firm performance: Evidence from GCC countries. *Research in International Business and Finance, 44*, 394–410. https://doi.org/10.1016/j.ribaf.2017.07.110

Pereira, V., Neal, M., Temouri, Y., & Qureshi, W. (Eds.). (2020). *Human capital in the middle east: A UAE perspective*. Springer Nature. https://doi.org/10.1007/978-3-030-42211-0

Qureshi, W. (2020). The role of human capital in the implementation of healthcare innovation in the UAE. In V. Pereira, M. Neal, Y. Temouri, & W. Qureshi. (Eds.), *Human capital in the middle east*. Palgrave Studies in Global Human Capital Management. Palgrave Macmillan. https://doi.org/10.1007/978-3-030-42211-0_11

Rod, I. (2019). TMT diversity and innovation ambidexterity in family firms: The mediating role of open innovation breadth. *Journal of Family Business Management, 9*(4), 377–392. https://doi.org/10.1108/JFBM-09-2023-0157

Rondeau, K. V., Williams, E. S., & Wagar, T. H. (2009). Developing human capital: What is the impact on nurse turnover? *Journal of Nursing Management, 17*(6), 739–748. https://doi.org/10.1111/j.1365-2834.2009.00988.x

Sheikh, J. I., Cheema, S., Chaabna, K., Lowenfels, A. B., & Mamtani, R. (2019). Capacity building in healthcare professions within the Gulf Cooperation Council countries: Paving the way forward. *BMC Medical Education, 19*(1), 1–10. https://doi.org/10.1186/s12909-019-1513-2

Singh, R., Khan, S., & Dsilva, J. (2021). A framework for assessment of critical factor for circular economy practice implementation|emerald insight. *Journal of Modelling in Management, 18*(5), 1476–1497. https://doi.org/10.1108/JM2-06-2021-0145

Singh, R., Khan, S., & Centobelli, P. (2022). Investigating the Interplay between social performance and organisational factors supporting circular economy practices. *Sustainability, 14*(24), 16781. https://doi.org/10.3390/su142416781

Tawk, C. J., & Daw, D. N. (2024). Digital transformation of human capital management in the Gulf Cooperation Council. *Corporate Ownership and Control, 21*(3), 89–102. https://doi.org/10.22495/cocv21i3art10

Tripathi, V., Singh, R., Dsilva, J., & Singh, S. (2021). *A review on Sustainable HRM: A study evaluating sustainability for organisational development*. Retrieved September 30, 2024, from ResearchGate website: https://www.researchgate.net/publication/355771746_A_review_on_Sustainable_HRM_A_study_evaluating_Sustainability_for_organisational_development

Wolfson, M. A., & Mathieu, J. E. (2018). Sprinting to the finish: Toward a theory of human capital resource complementarity. *Journal of Applied Psychology, 103*(10), 1145–1165. https://doi.org/10.1037/apl0000323

World Bank. (2018). The human capital project. https://www.worldbank.org/en/publication/human-capital

Yang, Z. (2023). Human capital space: A spatial perspective of the dynamics of people and economic relationships. *Humanities & Social Sciences Communications, 10*(1), 1–10. https://doi.org/10.1057/s41599-023-01639-5

Yee, W. M. T., & Rahman, S. (2019). Empowerment for economic and human capital development through education. *ASEAN Post-50: Emerging Issues and Challenges*, 81–99. https://doi.org/10.1007/978-981-13-8043-3_5

Zakaria, N., Zakaria, N., Alnobani, O., Almalki, M., El-Hassan, O., Alhefzi, M., Househ, M., & Jamal, A. (2022). Unlocking the eHealth professionals' career pathways: A case of Gulf Cooperation Council countries. *International Journal of Medical Informatics, 166*, 104914. https://doi.org/10.1016/j.ijmedinf.2022.104914

Part III
Technologies Driven Circular Economy for Different Sectors

Chapter 12
Transitioning to Digital Circular Economy Practices: A Case Study on SMEs

Joy Eghonghon Akahome

Abstract Integrating digital technologies in the practice of CE has been transformed which led to the emergence of the digital circular economy (DCE), which balances digital tools to improve circularity. Small and medium enterprises (SMEs) are necessary for international economic growth and face specific challenges and opportunities in the adoption of DCE practices. This study explores the SMEs transition to digital circular economy practices with the use of a qualitative case study approach that focused on the pure (sachet) water industry in Nigeria. The methods for collecting data consist of semi-structured interviews, observations, and document analysis to understand the experiences, challenges, and strategies of SMEs in their transition to circular models. The findings reveal that while these SMEs recognize the benefits of digital circular practices such as cost savings, improved brand reputation, and compliance with regulatory bodies, they face significant challenges of lack of funds, experience and limitation of supply chain and digital track system. Despite these challenges, some localized strategies were implemented such as reducing waste and partnering with local communities have encouraged SMEs to make progress toward circularity. The study highlights the importance of building a supportive policies, having access finance, and getting initiatives for capacity-building to facilitate the global adoption of DCE practices among SMEs. These information contributes to the growing body of knowledge on the implementation of digital circular economy in developing countries and offers some practical recommendations for policymakers, industry practitioners, and researchers whose aim is to foster technological sustainable business practices.

Keywords Digital circular economy (DCE) · Small and medium enterprises (SMEs) · Sustainability · Qualitative case study Nigeria · Sachet water industry

---

J. E. Akahome (✉)
Department of Entrepreneurship and Marketing, Federal University Otuoke, Bayelsa State, Yenagoa, Nigeria
e-mail: akahomejj@fuotuoke.edu.ng; jeakah@utu.fi

Turku School of Economics, University of Turku, Turku, Finland

R. Singh and V. Kumar (eds.), *Sustainable Innovations and Digital Circular Economy*,
https://doi.org/10.1007/978-981-96-1064-8_12

12.1 Introduction

12.1.1 Background and Context

The digital circular economy is viewed as a business strategy that puts sustainability first by designing products and systems to reduce waste, elongate product life cycles, and reuse materials. As opposed to traditional linear economy that follows a "take-make-dispose" approach, the circular economy majorly works on the principles of reducing, reusing, and recycling method. The fundamental goal is creating a closed-loop system where resources can be continuously recycled back into manufacturing thereby, minimizing the need for new raw materials and having a reduction of environmental impact (Ellen MacArthur Foundation, 2013). The shift from traditional to a circular economy is greatly seen as relevant to address the global sustainability challenges. In the traditional or conventional model, resources are extracted, products are manufactured, used, and discarded as waste. This traditional approach has led to depletion of resources and degradation of the environment, which lead to significant waste generation. In comparison with the linear economy, CE aims to separate economic growth from resource consumption by keeping use the resources for a long time to reduce the environmental footprint and foster economic resilience (Ghisellini et al., 2016). Consequently, there has been a rise of digital technologies that have facilitated a new phase of practice in circular economy and led to the concept of the digital circular economy (DCE). The integration of digital tools such as blockchain, the Internet of Things (IoT), and data analytics, help businesses are better equipped in tracking resources, optimizing operations, and promote sustainable innovation.

Small and medium enterprises (SMEs) play an important role in the global economy by contributing significantly to the growth of employment, innovation, and the economic. According to the World Bank (2020), SMEs has represented about 90% of businesses and more than 50% of employment worldwide. However on daily basis SMEs face challenges in the adoption of sustainable practices because of lack of resources, expertise and also funding. In a traditional economy, SMEs are majorly vulnerable to lack of resource scarcity, cost increase and issues of waste management which negatively affects their long-term viability (Johnson & Schaltegger, 2016).

It is easy for SMEs to successfully transition into a circular economy because of their adaptability, innovative capacity, and good relationships with local markets. Their adaptability allows them to experiment with new business models and immediately respond to changing market demands. Furthermore, SMEs can act as key catalyst for innovation by creating new products, processes, and services that are in line with the principles of circular economy. However, the transitioning from traditional to a circular economy poses several challenges for SMEs, which includes the need to redesign products, reconfigure supply chains, and regenerate or recreate new business strategies (Rizos et al., 2016). In presenting these challenges and opportunities, there is an urgent need for case studies that can explore how SMEs successfully implement circular economy practices. Such studies can provide useful information into the promotion and challenges of adopting circular economy in SMEs and offer practical advice to overcome the challenges. Scholars can identify some best ideas, practices, strategies and structure when they examine real-life examples that other SMEs can repeat, use, or adapt (Masi et al., 2017), while Oliveira et al. (2021) provide a transition to a society that is sustainable and provides theoretical framework for an integrated assessment methods in the circular economy. The SME case study will offer empirical evidence and examples of these principles. The case study provides examples of how circular economy practices can be implemented, the challenges faced, and the outcomes achieved for focusing on a specific SME in Nigeria. This empirical data presents an extension or validation of the theoretical models proposed by Oliveira et al.'s study where it emphasizes a global perspective on the shift to a circular economy. This SME case study will introduce a local and context perspective by exploring the transition of circular economy in Nigeria's unique socio-economic and regulatory environment. This localization extends Oliveira et al. (2021) study by showcasing how the international frameworks for sustainability and circular economy can be adapted to fit local contexts in developing economies.

The study will be guided by the following research questions:

R1 What are the strategies SMEs use in transitioning to a digital circular economy?
R2 What key drivers motivate SMEs to adopt digital circular economy practices?
R3 What barriers are faced by SMEs in the implementation of circular economy practices?
R4 What are the environmental, economic, and social outcomes of adopting circular economy practices in SMEs?

The following sections will delve into the literature review, methodology, the specific case of Nigerian sachet water SMEs, discussion, and offering understanding into how these enterprises can navigate the complex process of adopting circular economy principles.

12.2 Literature Review

12.2.1 Overview of Circular Economy: Concept and Principles

The circular economy (CE) is a comprehensive approach that is designed for economic development in order to benefit businesses, society, and the environment. In comparison with the traditional linear economy, resources are extracted, used, and disposed of where the circular economy is regenerative by design. It aims to separate economic growth from resource consumption by keeping products, materials, and resources in use for a long time. The core principles of the circular economy include the following.

Designing Out Waste and Pollution: According to (Ellen MacArthur Foundation, 2013), the CE aim is to eliminate waste through intentional design and innovative solutions by ensuring that products that are produced does not generate waste or pollution.

Keeping Products, Materials and Resources in Use: This principle involves extension of the Product lifecycle through the reusing, repairing, remaufacturimg, and recycling strategies to reduce the need for new materials (Ghisellini et al., 2016).

Regenerating Natural Systems: Circular economy practices focus on returning nutrients or materials that are valuable back to the ecosystem and improve the natural systems rather than destroying them (Geissdoerfer et al., 2017).

The circular economy is important to sustainability because it addresses the environmental challenges, which is affected by the traditional linear economy such as depletion of resources, degradation of the environment, and generation of waste. Following up and recycling on resource use, the CE can reduce the environmental impact of producing and consuming to promote resource efficiency and contribute to climate change mitigation (Kirchherr et al., 2017). Additionally, the CE fosters innovation in designing product, managing supply chain, and modeling businesses to lead to a more sustainable practice across industries (Prieto-Sandoval et al., 2018).

12.2.2 Digital Circular Economy in SMEs

The **Digital Circular Economy (DCE)** provides opportunities for small and medium enterprises (SMEs), which enables them to improve the sustainability, optimization of resource use, and reducing waste through the combination of digital technologies. SMEs is a relevant contributor to global economic development which faces challenges related to limited resources, but the adoption of digital tools can assist them in overcoming these barriers and achieve circularity.

Resource Optimization and Waste Reduction: SMEs can support **Internet of Things (IoT)** devices in monitoring the use of resources and identifying inefficiencies in the process of production. IoT sensors allow real-time raw materials tracking, which allows the SMEs in reducing waste and optimizing resource consumption. For example, in the food industry, IoT is applied to allow SMEs in monitoring the shelf life of products and preventing waste by optimizing inventory management (Zhu et al., 2022). In addition, **big data** analytics enables businesses in analyzing waste patterns and implementing waste reduction strategies.

Improved Supply Chain Transparency: SMEs use **Blockchain** technology to improve the transparency and traceability of supply chain. Blockchain provides a ledger that is secure and immutable, which can record every transaction and material flow in the supply chain to make it easier for SMEs to track the origin, movement, and environmental impact of products and materials. This can assist in ensuring sustainability standard compliance and allows SMEs to engage with consumers that are environmentally conscious (Upadhyay et al., 2021).

Product Lifecycle Extension and Predictive Maintenance: Applying **predictive maintenance** through the use of digital technologies for SMEs such as **machine learning** and IoT is creative. To monitor the equipment and products conditions in real time will help businesses to predict when maintenance is needed, thereby extending the lifecycle of their machinery and products (Antikainen & Valkokari, 2016). This can reduce the downtime, operational costs, and also minimize the need for extracting new resources that contribute to the principles of circular economy.

Circular Business Models: The DCE allows SMEs in exploring new business models for example, **product-as-a-service (PaaS)**, instead of selling products. SMEs provide services related to the use of those products. This model also promotes long life of the products, as businesses retain its responsibility for maintaining, repairing, and recycling of the products they provide. **Digitally enabled platforms** make it easier for SMEs to implement such models, reduce waste and encourage resource reuse (Bressanelli et al., 2020).

Collaboration and Knowledge Sharing: SMEs benefit from **digital platforms** that promotes how companies collaborate and share knowledge. These digital platforms enable SMEs to share their best practices, pool resources, and engage in innovation that is collaborative, making it easier to adopt the principles of circular economy. For example, the digital marketplaces provide SMEs the opportunity to exchange waste materials that can be repurposed by other businesses and promote a circular flow of resources (Gregson et al., 2015).

Customer Engagement and Circularity: Digital technologies also provide SMEs with the tools to engage consumers in the practice of circular economy. Through **digital communication channels and mobile apps**, SMEs can educate consumers on the environmental benefits of their products and services, promoting recycling programs, and incentivizing sustainable purchasing behaviors. This direct engagement assists SMEs in building strong bond with environmentally conscious consumers to improve their loyalty and reputation of the brand (Kääriäinen & Rueda Mejia, 2020).

This is to conclude that adopting the practices of digital circular economy can offer SMEs unique opportunities in sustainable innovation and operation. Utilizing digital technologies such as IoT, blockchain, and data analytics will enable SMEs to optimize resource use, reduce waste, and create new business models that can contribute to an economy that is more circular and sustainable.

12.2.3 Application Across Industries

The circular economy is being applied across various industries, from manufacturing to food production, construction, and fashion. For example, in the manufacturing sector, companies are adopting circular strategies like remanufacturing and product-as-a-service models to reduce waste and extend product lifecycles (Bocken et al.,

2016). In the fashion industry, many brands are exploring circular business models of clothing rentals, resaling, and textile recycling to minimize waste and reduce environmental impact (EMF, 2017). The construction industry is also embracing circular principles by focusing on resource efficiency, material reuse, and sustainable building practices (Pomponi & Moncaster, 2017).

12.2.4 Circular Economy in SMEs

A. **Circular Economy Practices**

Small and medium enterprises (SMEs) play an important role in the transition to a circular economy due to their flexibility, innovative capacity, and significant economic presence. However, the adoption of circular economy practices among SMEs varies widely depending on factors such as industry, size, location, and access to resources. Research indicates that SMEs often engage with circular practices through initiatives like waste reduction, resource efficiency, and eco-design, but they face challenges such as limited funds, resources, lack of expertise, and pressures from the markets (Rizos et al., 2016).

B. **Models and Frameworks for SMEs in the Circular Economy**

There are several models and frameworks that have been proposed to guide SMEs in their transition to a circular economy.

The circular business model canvas (CBMC): This model was adapted from the traditional business model canvas. The CBMC has helped the SMEs to map out circular economy opportunities by paying more attention on elements such as resource loops, product life extension, and creation of value through sustainability (Lewandowski, 2016).

The ReSOLVE Framework: This model was developed by the Ellen MacArthur Foundation. The ReSOLVE framework provides a strategic approach for businesses especially the SMEs to implement practices of circular economy. It focuses on six actions: Regeneration, sharing, optimization, looping, virtualization, and exchanges (EMF, 2015).

Eco-Innovation Framework: This framework emphasizes the inclusion of the consideration of environmental practices into innovation processes. This encourages SMEs to develop products, processes, and services that can minimize environmental impact while creating economic value (Klewitz & Hansen, 2014).

12.2.5 *Challenges and Drivers for SMEs in Transitioning to a Circular Economy*

SMEs face some specific challenges in the practical adoption of circular economy such as lack of capital, technical knowledge, and the need to collaborate with partners of supply chain. However, SMEs experience some opportunities that makes them adaptable and build innovative potential to create niche markets, differentiate themselves from competitors, and improve their sustainability credentials (Johnson & Schaltegger, 2016). When CE is successfully implemented in SMEs, it can lead to significant environmental and economic benefits that includes cost savings, new revenue streams, and improved brand reputation (Masi et al., 2017).

A. **Challenges for SMEs**

1. **High Initial Investment in Digital Technologies**: To implement digital solutions such as IoT, blockchain, and data analytics requires huge initial investment, which is specifically problematic for SMEs with lack of financial resources. There is great cost to acquire necessary technology infrastructure, training staff, and maintaining these systems, which is a significant barrier to entry for many businesses (Upadhyay et al., 2021). In addition to this, combining legacy systems with new digital platforms can further make the process complicated. One noticeable challenge that small and medium-sized enterprises (SMEs) face in transitioning to a circular economy is lack of funding because the process of implementing the practice of circular economy requires large investment in new technologies, processes, and training. Typically, SME is operated on tighter budgets and usually have limited access to capital in comparison with larger enterprises. Large enterprise may struggle to secure the necessary funding (Rizos et al., 2016) due to the high costs of associating it with redesigning products, reconfiguring supply chains, and adopting new business models, which can be less allowed for SMEs with small profit margins (Agyemang et al., 2019).
2. **Lack of Expertise and Knowledge in Digital Skills**: The second major challenge is the lack of technical know-how and expertise that regards the principles and practices of circular economy. Lack of skilled labor in specific fields such as data analytics, IoT, and blockchain can prevent the adoption of digital circular practices (Bressanelli et al., 2020). To transition into the DCE will require knowledge and skills that are specifically related to digital technologies, which are lacking in many SMEs. SMEs often face challenges in attracting and retaining talent with the necessary expertise to implement these innovations. A lot of SMEs lack the necessary technical knowledge or experience to effectively implement circular strategies. This gap in knowledge gap can prevent the realization of opportunities for resource efficiency, waste reduction, and product life extension. In addition, SMEs usually lack access to networks and information that could provide guidance and support for transitioning to a circular economy (Dey et al., 2020). It is factual that without

proper knowledge and skills, SMEs will find it a challenge to innovate and adopt sustainable practices (Ranta et al., 2018).

3. **Supply Chain Issues**: The challenge of supply chain can create significant barriers for SMEs because the transition to a circular economy will require changes to existing supply chains, which include the establishment of new partnerships, the securing of sustainable inputs, and developing closed-loop systems for product that are returned and recycled. SMEs may face challenges in influencing supply chains due to their relatively small size and financial capacity. In addition, the lack of agreement between supply chain partners can increase the complication in the implementation of circular practices (Govindan & Hasanagic, 2018). Also, most SMEs may find it a challenge to get suppliers or partners who are willing to engage in circular economy practices which can further limit their ability to transition. Supply chains complexity for global businesses presents a significant challenge in adopting circular practices. Digital tools such as blockchain can improve supply chain transparency, but the implementation of these tools across different and often fragmented supply chains can be challenging, especially when multiple stakeholders and suppliers are involved (Antikainen & Valkokari, 2016).
4. **Regulatory Barriers and Compliance Issues**: Developing countries lack clarity of the regulatory frameworks around the practice of digital circular economy, which can create uncertainty for businesses that look at transitioning to DCE models. Complying with existing environmental and digital regulations can also be complicated which may lead to legal challenges for businesses that are not fully aware of the requirements for data protection, resource recycling, and waste management (Zhu et al., 2022).
5. **Cultural Resistance to Change**: For businesses to take the step internally to change is a problem. Many businesses are accustomed to linear production models and may be hesitant to adopt circular practices due to uncertainty about their effectiveness and long-term benefits. Also, it will take some time to convince stakeholders to invest in circularity and digital innovation, which often requires a cultural shift within the organization (Gregson et al., 2015).

B. **Drivers for SMEs**

1. **Cost Savings and Resource Efficiency**

A key motive for adopting DCE is to potentially save costs through an improved resource efficiency. By supporting digital technologies to optimize production processes and reduce waste, businesses can reduce their operating costs. For example, IoT-enabled sensors can assist companies to monitor resource consumption in real time, which may lead to reducing use of energy and material waste (Zhu et al., 2022). This is not only helpful for businesses to cut costs but also to improve their environmental impact.

2. **Improved Brand Reputation and Competitive Advantage**

Businesses that learn to adopt the practice of circular economy can differentiate themselves from competitors by demonstrating their commitment to sustainability. Consumers are desperately seeking eco-friendly and socially responsible brands, and companies that know well to adopt DCE models can improve their reputation and customer loyalty (Bressanelli et al., 2020). Furthermore, businesses that are leading in sustainability are viewed as being more innovative, which provides them with a competitive edge in the marketplace.

3. **Compliance with Regulations and Future-Proofing**

Governments around the world are introducing strict environmental regulations that adopts DCE practices that can be of help to businesses that choose to stay ahead of compliance requirements. To implement circular models and use digital tools in tracking the use of resources and waste Management can ensure that businesses are well prepared for future regulatory changes to reduce the risk of non-compliance penalties (Lacy & Rutqvist, 2015).

4. **Innovation and Business Resilience**

Incorporating digital technologies with circular economy principles can encourage innovation, which allows businesses to develop new products, services, and business models. Such circular models include product-as-a-service or shared economy platforms that can promote creativity and explore alternative revenue streams (Bressanelli et al., 2020). Also, businesses adopting DCE practices are more resilient to the shortage of resources and market disruptions as they rely on renewable and reusable materials rather than finished resources.

5. **Consumer Demand**

An important driver for SMEs to adopt the practice of circular economy is the growing consumer demand for sustainable products and services. As the increase in awareness on environmental issues becomes relevant, consumers are becoming more conscious of their environment and how their purchase decisions can affect the environment. This shift in consumer behavior will present an opportunity for SMEs to differentiate themselves by offering environmentally friendly products and services that aligns with the principles of circular economy. In response to consumer demand for sustainability, SMEs can improve their market competitiveness and build brand loyalty (Moktadir et al., 2018).

6. **Regulatory Pressure**

The pressure from regulatory bodies can be a key driver for SMEs to transition to a circular economy. The government and regulatory bodies can increase the implementation of policies and regulations that are aimed at promoting sustainability and reducing waste. These regulations may include some requirements for extending producer's responsibility, targeting waste reduction, and providing incentives for recycling and resource efficiency. SMEs that choose to adopt practices of circular economy will not only comply with these regulations but also benefit from government incentives so as to avoid potential penalties (Kirchherr et al., 2017). Furthermore, keeping up with the regulatory trends can give SMEs a competitive edge in the market.

7. **Competitive Advantage and Access to New Markets and Investment Opportunities**

To transition to DCE can explore new business opportunities such as access to **green finance** and **investors** impact can increase interest in the support of sustainable enterprises. The adoption of the practice on circular economy can provide SMEs with a competitive advantage by giving them the opportunity to be innovative, reducing costs, and improving their reputation. By engaging in the implementation of resource-efficient processes and waste minimization, SMEs can save cost and improve on their operational efficiency. In addition, circular business models, such as product-as-a-service or recycling, can create new streams of revenue and increase the engagement of customers. Furthermore, the demonstration of commitment to sustainability can improve the reputation of SMEs and attract environmentally conscious consumers and business partners (Prieto-Sandoval et al., 2018). In a market that is constantly competitive, these advantages can help SMEs to be unique and thrive. Therefore, transitioning to a circular economy presents both challenges and opportunities for SMEs. While lack of funding, lack of experience and knowledge, and issues in the supply chain can pose some significant challenges while consumer demand, pressure in regulatory bodies, and competitive advantage potentials are powerful advantages that can motivate SMEs to adopt the practices of circular economy. Knowing these facts on barriers and drivers is important for the development of effective strategies that can support SMEs in their journey to achieve sustainability.

12.3 Methodology

12.3.1 Research Design

A qualitative case study methodology is used in the study because it works well for the examination of complex phenomena that exist in the context of real-world situations. A qualitative case study is appropriate because it offers information that is rich and comprehensive which is difficult to realize through quantitative methods. This study enables an in-depth examination of the adoption of circular economy practices that are within the SMEs (Yin, 2018). in addressing the "what" questions instead of "how" or "why" in qualitative research describes focusing on understanding the nature, traits, and experiences that are related to a specific phenomenon. In examining what occurs within a particular context, this method is useful because it offers a comprehensive account that can guides future research (Stake,1995; Creswell & Poth, 2016).

For example, the qualitative case study might seek to address questions in the context of "what". With this focus, the investigation becomes less about the processes (the "how") or the underlying causes ("why") and more about the particular behaviors and results seen inside the SMEs (Yin, 2018). In conducting an exploratory research, this method works best because of its intention to be descriptive in nature and identify patterns, practices, or phenomena without getting too deep into processes or causation.

12.3.2 Case Selection

The case study contains a representative sample of SMEs that are involved in the production of pure (sachet) water in Nigeria, the case selection The procedure of the case selected s important due to the fact that the SMEs are chosen based on the following criteria.

Industry Type: Due to the widespread of pure (sachet) water consumption across Nigeria and how it has a negative environmental impact on plastic waste, it is a significant sector in Nigeria. The study focuses on SMEs in this industry.

Size: Most SMEs in Nigeria are known to have fewer than 200 employees and a lack of market reach in the Nigerian context (Osotimehin et al., 2012).

Location: In capturing a variety of context in the environmental and regulatory aspects, the SMEs chosen for the study are spread throughout Nigeria.

Degree of Adoption of the Circular Economy: The study encapsulates SMEs at different phases of the adoption process from starting the investigation to completing the execution. This makes it important to compare the elements that promote or impedes their development (Rizos et al., 2016).

12.3.3 Data Collection

In this study, several qualitative methods were employed in the process of collecting data in order to guarantee a thorough understanding of adopting SMEs for the practice of circular economy.

Interviews: In gaining understanding into the experiences, drivers, and challenges of the key stakeholders such as SME owners, managers, and employees, a semi-structured interview will be conducted. According to Rubin and Rubin (2012), interviews conducted in a study are necessary for gathering individual mindset behind their corporate choices.

Observations: There is an observation on how the operational procedures of the SMEs handle resources, waste products, and engage in recycling. In verifying information that are gathered from interviews and current data on how CE are applied will be provided by observations (Creswell & Poth, 2016).

Document Analysis: The comprehension of the official strategies and policies can direct the SMEs' transition to the practice of circular economy using relevant document such as examining company reports, analyzing business plans, and providing environmental audits. The offering of accurate proof of the SMEs commitments and practices in the document analysis can improve the interview and observational data (Bowen, 2009).

12.4 Data Analysis

The data analysis process involves organized coding and thematic analysis for the identification of patterns and themes that are related to the transition to a circular economy.

Coding :The coding combined the inductive and deductive methods to code the documents contents, notes during observation, and transcripts from the interview. While deductive coding is informed by established previous theoretical frameworks on circular economy practices, the inductive coding allows themes to be generated directly from the data (Braun & Clarke, 2006).

Thematic Analysis: By following up on the data coding, a thematic analysis is carried out to describe the main themes and sub-themes that best explain the elements that affects the SMEs to shift to a circular economy. Data interpretation is offered by thematic analysis to connect the conclusions to the research questions and goal of the research (Nowell et al., 2017).

Triangulation: Data collected from observations, interviews, and documents are cross checked to increase the validity of the results. Additionally, the use of multiple data sources is done to support the themes that are identified in the process to increase the validity of the conclusions reached (Denzin, 2012).

12.5 Case Study Results

12.5.1 Case Studies Presentations

This research case study focuses on Small and Medium-sized Enterprises (SMEs) in Nigeria's pure (sachet) water sector. This industry contributes to the Nigerian's economy by providing millions of people with affordable drinking water. However, it presents some environmental issues in the aspect of plastic waste. The chosen SMEs is different in terms of size, location, and level of adoption of the circular economy. For example, the first SME in Lagos is a medium-sized business that has been operating for more than ten years, and another SME is smaller with its headquarters in Ibadan. These SMEs has its little focus on the circular economy that ranges from having few knowledge on the practice of sustainable to having already implemented some fundamental waste management techniques.

A. **Background of SMEs**:

SME 1 (Lagos): The SME is a medium-sized sachet water producer with 72 employees, and this company had few to none engagement with practice of circular economy but faced increasing pressure from both regulatory bodies and consumers to address its plastic waste.

SME 2 (Ibadan): The small-sized SME is about 20 employees, and this SME had already started its experimentation with recycling initiatives but lacked a skilled strategy for implementing its circular economy principles across its operations.

B. **Implementation of Circular Economy Practices**

The transition to circular economy practices involves some key initiatives across the selected SMEs.

Waste Reduction:

SME 1: In terms of reducing waste, this SME implemented a waste segregation system that allowed collecting and sorting of plastic waste directly at the manufacturing site. The waste was later on sent to recycling facilities thereby reducing the overall waste that was sent to landfills (Rizos et al., 2016).

SME 2: This SME partnered with local artisans to reuse waste materials into reusable products such as environmentally friendly bags that are sold in local markets (Moktadir et al., 2018).

Product Life Extension:

SME 1 started a program and event that encouraged customers to return sachet water plastic that has been used and getting paid in exchange for discounts on future purchases. This initiative not only extended the life of the packaging materials but also fostered customer loyalty (Prieto-Sandoval et al., 2018).

SME 2 is more focused on redesigning its package to make it acceptable, durable, and reusable thereby extending the life cycle of the products and reducing the rate of frequent purchases (Geissdoerfer et al., 2017).

Resource Recovery:

Both SMEs adopted the practice of resource recovery where waste materials were collected and processed for reuse. SME 1 created a small-scale facility to process and convert plastic waste into raw materials that could be resold or used within the homes for packaging (Kirchherr et al., 2017).

Challenges and Solutions:

The SMEs have faced many challenges during their transition to the practice of circular economy using developed unique solutions.

Financial Constraints:

The two SMEs struggled with getting some necessary funding to implement the practice of circular economy. SME 1 addressed this issue when it applied for government grants and NGO partnerships that supported sustainable business initiatives (Agyemang et al., 2019).

SME 2 increased community support by crowdsourcing funds from environmentally conscious consumers and local businesses which enabled them to finance their waste reduction initiatives (Govindan & Hasanagic, 2018).

Lack of Expertise:

SME 1 At first, they lacked the skills and experience to effectively implement the practice of circular economy. To improve, they paid more attention in training of the staff by introducing programs they can learn from and collaborate with other environmental specialist who can provide the necessary expertise (Dey et al., 2020).

SME 2 engaged with some local universities to tap into research and innovation resources, which aided them in designing more efficient recycling processes (Johnson & Schaltegger, 2016).

Supply Chain Challenges:

SME 1 faced some problems in convincing partners of supply chain to participate in their circular economy initiatives. They addressed this by offering incentives, such as discounts and marketing support for suppliers who adopted sustainable practices (Ranta et al., 2018).

SME 2 facilitated its supply chain by focusing on local suppliers who shared their commitment to sustainability thereby reducing complex measures in logistics to foster stronger partnerships (Masi et al., 2017).

Outcomes and Impact:

Implementing the practice of circular economy had a positive impact on both SMEs in terms of business performance, sustainability, and competitiveness:

Business Performance:

SME 1 did report a 15% reduction in its operational costs because of expenses on decreased waste disposal and increase on revenue from the sale of materials that have been recycled. In addition, the program on waste return improved the customers' retention and brand loyalty rate (Prieto-Sandoval et al., 2018).

SME 2 saw a 20% increase in product sales that drives growing consumer demand for eco-friendly products. Partnering with local artisans also created new streams of revenue and boosted profitability (Moktadir et al., 2018).

Sustainability:

SME 2's reuse strategies on reduced waste was generated by 30%, while SME 1's waste segregation and efforts on recycling diverted about 10 tons of plastic waste from landfills annually (Geissdoerfer et al., 2017). By allowing the SMEs to recycle up to 80% of the materials they used in manufacturing, the resource recovery initiatives also assisted them in achieving their sustainability goals (Kirchherr et al., 2017).

Competitiveness:

Adopting the practice of circular economy for both SMEs improved their market positioning. SME 1 is committed themselves to sustainability that attracted environmentally conscious consumers, which led to a 25% increase in market share. SME 2's innovative packaging design and waste reuse strategy improved its brand image, which made it a preferred choice among eco-conscious customers (Rizos et al., 2016).

12.6 Discussion

The case study findings on the transition of SMEs in Nigeria's pure (sachet) water industry to circular economy practice showed some support and disagreement with existing literature. The identification of the challenges in supply chain, lack of expertise, and limited funding as major challenges to the adoption of circular economy practices is consistent with existing literature. The literature has extensively documented these challenges especially in studies that emphasizes SMEs' resource constraints when compared to larger enterprises (Agyemang et al., 2019; Rizos et al., 2016). Furthermore, the adoption of the practice of circular economy is influenced by consumer demand, regulatory pressure, and competitive advantage. These factors are consistent with previous research that outlines the importance of market opportunities and external pressures in encouraging small and medium-sized enterprises to transition (Kirchherr et al., 2017; Moktadir et al., 2018). However, certain differences were noted. The case studies demonstrated that localized supply chains and community-based partnerships would reduce the difficulties SMEs face as a result of rigid and complex supply chains in spite of the fact that previous research has frequently highlighted these challenges (Govindan & Hasanagic, 2018). For example, the smaller SME in Ibadan did well by effectively using local partnerships to improve its circular practices. Such strategy is not well emphasized in larger research on global supply chains. This implies that SMEs that operate in more local, smaller environments which may have a competitive advantage over those in larger more international markets when it comes to implementing circular practices. Another discrepancy is related to the level of innovation in resource recovery. While existing literature outlines the huge investment required for advanced resource recovery technologies (Geissdoerfer et al., 2017), the SMEs in the case studies demonstrated that smaller-scale, community-driven recovery initiatives can be feasible and effective. This finding suggests that SMEs do not always need high-tech solutions to achieve circularity, which contrasts with the emphasis on technological solutions found in much of the existing literature.

12.7 Key Insights and Implications

The findings from the case studies offer several key insights for SMEs, policymakers, and industry practitioners.

Localized Solutions for Circularity: The success of having localized community-driven strategies in the case study describes the potential for SMEs to increase their partnership with local markets and communities in order to effectively implement the practices of circular economy. This information suggests that policymakers and industry practitioners should support the development of a localized circular economy frameworks that are tailored to meet the specific needs and capabilities of SMEs.

Incremental Implementation: SME gradual adoption of circular practices starts with low-cost initiatives like reducing waste and extending the product life cycle which illustrates that incremental implementation can be a possible strategy for SMEs with lack of resources. This approach will allow SMEs to build their capacity over time while they gradually realize the benefits of circular economy practices.

Role of External Support: Having external support is of importance, such as government grants, NGO partnerships, and university collaborations were evident in the case studies. This findings has reinforced the need for policies and programs that provide SMEs with access to funding, experience, and networks that can facilitate their transition to a circular economy.

Market Differentiation and Competitive Advantage: The case studies showed that SMEs can gain a competitive edge by adopting the practice of circular economy especially in areas where consumers are more concerned about sustainability. This understanding outlines the potential for SME to build CE that will serve as a strategic tool in differentiating themselves and building stronger customer relationships.

12.8 Theoretical Contributions

The case study findings contributed to the existing theoretical framework on the practice of digital circular economy within the SMEs by offering empirical evidence on the feasibility of low-cost, community-driven circular solutions. The research expands the understanding of how SMEs can overcome challenges through localized strategies and incremental adoption which challenges the prevailing notion that advanced technology and significant investment are prerequisites for successful transition to circular economy (Bocken et al., 2016; Prieto-Sandoval et al., 2018). Moreover, the study provides relevant information into the role of local supply chains and community engagement to enable the practice of circular economy that suggest how the circular economy framework can be adapted to suit different contexts especially in developing countries. This adds to the body of knowledge on how contextual factors, such as local market dynamics and community networks, can influence SMEs adoption of circular economy practices (Ranta et al., 2018). Therefore, the study highlights the need for a much better and context-sensitive approach to circular economy theory—one that can recognize the diversity of SMEs and the varied environments in which they operate. Furthermore, this study calls for a more comprehensive theoretical framework that covers SME experiences that demonstrates how CE can be achieved through localized, low-cost techniques and challenges the dominance of high-technological and capital intensive models in the discussion of CE.

12.9 Conclusion

The study explored SMEs transition of Nigeria's pure (sachet) water industry to practice of circular economy. The conclusions of the key findings suggest lack of funding, lack of expertise, and challenges of supply chain were significant issues in adopting circular practices. However, consumer demand, regulatory pressure, and competitive advantage served as strong motivators for the transition. The research also highlighted the importance of localized, community-driven solutions and incremental implementation strategies which allowed SMEs to overcome these barriers effectively. In addition, the study demonstrated that adopting circular economy practices led to improved business performance, sustainability, and greater market competitiveness for the SMEs involved. These findings addressed the research questions by providing insights into the strategies SMEs use to transition to a circular economy, the challenges they face, and the outcomes of their efforts.

12.10 Practical Recommendations

The research examines SMEs in Nigeria that tries to transition to a circular economy and adopts few number of practical recommendations: Adopting Increased Implementation Strategies: firstly, the SMEs should focus on low-cost, high-impact projects by extending the life of their products and reducing waste. In the early stage, SMEs can develop their capability and experience with this gradual approach, which can create a better situation for more extensive reforms (Rizos et al., 2016). Leverage Local Partnerships and Community Engagement: having a close working relationship with local stakeholders such as suppliers, businessmen, and customers, SMEs can improve their efforts related to the circular economy. These collaborations may offer more funds, expertise, and market prospects which can improve the feasibility and significance of the transition (Johnson & Schaltegger, 2016). Seek External Support: in focusing on the sustainability activities, SMEs can look into grants from the government, NGOs and financial institutions. Furthermore, SMEs can obtain the knowledge required to successfully implement circular economy methods by collaborating with universities and research centres (Agyemang et al., 2019). Emphasis on Market Differentiation: SMEs can stand out in the marketplace and draw in environmentally conscientious customers by emphasizing their dedication to sustainability. According to Moktadir et al. (2018), this can increase brand loyalty and provide new revenue streams.

12.11 Limitations and Future Research

Despite the limitations found in the study, it has offered useful information about the adoption of circular economy principles in Nigeria's pure water sector. One of the limitations was that the case study scope was restricted to a specific industry and geographical area, which may not accurately affect the diversity of SMEs in other industries or nations. Secondly, the results are based on a small sample of SMEs where the qualitative nature of the study may introduce some potential biases in the collection and interpretation of data. Future studies should expand the scope of the study to include SMEs from different industries and other geographical areas in order to overcome these limitations. Studies that compare various industries could offer a more comprehensive understanding of how various contexts affect the adoption of circular economy practices. Furthermore, quantitative research could support the qualitative findings by offering statistical proof of the results, barriers, and drivers that have been identified.

Finally, the circular economy offers relevant potential for SMEs to improve their sustainability, resilience, and competitiveness. By adopting circular practices, SMEs can contribute internationally to sustainability goals while benefiting from costs reduction, increased efficiency and improved market positioning. However, the idea of transitioning requires careful planning, strong partnerships, and willingness to be innovative. The findings of this study outlines the importance of localized, incremental approaches that combines with the unique challenges and opportunities faced by SMEs. As the circular economy continues to gain recognition worldwide, the information gained from this research can guide SMEs, policymakers, and industry practitioners in fostering a more sustainable and resilient future.

References

Antikainen, M., & Valkokari, K. (2016). A framework for sustainable circular business model innovation. *Journal of Cleaner Production, 30*, 40–46.

Agyemang, M., Kusi-Sarpong, S., Mani, V., & Kusi, S. (2019). Drivers and barriers to circular economy implementation: An explorative study in Pakistan's automobile industry. *Management of Environmental Quality: An International Journal, 30*(3), 594–613.

Bocken, N. M. P., de Pauw, I., Bakker, C., & van der Grinten, B. (2016). Product design and business model strategies for a circular economy. *Journal of Industrial and Production Engineering, 33*(5), 308–320.

Bowen, G. A. (2009). Document analysis as a qualitative research method. *Qualitative Research Journal, 9*(2), 27–40.

Braun, V., & Clarke, V. (2006). Using thematic analysis in psychology. *Qualitative Research in Psychology, 3*(2), 77–101.

Bressanelli, G., Saccani, N., Perona, M., & Baccanelli, I. (2020). Towards circular economy in the household appliance industry: An overview of cases. *Resources, 9*(11), 1–23. https://doi.org/10.3390/resources9110128

Creswell, J. W., & Poth, C. N. (2016). *Qualitative inquiry and research design: Choosing among five approaches* (4th ed.). Sage Publications.

Denzin, N. K. (2012). Triangulation 2.0. *Journal of Mixed Methods Research, 6*(2), 80–88.

Dey, P. K., Malesios, C., De, D., Budhwar, P., Chowdhury, S., & Cheffi, W. (2020). Circular economy to enhance sustainability of small and medium-sized enterprises. *Business Strategy and the Environment, 30*(3), 143–161.

Ellen MacArthur Foundation (EMF). (2013). *Towards the circular economy: Economic and business rationale for an accelerated transition.* Ellen MacArthur Foundation.

Ellen MacArthur Foundation. (2015). *Towards a circular economy: Business rationale for an accelerated transition.* Retrieved from https://www.ellenmacarthurfoundation.org/towards-a-circular-economy-business-rationale-for-anaccelerated-transition

Ellen MacArthur Foundation. (2017). *A new textiles economy: Redesigning fashion's future.* Retrieved from https://www.ellenmacarthurfoundation.org/a-new-textiles-economy

Geissdoerfer, M., Savaget, P., Bocken, N. M. P., & Hultink, E. J. (2017). The circular economy—a new sustainability paradigm? *Journal of Cleaner Production, 143*, 757–768.

Ghisellini, P., Cialani, C., & Ulgiati, S. (2016). A review on circular economy: The expected transition to a balanced interplay of environmental and economic systems. *Journal of Cleaner Production, 114*, 11–32.

Govindan, K., & Hasanagic, M. (2018). A systematic review on drivers, barriers, and practices towards circular economy: A supply chain perspective. *International Journal of Production Research, 56*(1–2), 278–311.

Gregson, N., Crang, M., Fuller, S., & Holmes, H. (2015). Interrogating the circular economy: The moral economy of resource recovery in the EU. *Economy and Society, 44*(2), 218–243.

Johnson, M. P., & Schaltegger, S. (2016). Two decades of sustainability management tools for SMEs: How far have we come? *Journal of Small Business Management, 54*(2), 481–505.

Kääriäinen, P., & Rueda Mejía, S. (2020). Experimental wood-based materials. In *Emerging materials & technologies: New approaches in design teaching methods in four exemplified areas* (pp. 29–51). Franco Angeli Design International s.r.l.

Kirchherr, J., Reike, D., & Hekkert, M. (2017). Conceptualizing the circular economy: An analysis of 114 definitions. *Resources, Conservation and Recycling, 127*, 221–232.

Klewitz, J., & Hansen, E. G. (2014). Sustainability-oriented innovation of SMEs: A systematic review. *Journal of Cleaner Production, 65*, 57–75.

Lacy, P., & Rutqvist, J. (2015). *Waste to wealth: The circular economy advantage.* Palgrave Macmillan.

Lewandowski, M. (2016). Designing the business models for circular economy—towards the conceptual framework. *Sustainability, 8*(1), 43.

Masi, D., Day, S., & Godsell, J. (2017). Supply chain configurations in the circular economy: A systematic literature review. *Sustainability, 9*(9), 1602.

Moktadir, M. A., Rahman, T., Rahman, H., Ali, S. M., & Paul, S. K. (2018). Drivers to sustainable manufacturing practices and circular economy: A perspective of leather industries in Bangladesh. *Journal of Cleaner Production, 174*, 1366–1380.

Nowell, L. S., Norris, J. M., White, D. E., & Moules, N. J. (2017). Thematic analysis: Striving to meet the trustworthiness criteria. *International Journal of Qualitative Methods, 16*(1), 1–13.

Oliveira, M., Miguel, M., Kevin van Langen, S., Ncube, A., Zucaro, A., Fiorentino, G., Passaro, R., Santagata, R., Coleman, N., Lowe, B., Ulgiati, S., & Genovese, A. (2021). Circular economy and the transition to a sustainable society: Integrated assessment methods for a new paradigm. *Circular Economy and Sustainability, 1*(99–113). https://doi.org/10.1007/s43615-021-00019-y

Osotimehin, K. O., Jegede, C. A., Akinlabi, B. H., & Olajide, O. T. (2012). An evaluation of the challenges and prospects of micro and small scale enterprises development in Nigeria. *American International Journal of Contemporary Research, 2*(4), 174–185.

Pomponi, F., & Moncaster, A. (2017). Circular economy for the built environment: A research framework. *Journal of Cleaner Production, 143*, 710–718.

Prieto-Sandoval, V., Jaca, C., & Ormazabal, M. (2018). Towards a consensus on the circular economy. *Journal of Cleaner Production, 179*, 605–615.

Ranta, V., Aarikka-Stenroos, L., Ritala, P., & Mäkinen, S. J. (2018). Creating value in the circular economy: A structured multiple-case analysis of business models. *Journal of Cleaner Production, 201*, 988–1000.

Rizos, V., Behrens, A., Kafyeke, T., Hirschnitz-Garbers, M., & Ioannou, A. (2016). The circular economy: Barriers and opportunities for SMEs. *CEPS working document, 412.*

Rubin, H. J., & Rubin, I. S. (2012). *Qualitative interviewing: The art of hearing data* (3rd ed.). Sage Publications.

Upadhyay, A., Laing, T., Kumar, V., & Dora, M. (2021). Exploring barriers to the circular economy: Small and medium-sized enterprises in the European Union. *Business Strategy and the Environment, 30*(1), 195–203.

World Bank. (2020). *Small and medium enterprises (SMEs) Finance.* Retrieved from https://www.worldbank.org/en/topic/smefinance

Yin, R. K. (2018). *Case study research and applications: Design and methods* (6th ed.). Sage Publications.

Zhu, B., Nguyen, M., Siri, N. S., & Malik, A. (2022). Towards a transformative model of circular economy for SMEs. *Journal of Business Research, 144*, 545–555. https://doi.org/10.1016/j.jbusres.2022.01.093

Chapter 13
Enhancing Role of Innovations in Shaping a Digital Circular Economy

Vishal Jain and Archan Mitra

Abstract This chapter investigates the role of digital technologies in enabling the transition towards a circular economy, a model focused on sustainability by emphasising resource efficiency, waste reduction, and environmental regeneration. The study aims to bridge the gap between traditional linear economic practices and the emerging digital circular economy. The chapter employs a comprehensive literature review, examining the integration of digital tools such as the Internet of Things (IoT), artificial intelligence (AI), blockchain, and big data analytics in various industries, including manufacturing, agriculture, and energy. The study also explores case studies that highlight practical implementations of these technologies in promoting sustainable practices. The findings reveal that digital technologies significantly enhance resource optimisation, increase transparency in supply chains, and foster sustainable business models. However, challenges such as the high costs of digital infrastructure, the complexity of transitioning business models, and the environmental impacts of digital technologies themselves remain barriers to widespread adoption. This chapter contributes to the literature by providing a holistic view of how digital innovations can support the circular economy. It underscores the potential for digital technologies to drive sustainable growth while also identifying the challenges that need to be addressed for these technologies to realise their full potential in supporting a sustainable and equitable future. The study offers valuable insights for academics, corporations, and policymakers striving towards sustainability in the digital age.

Keywords Digital circular economy · Sustainable innovations · Internet of Things (IoT) · Blockchain technology · Resource efficiency

---

V. Jain
Department of Computer Science and Engineering, Sharda School of Engineering and Technology, Sharda University, Greater Noida, India

A. Mitra (✉)
Nitte Institute of Communication (NICO), NITTE University, Mangalore, India
e-mail: archan6644@gmail.com

R. Singh and V. Kumar (eds.), *Sustainable Innovations and Digital Circular Economy*,
https://doi.org/10.1007/978-981-96-1064-8_13

13.1 Introduction

Digitalisation and sustainability are transforming the global economy. The linear model of economics—represented by "take, make, dispose" strategy—have been extensively studied as environmental concerns develop. This worldview causes environmental waste, pollution, and greenhouse gas emissions. It depletes scarce and available resources. Circular economies are popular for sustainable solutions to these concerns. A circular economy redefines growth by emphasising social benefits. Waste and pollution control, resource reuse, and natural system rebuilding are its main priorities, according to the Ellen MacArthur Foundation (2013). Digital technology have transformed every element of modern life, from energy and manufacturing to communication and business. Big data analysis, blockchain, AI, and the IOT could improve circular economy ecosystem operations, according to Esposito et al. (2018). Businesses can use digital technologies to switch from linear to circular models, which will boost sustainability. These innovations enable monitoring and predictive maintenance of resource optimisation, and chain of supply transparency. The digital circular economy is increasingly shaped by innovative technologies, enabling sustainable resource management and reducing waste. According to Bonina et al. (2022), integrating digital innovations such as blockchain and IoT enhances transparency and traceability in supply chains, promoting sustainability. Additionally, digital twins and data analytics, as outlined by Rachna et al. (2021), play a pivotal role in optimising resource usage and fostering a more resilient circular economy model.

This study examines how digital technologies might help implement circular economy ideas. The study will examine the underlying discoveries that caused this change and the pros and cons of incorporating digital solutions into sustainable practices. Its goal is to contribute to the expanding corpus of research on digitisation and sustainability and provide useful information to academics, corporations, and governments.

Unlike linear economy, that assumes resources are abundant, cheap, and disposable, the economy which is circular is a change in paradigm. However, Stahel (2016) suggests designing a economy which is circular to be regenerative and restorative. The company distinguishes biological cycles from technical cycles and strives to maximise resource, component, and product value. Technology reduces waste and virgin resource use by generating items that may be recycled, remanufactured, refurbished, and reused. Composting is a biological cycle process that returns materials to the biosphere and regenerates natural systems, according to Bocken et al. (2016). According to Geissdoerfer et al. (2017), the circular economy relies on lowering resource input and waste, extending product life cycles, using renewable energy, and separating economic growth from resource consumption. These are only a few circular economy ideas. These concepts are congruent with sustainable development's general goals of meeting current needs without compromising future needs (United Nations, 1987). Sustainable development goals align with these ideals.

Digital technologies make the circular economy transition possible and accelerate it. These technologies can help commercial firms and public administrations globally embrace circular economy principles by providing resource management tools and platforms. According to Porter and Heppelmann (2014), the IOT allows tracking of resources through real-time and goods throughout their life cycles. This improves resource optimisation and waste reduction. Similar to this, AI can analyse massive data to find patterns and insights that might improve decision-making and productivity (McKinsey Global Institute, 2017).

Blockchain technology can improve supply chain traceability and transparency by providing a decentralised, irreversible record. The use acquisition and goods disposal are environmentally friendly (Saberi et al., 2019). According to Antikainen et al. (2018), digital markets and platforms can let people share, rent, and trade commodities and resources, supporting the economy in circular format recycling and reuse principles.

These technologies broader digitisation trend may change business structures and value chains. For instance, the "as-a-service" model, which sells items as services, is growing in popularity. This model extends product life cycles and reduces waste. This model follows circular economy concepts, according to Tukker (2015). Digital technology may enable new kinds of creativity and cooperation, such as open-source design and co-creation, which could encourage more circular and sustainable business practises (Lieder & Rashid, 2016).

13.1.1 Sustainable Innovations in the Digital Circular Economy

Without sustainable technologies, a digital circular economy cannot succeed. These innovations may include new goods, materials, business strategies, or workflows. They might also be anything in between. Niero and Olsen (2016) suggest using eco-friendly packaging and biodegradable materials to reduce product environmental impact and promote product circularity. Modularity and easy disassembly can also simplify product recycling, refurbishing, and reuse (Bocken et al., 2016). Innovative features include modularity and disassembly. Digitalisation and environmentally responsible practices are often needed to build sustainable technologies. Circular economies aim to reduce the link between economic growth and resource consumption. Intelligent networks and energy management systems can boost energy efficiency and lower emissions (European Commission, 2020). Digital platforms supporting product service models can lengthen PLC and reduce waste (Tukker, 2015). This follows circular economy concepts. Sustainable innovations are often used with a systematic approach in the digital circular economy. This strategy integrates many technologies and activities to meet sustainable goals. Blockchain technology for traceability, IoT sensors for product use, and AI algorithms for maintenance may be used in a circular supply chain (Esposito et al., 2018). This is done inside a reuse-and-recycle business model. These integrated approaches are needed to maximise the digital circular economy's potential and achieve long-term sustainability.

13.1.2 Challenges and Opportunities

Despite its potential to improve sustainability, the digital circular economy has many drawbacks. Kumar et al. (2018) state that SMEs and companies in developing regions face huge investments in digital infrastructure and technology. Digital technology integration into business models can be difficult and require major organisational change, which stakeholders may oppose (Geissdoerfer et al., 2017). Another fear is that poorly managed digital technologies could worsen the environment. Belkhir and Elmeligi (2018) showed that blockchain networks and data centres may use a lot of electricity, increasing carbon emissions. Since this is the case, digital solutions must be created and executed to minimise environmental effect and support sustainability goals. Despite obstacles, the digital circular economy has great potential. Circular processes can boost productivity, cut costs, and generate new revenue, according to Antikainen et al. (2018). Policymakers are given a framework to attain (SDGs) and handle global issues like waste management, resource shortages, and climate change (United Nations, 2015). Lieber and Rashid (2016) say the digital circular economy offers innovation and entrepreneurship in digital platforms, sustainable commodities, and circular business models. These opportunities are abundant in the digital economy. The digital circular economy may inspire creative sustainability solutions. We may achieve this power by encouraging cooperation and information exchange among governments, corporations, and civil society.

The convergence of the digital circular economy and green technology is crucial to research and practice. Digital technology can help businesses and governments implement the circular economy and reap economic and environmental benefits. To fully profit from the digital circular economy, digitisation and sustainability concerns must be addressed. These concerns include funding, organisational restructuring, and environmental management. As the global economy changes, the digital circular economy will become more vital for building a sustainable future.

13.2 Literature Review

Sustainable innovations and the digital circular economy have become important topics of study due to global challenges like resource shortages, environmental degradation, and climate change. This literature study examines sustainable innovations, their convergence, and how digital technology advances circular economy practices. It uses many studies and assessments to get a complete picture of existing knowledge, showing trends, gaps, and future directions.

13.2.1 Sustainable Innovations in the Digital Age

Recycling, reuse, and resource efficiency are the biggest hurdles when systems switch from linear production models to circular economies. Environmentally friendly technologies are used now. The authors Jain and Mitra (2024a, b, c, d) emphasise biomimicry in bio-inspired optimisation. In addition, they show how natural processes might influence the development of environmentally friendly technologies. According to their research, biomimicry can inspire sustainable innovation in industry and renewable energy. Technology and manufacturing are examples. Biomimicry-based energy collectors improve efficiency and help achieve sustainability goals (Jain & Mitra, 2024a, b, c, d). Because biomimicry mimics biological processes. Because they reduce reliance on non-renewable resources.

Combining environmentally friendly innovations and digital technologies can increase the effect of such breakthroughs. Digital technologies and platforms enable real-time monitoring, predictive analytics, and resource optimisation, making circular economy implementation easier. Das and Mitra's (2024a, b) research on ethical marketing in the big data era shows how digital advances might promote sustainability. This research is a great example of how digital advances may improve sustainability. Businesses may ensure their marketing efforts are ethical and environmentally friendly by developing rigorous data governance rules. This method would promote ethical production and consumption, which would increase consumer trust and progress the circular economy.

13.2.2 Case Studies and Practical Implementations

Several case studies within the DCE setting exemplify the implementation of environmentally appropriate technologies. Jain and Mitra (2024a, b, c, d) give case studies that demonstrate the implementation of biomimicry in renewable energy applications. These case studies illustrate how we might derive inspiration from nature to develop energy systems that are both more efficient and sustainable. These case studies offer useful insights into how biomimicry might potentially drive innovation in several industries, including manufacturing and waste management. Mitra and Das (2024) conducted study on digital celebrities and the sustainability of celebrity on over-the-top (OTT) platforms, offering a different viewpoint on the connection between digital technologies and sustainability. This study was carried out on the subject of digital celebrities themselves. The objective is to examine the impact of the emergence of digital celebrities on societal trends and the media sector, specifically in terms of consumption patterns and environmental sustainability. The findings indicate that digital platforms can both enable the creation and distribution of eco-friendly content and contribute to the faster obsolescence of digital assets, leading to an increase in electronic waste generation. Khan et al. (2019) study examines the impact of community education and training on disaster preparedness and mitigation in the field of disaster management. Their analysis of the flash flood that took place in Uttarakhand in 2013 emphasises the significance of digital technologies in promoting community resilience and facilitating environmentally conscious approaches to disaster management. This case study aims to demonstrate how digital technology can enhance resource efficiency and minimise the environmental impact of disaster response initiatives, thereby contributing to the circular economy.

13.2.3 Gaps and Future Directions

Although it offers a strong basis for comprehending the significance of digital technology and sustainable technologies in the circular economy, there are still some gaps in the current research. Further investigation is crucial in this critical domain, specifically focussing on the scalability of these breakthroughs. A significant proportion of case studies and analyses primarily concentrate on smaller-scale implementations. However, it remains unclear how these strategies might be successfully scaled up to yield broader advantages. Furthermore, there is a valid cause for apprehension regarding the impact of digital technology on the environment, specifically in terms of the substantial energy consumption and the generation of electronic waste. This issue necessitates further investigation, as highlighted by Belkhir and Elmeligi (2018). Furthermore, there exists a void in the existing body of knowledge that could be addressed by undertaking novel research that establishes linkages across the domains of digital technology, sustainability, and circular economy within various academic departments. Despite the increasing awareness of the interrelatedness of many domains, most of the research produced so far remains fragmented and fails to properly include concepts from diverse fields. In order to address this gap, it is crucial for researchers, policymakers, and industry stakeholders to collaborate and develop comprehensive and holistic approaches to sustainable innovation. Ultimately, further investigation is necessary to explore how rules and regulations may effectively facilitate the advancement of the digital circular economy. Although there is ongoing research on the economic and technological components of the circular economy, there remains a dearth of understanding regarding its impact on policy. Future research should investigate the impact of government legislation on supporting environmentally responsible innovation and assuring the adoption of digital technologies in accordance with the concepts of the circular economy.

The existing body of literature on sustainable innovations and the digital economy presents compelling evidence that digital technologies play a crucial role in facilitating and disseminating the adoption of circular economy practices. This is evidenced by the fact that digital technologies play a pivotal role. Despite considerable advancements in harnessing the capabilities of these technologies, there are several obstacles and deficiencies that must be addressed. Future research should prioritise addressing these shortcomings, especially in terms of scalability, interdisciplinary collaboration, and the impact of legislation on promoting sustainability. By implementing this action, there is a potential for digital technologies and environmentally friendly solutions to fully realise the shift to a circular economy.

13.3 Case Study 1: Sustainable Innovations in Manufacturing

13.3.1 Analysis of How Digital Tools Are Used to Reduce Waste and Enhance Resource Efficiency

The manufacturing industry has been in the forefront of utilising digital tools to create sustainable and resource-efficient solutions. The implementation of digital technologies such as the Internet of Things (IoT), artificial intelligence (AI), and big data analytics has facilitated the transformation of traditional industrial processes into circular and environmentally sustainable models. These technologies aid companies in diminishing their environmental footprint, optimising resource utilisation, and minimising waste through enhanced manufacturing efficiency and precision. IoT devices are frequently utilised to monitor and control manufacturing operations in real time. This facilitates the speedy identification of inefficiencies and waste for companies. Smart Internet of Things sensors offer data for the improvement of the production process. These insights are acquired by the measurement of energy use, material utilisation, and machine efficiency. This level of control enhances operational efficiency and minimises manufacturing waste. AI algorithms facilitate predictive maintenance by accurately forecasting equipment malfunctions. This leads to a decrease in the amount of time that a system is not functioning and a decrease in the need for new parts, which in turn improves the efficiency of resources (Mitra et al., 2024). Big data analytics is essential for evaluating enormous datasets derived from IoT devices and other digital technologies. These analytics assist producers in identifying patterns and trends that can be leveraged to decrease energy consumption and minimise material wastage. Big data enables companies to maximise the efficiency of resource utilisation throughout the entire process, from acquiring raw materials to assembling the final product (Jain & Mitra, 2023).

13.3.2 Examples of Companies Adopting Digital Circular Economy Practices in Manufacturing

The efficacy of digital tools in promoting sustainability in the manufacturing sector has been evidenced by the successful implementation of digital circular economy practices by numerous manufacturing firms in recent years. Siemens exemplifies this phenomenon, as it holds a significant position as a worldwide frontrunner in the production of industrial goods. Siemens has initiated a comprehensive digitalisation plan that integrates technologies like the Internet of Things (IoT), artificial intelligence (AI), and digital twins to enhance resource efficiency and reduce waste in its industrial operations. The notion of "Digital Enterprise" allows for the virtual simulation of the entire production process, facilitating real-time monitoring and improvement. Siemens use digital twins, virtual representations of real assets, to simulate different production scenarios and identify the most efficient and sustainable solutions (Siemens, 2023). Siemens is able to minimise waste and preserve resources through this process.

The effectiveness of digital tools in promoting sustainability in the manufacturing sector has been demonstrated by the successful adoption of digital circular economy practices by many manufacturing enterprises in recent years. Siemens is a prominent global leader in the manufacturing of industrial goods, illustrating this tendency. Siemens has implemented a thorough digitalisation strategy that incorporates technologies such as the Internet of Things (IoT), artificial intelligence (AI), and digital twins to improve resource efficiency and minimise waste in its industrial operations. The concept of a "Digital Enterprise" enables the virtual replication of the complete manufacturing process, enabling real-time monitoring and enhancement. Siemens utilises digital twins, which are virtual representations of actual assets, to simulate various production scenarios and determine the most optimal and environmentally friendly solutions (Siemens, 2023). Siemens can effectively reduce waste and conserve resources with this approach.

In recent years, various manufacturing companies have successfully implemented digital circular economy practices, which is evidence of the efficacy of digital tools in promoting sustainability in the manufacturing sector. This success has been demonstrated by the effective deployment of digital technologies. Siemens is a prime example of this phenomena because it is a big player in the manufacture of industrial goods all over the world and occupies a leading position in the competition. For the purpose of improving resource efficiency and cutting down on waste in its manufacturing operations, Siemens has begun a comprehensive digitalisation plan that incorporates technologies such as the Internet of Things (IoT), artificial intelligence (AI), and digital twins. The concept of "Digital Enterprise" makes it possible to simulate the entire manufacturing process virtually, which makes it easier to monitor and optimise the process in real time. According to Siemens (2023), the company uses digital twins, which are virtual representations of real assets, to simulate various production scenarios and determine the solutions that are both the most efficient and the most sustainable at the same time. Using this method, Siemens is able to reduce the amount of trash produced and preserve the resources available.

13.4 Case Study 2: Digital Circular Economy in the Agricultural Sector

13.4.1 Exploration of Sustainable Agricultural Practices Enabled by Digital Technologies

In order to implement ecologically responsible practices that foster circularity and decrease the impact on the environment, the use of digital technology has become increasingly widespread in the agricultural industry. They are being used to implement these practices. Examples of digital tools that are revolutionising traditional agricultural techniques include blockchain, Internet of Things, and precision agriculture. These tools are maximising the utilisation of resources, increasing production, and reducing the amount of resources that are wasted.

Precision agriculture provides farmers with the ability to monitor the health of their crops, the conditions of the soil, and the weather patterns in real time. This is accomplished through the utilisation of sensors connected to the Internet of Things (IoT), drones, and satellite photography. With the help of precise agriculture, farmers are able to apply the necessary amount of water, fertilisers, and pesticides at the appropriate time. This is made possible by the availability of correct data on the variables in question. The result of this is that farmers are able to eliminate waste and lessen the negative effects that agricultural practices have on the surrounding ecosystem. Internet of Things sensors, for example, are able to monitor the amount of moisture that is present in the soil and activate irrigation systems only when it is required. This allows for the reduction of the amount of water that is utilised and the prevention of over-irrigation (Jain & Mitra, 2024a, b, c, d).

Furthermore, blockchain technology is playing a significant role in the development of environmentally friendly agricultural practices, which is a significant development. Through the creation of a distributed and transparent ledger within the network, blockchain technology makes it possible to monitor the journey of agricultural products from the farm to the consumer's table. As a result of this traceability, products are assured to be procured and processed in a sustainable manner, which contributes to the promotion of circularity in the agricultural sector. The deployment of blockchain technology, for instance, can be used to monitor the application of organic pesticides and fertilisers, which will ensure that agricultural practices are in conformity with the requirements for sustainability. According to Sachdeva and Mitra (2023), blockchain technology has the potential to make it easier for farmers to share information and resources with one another. This would result in increased collaboration and a reduction in the amount of work that is repeated over and over again.

13.4.2 Impact of Digital Innovations on Reducing Environmental Impact and Promoting Circularity in Agriculture

A considerable influence has been made by digital advances in agriculture, which have contributed to the reduction of environmental damage and the promotion of circularity. In example, the decrease of resource waste, notably in terms of water, energy, and fertilisers, is one of the most important advantages brought about by these technologies. Through the optimisation of resource utilisation, digital technologies contribute to the reduction of the carbon footprint of agricultural operations and the overall influence that farming has on the environment (environmental impact). When it comes to water use in agriculture, for instance, the implementation of Internet of Things (IoT) sensors and precision agriculture techniques has resulted in significant reductions. Farmers in areas where water shortage is a serious concern have been able to conserve water resources while maintaining or even increasing agricultural yields thanks to the technologies that have been developed. By optimising the use of fertilisers and pesticides, precision agriculture minimises the amount of toxic chemicals that run off into water bodies, hence reducing the negative impact that farming methods have on the environment (Mitra & Das, 2023).

One more significant result of digital advancements is the encouragement of circularity in agricultural practices of all kinds. The implementation of circular supply networks in agriculture has been made possible by the development of digital platforms and blockchain technology. These supply chains aim to reduce waste while maximising the reuse and recycling of resources. For instance, agricultural waste such as crop leftovers and animal manure can be transformed into bioenergy or organic fertilisers. This eliminates the requirement for synthetic inputs and brings the production cycle of agriculture full circle. Furthermore, digital platforms that connect farmers with consumers and other stakeholders encourage the exchange of resources and expertise, which in turn fosters an agricultural ecosystem that is more collaborative and sustainable (Sachdeva et al., 2024).

13.5 Case Study 3: Circular Economy in the Energy Sector

13.5.1 Examination of Sustainable Energy Innovations and Their Digital Integration

There have been considerable breakthroughs in environmentally friendly developments within the energy sector, which have been driven in large part by the use of digital technologies. The use of digital tools that optimise energy output, distribution, and consumption is becoming increasingly prevalent in the process of harnessing renewable energy sources such as solar, wind, and hydropower. This move to a circular economy in the energy sector, where the emphasis is placed on lowering reliance on non-renewable resources and minimising waste, is made possible by these technologies, which are essential for the transformation.

The creation of smart grids is one of the most important digital advances that have occurred in the space of energy. Using Internet of Things (IoT) sensors, artificial intelligence (AI), and big data analytics, smart grids are able to monitor and manage energy flows in real time, thereby optimising the equilibrium between energy supply and demand. Because of this optimisation, energy waste is reduced, and the efficiency of energy systems is improved. As a result, it is possible to include a greater proportion of renewable energy sources into the grid. In addition, smart grids make it possible to decentralise the production of energy. This enables users to generate their own energy through the use of solar panels or wind turbines, and then feed any extra energy back into the grid (European Commission, 2020).

Optimisation of the operation and maintenance of renewable energy systems is also being accomplished through the use of artificial intelligence and machine learning techniques. These technologies have the ability to forecast the breakdown of equipment, optimise energy output based on the circumstances of the weather, and ensure that energy storage systems are operating at their maximum efficiency. According to Mitra and Lohochoudhury (2019), artificial intelligence has the capability to analyse past weather data and make predictions regarding the ideal times for solar panel operation. This allows for both the maximisation of energy generation and the reduction of the requirement for backup power from non-renewable sources.

13.5.2 Role of Digital Platforms in Managing and Optimising Renewable Energy Resources

In order to facilitate the incorporation of renewable energy resources into the larger energy system, digital platforms play a crucial role in the management and optimisation of these resources. Through the provision of the appropriate tools and infrastructure, these platforms enable energy producers, consumers, and grid operators to collaborate and coordinate their operations, thereby contributing to the development of an energy system that is both more efficient and more sustainable.

The utilisation of blockchain technology for the purpose of managing energy transactions is an example of a digital platform that is utilised in the energy sector. The use of blockchain technology enables the formation of decentralised energy markets by providing a method that is both transparent and safe for recording and verifying any energy transactions. Through the use of these markets, users are able to buy and sell extra energy directly with one another, so avoiding the influence of conventional energy utilities and lowering the requirement for large-scale energy infrastructure. This decentralisation encourages the utilisation of renewable energy sources since it makes it possible to directly integrate distributed energy resources like rooftop solar panels and small-scale wind turbines (Saberi et al., 2019). Decentralisation also makes it possible to integrate these resources.

As another illustration, the utilisation of digital platforms to optimise the storage and distribution of energy is an example. Batteries and other energy storage devices are extremely important in maintaining a healthy equilibrium between the supply and demand of energy, particularly in the context of integrating intermittent renewable energy sources such as solar and wind. The ability of digital platforms to monitor the levels of energy storage in real time allows for the optimisation of the charging and discharging of batteries, which in turn ensures a consistent supply of energy. These platforms also make it possible to aggregate distributed energy resources, which enables these resources to be utilised collectively for the purpose of providing grid services such as frequency regulation and peak load shaving (Belkhir & Elmeligi, 2018).

It can be concluded that digital technologies are driving significant breakthroughs in the manufacturing, agriculture, and energy sectors. These innovations are also boosting the adoption of circular economy principles and enhancing sustainability. These case studies demonstrate the potential of digital tools to lessen waste, maximise the utilisation of resources, and incorporate renewable energy sources, all of which contribute to an economy that is more sustainable and circular.

13.6 Discussion

The transition towards a digital circular economy, as evidenced by the case studies in manufacturing, agriculture, and the energy sector, is not only a technological shift but also a profound transformation in how industries approach sustainability and resource management. This discussion delves into the implications of these findings, exploring how digital tools and sustainable innovations are reshaping industry practices, the challenges faced in implementing these changes, and the broader impact on society and the environment.

13.6.1 Synthesis of Findings

The case studies demonstrate that digital tools played a significant role in aiding the circular economy by optimising the use of resources, decreasing waste, and promoting sustainable practices across a variety of industries. With the integration of Internet of Things (IoT), artificial intelligence (AI), and big data analytics, manufacturing organisations have been able to monitor and optimise production processes in real time, which has led to considerable reductions in the amount of resources that are wasted and the amount of energy that is consumed. The implementation of digital twins, which can be observed in Siemens' operations, is an illustration of how virtual simulations may forecast and prevent inefficiencies, hence prolonging the lives of items and machinery (Siemens, 2023). By adhering to the ideas of the circular economy, which emphasise the need of preserving the highest possible utility of products and resources for the longest period of time possible, this strategy is in harmony with those ideals.

Both blockchain technology and precision agriculture have emerged as powerful instruments for improving sustainability in the agricultural sector: precision agriculture and blockchain technology. The use of precision agriculture enables the focused application of resources such as water and fertilisers, hence reducing the amount of waste produced and the negative impact on the environment. Blockchain technology, on the other hand, ensures transparency and traceability in supply chains. This helps to promote sustainable farming practices by providing customers and regulators with the capacity to verify the origins of agricultural products as well as their sustainability credentials (Sachdeva & Mitra, 2023). These technologies not only enhance productivity, but they also cultivate trust and accountability, both of which are necessary for the widespread adoption of environmentally responsible agricultural methods.

Innovations in digital technology have also been beneficial to the energy sector, particularly in the management and optimisation of renewable energy supplies. Smart grids and energy management systems that are driven by artificial intelligence make it possible to integrate renewable energy sources into the grid in a manner that is both more efficient and reliable. This helps to reduce reliance on fossil fuels and minimise energy waste. As can be observed in the evolving models of peer-to-peer energy trading (European Commission, 2020), the utilisation of digital platforms to manage energy transactions and storage provides additional support for the circular economy. This is accomplished by facilitating the decentralised production and consumption of energy. The potential of digital technology to transform energy systems into infrastructures that are more sustainable and resilient is brought to light by these breakthroughs.

13.6.2 Challenges and Barriers

Although there is a tremendous potential for digital technologies to drive sustainable behaviours, there are a number of obstacles and constraints that need to be solved before the benefits of these technologies can be completely realised. One of the most major obstacles is the substantial amount of money that must be invested in developing digital infrastructure and technology. It is possible that the expense of adopting advanced digital technologies is prohibitive for a significant number of small and medium-sized organisations (SMEs) as well as businesses located in developing nations. Due to the existence of this cost barrier, the general adoption of these technologies is restricted, particularly in industries that are already experiencing resource constraints. With regard to the implementation of digital twins and complex Internet of Things systems, for instance, while large firms such as Siemens and General Electric are able to afford to do so, smaller manufacturers may have difficulty justifying the initial expenses, despite the fact that the long-term benefits are substantial (General Electric, 2023).

One further key obstacle is the difficulty of incorporating digital technology into the business models and procedures that are already in place of the company. In order to successfully transition to a digital circular economy, organisations frequently need to undergo a fundamental rethinking of their operations, ranging from the management of supply chains to the engagement of customers. There is a possibility that stakeholders who are accustomed to conventional linear models of production and consumption would express opposition to this shift. In addition, the incorporation of digital tools necessitates the presence of a competent labour force that is able to successfully manage and utilise these technologies. According to Mitra et al. (2024), the lack of digital literacy and technical skill among workers can be a barrier to the adoption of sustainable innovations. This is especially true in industries that have been sluggish to embrace digital transformation.

There is also a substantial obstacle presented by environmental issues that are associated with the implementation of digital technologies. Although digital technologies have the potential to improve sustainability, if they are not handled appropriately, they can also contribute to the damage of the environment. There is a possibility that the manufacturing, utilisation, and disposal of Internet of Things (IoT) devices, data centres, and blockchain networks can result in significant amounts of energy consumption and electronic waste, which can counteract some of the environmental benefits that these technologies are expected to offer. According to Belkhir and Elmeligi (2018), this paradox highlights the importance of giving careful consideration to the environmental footprint that digital innovations leave behind and developing strategies to mitigate the impact that these innovations have. Some examples of such strategies include the utilisation of renewable energy sources to power data centres and the promotion of recycling electronic devices.

13.6.3 Opportunities and Future Directions

There are various potential for innovation, growth, and sustainability that are presented by the digital circular economy, despite the challenges that it offers. The potential for digital technology to generate new business models that are in accordance with the principles of the circular economy is one of the most significant opportunities that exists today. Take, for instance, the proliferation of product-as-a-service models, which are made possible by digital platforms. These models enable businesses to keep ownership of their products while simultaneously generating income through service agreements rather than through one-time sales. Through the use of this approach, manufacturers are encouraged to build goods that are long-lasting, repairable, and upgradeable, hence decreasing waste and prolonging the lifecycles of products. It is expected that organisations that implement such approaches will gain a competitive edge (Tukker, 2015). This is because consumers are increasingly placing a higher priority on sustainability. One further possibility resides in the potential for digital platforms to encourage collaboration and the exchange of information among the various parties involved in the circular economy environment. Blockchain technology and the Internet of Things are examples of digital tools that not only enhance transparency and traceability but also make it easier for supply chains to share information and resources with one another. It is possible that this partnership will result in the creation of novel solutions to complicated sustainability challenges. Some examples of these challenges include minimising the carbon footprint of supply chains and making the most of utilisation of renewable energy resources. Digital platforms have the potential to expedite the transition to a circular economy and drive systemic change across industries (Saberi et al., 2019). This is accomplished by utilising the collective intelligence of stakeholders using digital platforms.

Consumers are becoming more conscious of environmental and social issues, which gives a chance for firms to differentiate themselves through the implementation of sustainable practices. There has been an increase in the demand for products and services that are both sustainable and ethical as a result of customers becoming more knowledgeable about the impact that their shopping decisions have on the environment. The capacity of businesses to demonstrate their commitment to sustainability by utilising supply chains that are both transparent and traceable, with the assistance of digital technology, is likely to result in the development of stronger relationships with customers and an improvement in the reputation of their brand. According to Jain and Mitra (2024a, b, c, d), it is anticipated that this trend towards conscientious consumerism would foster additional innovation in environmentally responsible corporate practices and the implementation of principles of circular economies. In the future, there is a large potential for research that bridges the fields of digital technology, sustainability, and circular economy. This study might be conducted by researchers from many disciplines. The present body of research is mostly compartmentalised, with only a limited amount of integration of ideas across different fields. In spite of this, resolving the multifaceted problems associated with sustainability calls for an all-encompassing strategy that takes into account the dynamic relationship that exists between technology advancement, economic models, and the influence on the environment. It is recommended that future research concentrate on the development of comprehensive frameworks that integrate digital technologies with the principles of circular economies. These frameworks should provide enterprises, policymakers, and other stakeholders with insights that can be put into action. The potential of future technologies, such as artificial intelligence and blockchain, to drive sustainability in new and novel ways could also be investigated through this type of research (Lieder & Rashid, 2016).

13.6.4 Broader Implications

The broader ramifications of the digital circular economy extend beyond the scope of specific industries, having an effect on society as well as the environment on a global scale. The digital circular economy contributes to the achievement of several

Sustainable Development Goals (SDGs) set forth by the United Nations. These SDGs include responsible consumption and production (SDG 12), climate action (SDG 13), and sustainable cities and communities (SDG 11) (United Nations, 2015). The digital circular economy does this by promoting resource efficiency, reducing waste, and lowering environmental impact. Increasing economic resilience can also be accomplished through the implementation of digital circular economy principles. This is accomplished by decreasing reliance on limited resources and generating new opportunities for growth and innovation. Furthermore, the shift to a digital circular economy has the potential to address some of the most important environmental concerns of our day, such as the loss of biodiversity, the threat of climate change, and the shortage of resources. Changing away from linear models of production and consumption allows industries to lessen their impact on the environment by lowering their carbon footprint, minimising the depletion of resources, and fostering the regeneration of natural systems. For the purpose of establishing a future that is more sustainable and egalitarian, in which economic growth is decoupled from environmental deterioration, this shift in systemic understanding is vital.

On the other hand, the success of the digital circular economy is contingent upon the concerted efforts of consumers, corporations, and governments. One of the most important roles that policymakers play is in the process of developing the regulatory framework and incentives that are required to encourage the implementation of circular economy principles. Among these are the establishment of standards for the efficient use of resources, the promotion of the utilisation of renewable energy sources, and the encouragement of the development of digital infrastructure. While customers need to adopt more sustainable lifestyles and demand responsibility from the businesses they support, businesses need to take the initiative to innovate and implement sustainable practices. Furthermore, businesses need to take the lead in implementing sustainable practices.

13.7 Conclusion

The investigation of environmentally friendly technologies within the context of a digital circular economy indicates a transformational potential that spans across a variety of industries, such as the manufacturing industry, agriculture, and the energy sector. Not only has the incorporation of digital technologies like the Internet of Things (IoT), artificial intelligence (AI), blockchain, and big data analytics into these industries led to an increase in efficiency and a decrease in waste, but it has also opened the way for an economic model that is more sustainable and robust. Industries are increasingly able to dissociate economic growth from environmental deterioration as a result of their transition from linear to circular economic practices. This allows them to solve some of the most important concerns of our day, including as the depletion of resources, the escalation of climate change, and the pollution of the environment. Significant improvements in resource efficiency, waste reduction, and general sustainability have been brought about in the manufacturing industry as a result of the implementation of digital tools. In order to illustrate how digital twins, predictive maintenance, and sophisticated analytics can change conventional production processes into models that are more environmentally friendly and circular, companies such as Siemens and General Electric have demonstrated their capabilities. In a similar vein, digital technologies in agriculture, such as precision agriculture and blockchain, have made it possible to implement more environmentally friendly farming practices, hence lowering the impact that agriculture has on the environment and fostering circularity in the agricultural supply chain.

The incorporation of digital technology has also been beneficial to the energy sector. Smart grids, energy management systems driven by artificial intelligence, and digital platforms have all contributed to the effective utilisation and distribution of renewable energy resources. The transition to a more decentralised and sustainable energy system is supported by these developments. This system is one in which energy generation and consumption are optimised to eliminate waste and dependence on resources that do not replenish themselves or replenish themselves. The move to a digital circular economy, on the other hand, is not without its difficulties. One of the most significant obstacles to the widespread adoption of these technologies is the presence of significant financial barriers, particularly for small and medium-sized enterprises (SMEs) and firms located in developing nations. In addition, the incorporation of digital tools into pre-existing business models necessitates a substantial amount of organisational change as well as the acquisition of new skills and experience. The environmental concerns that are associated with the development and disposal of digital technologies also need to be properly handled in order to guarantee that the positive effects that digital advances have on the environment are not counterbalanced by the negative effects that they have.

The opportunities that are given by the digital circular economy are enormous, notwithstanding the problems that will be encountered. A large amount of opportunity exists for the development of new business models, an increase in collaboration, and the establishment of value chains that are sustainable. There is a high probability that the adoption of circular economy practices, which are backed by digital technology, will accelerate as firms, policymakers, and consumers become more aware of the significance of sustainability. The implementation of this change not only has the potential to improve economic resilience, but it also makes a contribution to the overarching objectives of sustainable development. This will ensure that future generations will be able to flourish in a world where economic progress is balanced with environmental stewardship. I would want to conclude that the digital circular economy is an important step towards a future that is both more sustainable and more equitable. By integrating digital advances and sustainable practices, industries have the ability to lessen their impact on the environment, maximise the utilisation of resources, and generate new opportunities for growth and innovation. The success of this transformation, on the other hand, is contingent upon the concerted efforts of all stakeholders, including enterprises, governments, and consumers, working together to overcome obstacles and make the most of opportunities given by the digital circular economy. The ideas of circularity and sustainability will play an increasingly crucial role in moulding the future of companies and society as a whole as the global economy continues to undergo further transformations.

References

Antikainen, M., Uusitalo, T., & Kivikyto-Reponen, P. (2018). Digitalization as an enabler of circular economy. *Procedia CIRP, 73*, 45–49. https://doi.org/10.1016/j.procir.2018.04.027

Belkhir, L., & Elmeligi, A. (2018). Assessing ICT global emissions footprint: Trends to 2040 and recommendations. *Journal of Cleaner Production, 177*, 448–463. https://doi.org/10.1016/j.jclepro.2017.12.239

Bocken, N. M., de Pauw, I., Bakker, C., & van der Grinten, B. (2016). Product design and business model strategies for a circular economy. *Journal of Industrial and Production Engineering, 33*(5), 308–320. https://doi.org/10.1080/21681015.2016.1172124

Bonina, C., Koskinen, K., Eaton, B., Gawer, A., & Badinelli, R. (2022). Governing data for the public interest: Challenges and opportunities for smart cities. *Journal of Modelling in Management, 17*(2), 460–475. https://doi.org/10.1108/JM2-06-2021-0145

Das, S., & Mitra, A. (2024). Advancing lightweight digital trust architectures in the internet of medical things: A multi-dimensional analysis. In *Lightweight digital trust architectures in the internet of medical things (IoMT)* (pp. 1–14).

Das, S., & Mitra, A. (2024). Balancing the scale: Ethical marketing in the age of big data—A comprehensive analysis of data governance standards and the role of effective technology. In *Ethical marketing through data governance standards and effective technology* (pp. 50–61).

Ellen MacArthur Foundation. (2013). *Towards the circular economy: Economic and business rationale for an accelerated transition.* Ellen MacArthur Foundation.

Esposito, M., Tse, T., & Soufani, K. (2018). Introducing a circular economy: New thinking with new managerial and policy implications. *California Management Review, 60*(3), 5–19. https://doi.org/10.1177/0008125618764691

European Commission. (2020). *A new circular economy action plan for a cleaner and more competitive Europe.* European Commission.

Galli, A., Piscitelli, M. S., Moscato, V., & Capozzoli, A. (2022). Bridging the gap between complexity and interpretability of a data analytics-based process for benchmarking energy performance of buildings. *Expert Systems with Applications, 199*, 117649. https://doi.org/10.1016/j.eswa.2022.117649

Geissdoerfer, M., Savaget, P., Bocken, N. M., & Hultink, E. J. (2017). The circular economy–A new sustainability paradigm? *Journal of Cleaner Production, 143*, 757–768. https://doi.org/10.1016/j.jclepro.2016.12.048

General Electric. (2023). *Predix: The industrial IoT platform for digital transformation.* Retrieved from https://www.ge.com/digital/predix

Jain, V., & Mitra, A. (2023). Development and application of machine learning algorithms for sentiment analysis in digital manufacturing: A pathway for enhanced customer feedback. In *Emerging technologies in digital manufacturing and smart factories* (pp. 26–38).

Jain, V., & Mitra, A. (2024a). A critical examination of ethical implications in AI-Driven consumer behaviour media discourse and environmental sustainability. In *Enhancing and predicting digital consumer behavior with AI* (pp. 1–16).

Jain, V., & Mitra, A. (2024b). Case studies and practical implementation of biomimicry in bio-inspired optimization. In *Bio-inspired intelligence for smart decision-making* (pp. 85–100).

Jain, V., & Mitra, A. (2024c). Culinary narratives and community revitalization: The role of media in promoting sustainable gastronomy tourism. In *Promoting sustainable gastronomy tourism and community development* (pp. 67–80).

Jain, V., & Mitra, A. (2024d). Energy harvesting system design and analysis using biomimicry: A sustainable approach for renewable energy applications. In *Emerging materials technologies and solutions for energy harvesting* (pp. 234–257).

Khan, A., Baharul Islam, K. M., & Mitra, A. (2019). Exploring the status of community information and training for disaster preparation and mitigation practices: An appraisal of 2013 flash flood in Uttarakhand. *International Journal of Emergency Management, 15*(2), 147–165.

Kumar, V., Subramanian, R., & Strandhagen, J. O. (2018). Circular economy in India: Current practices and future opportunities. *Journal of Industrial Ecology, 22*(1), 199–210. https://doi.org/10.1111/jiec.12741

Lieder, M., & Rashid, A. (2016). Towards circular economy implementation: A comprehensive review in context of manufacturing industry. *Journal of Cleaner Production, 115*, 36–51. https://doi.org/10.1016/j.jclepro.2015.12.042

McKinsey Global Institute. (2017). *Artificial intelligence: The next digital frontier?* McKinsey & Company.

Mitra, A., & Das, S. (2023). SME and environmental sustenance: Digital marketing in SMEs via data-driven strategies. In *Sustainability, green management, and performance of SMEs* (pp 315–332). https://doi.org/10.1515/9783111170022-017

Mitra, A., & Das, S. (2024). The rise of digital celebrities: Analyzing the impact and sustainability of stardom on OTT platforms. In *Exploring the impact of OTT media on global societies* (pp. 55–73).

Mitra, A., & Lohochoudhury, B. (2019). Identifying anthropogenic factors of groundwater pollution through students' opinion in rural West Bengal. *Anthropogenic Pollution, 3*(2), 49–59.

Mitra, A., Mitra, A., & Das, S. (2024). The impact of IoT on the environment: A descriptive study in India. In *intelligent decision support system for IoT-enabling technologies: Opportunities, challenges and applications* (pp. 35–58).

Niero, M., & Olsen, S. I. (2016). Circular economy: To be or not to be in a closed product loop? A life cycle assessment of aluminium cans with inclusion of alloying elements. *Resources, Conservation and Recycling, 114*, 18–31. https://doi.org/10.1016/j.resconrec.2016.06.023

Porter, M. E., & Heppelmann, J. E. (2014). How smart, connected products are transforming competition. *Harvard Business Review, 92*(11), 64–88.

Rachna, R., Akshay, J., & Bijit, S. (2021). Exploring the role of digital twins in sustainable circular economy. *Informatics, 5*(4), 62. https://doi.org/10.3390/informatics5040062

Saberi, S., Kouhizadeh, M., Sarkis, J., & Shen, L. (2019). Blockchain technology and its relationships to sustainable supply chain management. *International Journal of Production Research, 57*(7), 2117–2135. https://doi.org/10.1080/00207543.2018.1533261

Sachdeva, P., & Mitra, A. (2023). SME and environmental sustenance: Digital marketing in SMEs via data-driven strategies. In *Sustainability, green management and performance of SMEs* (pp. 315–332).

Sachdeva, P., Kumar, M. D., & Mitra, A. (2024). Green discourse analysis on Twitter: Imperatives to green product management in sustainable cities (SDG11). In *Studies in systems, decision and control* (Vol. 487, pp. 245–256). https://doi.org/10.1007/978-3-031-35828-9_22

Siemens. (2023). *Digital enterprise: The digital transformation of industry.* Retrieved from https://www.siemens.com/global/en/home/company/topic-areas/digital-enterprise.html

Stahel, W. R. (2016). The circular economy. *Nature, 531*(7595), 435–438. https://doi.org/10.1038/531435a

Tukker, A. (2015). Product services for a resource-efficient and circular economy–a review. *Journal of Cleaner Production, 97*, 76–91. https://doi.org/10.1016/j.jclepro.2013.11.049

Unilever. (2023). Annual report and accounts. Retrieved January 28, 2024, from https://www.unilever.com/investors/annual-report-and-accounts/

United Nations. (1987). *Our common future.* United Nations.

United Nations. (2015). *Transforming our world: The 2030 agenda for sustainable development.* United Nations.

Chapter 14
Technology for Sustainability: Barriers to Digital Circular Economy Adoption

Seema Rani, Rubee Singh, and Akshay Singh

Abstract The circular economy is a system of managing production processes that aims at minimizing the use of resources and promote their reuse while considering implications for economy, society, and environment simultaneously. It is a kind of sustainable measure which can be adopted by the manufacturers that harvest natural resources and dispose of waste. These tactics may be optimized by using the digital technologies (DTs) such as blockchain or big data which is then known as digital circular economy (DCE). This technology-based CE facilitates failure prediction and preventive maintenance, techniques aimed at enhancing resource efficiency. Extant research remained concentrated on the initial phases of the CE transition, including the identification of opportunities and the building of business models. Discussions concerning the barriers in the way of implementing of CE, specifically the DCE, are infrequent. Nonetheless, to achieve success in the shift, identifying the barriers is as significant as conceptual and technological advancements. This research adopts a comprehensive perspective on identifying barriers related to DCE adoption. The integration of DTs create certain situations that hinder the adoption of DCE such as costs, data safety, limited knowledge, lack of policy support, organizational culture, change management, technological issues, etc. This subject has garnered significant academic interest; however, there consensus remains elusive because the findings from the publications are largely non-generalizable. Consequently, this research presents a review of prior investigations to establish a full understanding of the barriers to the DCE for offering implications to the organizations which are implementing DCE-based practices.

S. Rani
Department of Commerce, Kurukshetra University, Kurukshetra, India

R. Singh (✉)
Department of Management Studies, Kumaun University, Nainital, Uttarakhand, India
e-mail: dr.rubeerajput@gmail.com

Institute of Business Management, GLA University, Mathura, India

A. Singh
School of Management Studies, Indra Gandhi National Open University, New Delhi, India

R. Singh and V. Kumar (eds.), *Sustainable Innovations and Digital Circular Economy*,
https://doi.org/10.1007/978-981-96-1064-8_14

Keywords Digital circular economy · Barriers · Industry 4.0 · Digital technology · Sustainability

14.1 Introduction

The shift towards a circular economy (CE) has increasingly emerged as an important objective for organizations worldwide. The CE is nowadays is the most appropriate alternate for the "take–make–waste" approach of linear system, because it focuses on maintaining material utility, minimizing waste, and decreasing pollution (Singh et al., 2023). This system consequently substitutes the 'end-of-life' paradigm with the principles of reduction, reuse, recycling, and recovery. Organizations need transition to a CE-based system rather than the linear model of economy; yet, challenges such as organizational factors, data unavailability, and integration frequently hinder this transformation (Kandasamy et al., 2023; Trevisan et al., 2023). Therefore, scholars contend that the transformation of the CE is intrinsically linked to digitalization technologies (DTs) propounded by Industry 4.0 (I4.0) (Ajwani-Ramchandani et al., 2021; Chauhan et al., 2019), encompassing the proficient use of big data (BD), artificial intelligence (AI), blockchain technology (BT), and the Internet of Things (IoT). IoT technology gathers substantial quantities of real-time data based on circular practices (Brous et al., 2020). The CE-based business models are transforming through DTs. It made communication possible through internet among items which collect and generate data based on the use, scenario, as well as connections with various things and persons (Kortuem et al., 2010). Energy sector can be transformed with BT with its potential applications in the identification of consumption patterns, advanced metering, managing prepaid payment systems, the use of AI and machine learning and micro transactions, to facilitate the provision of distinctive energy products, along with enhanced value in marketing domains (Kouhizadeh et al., 2020). Consequently, scholars concur that the implementation of the CE is intrinsically associated with digitization, as it enhances the capabilities to forecast, track, and monitoring across the whole life cycle of product for enterprises (Chauhan et al., 2019).

Numerous studies have identified digitalization as a catalyst for the conversion to a CE for various reasons. These DTs can convert theoretical CE principles into viable and practical actions (Antikainen et al., 2018). From CEs, the implementation of innovative technologies yields quantifiable advantages. Technologies can enhance workers' skills and capacities, hence facilitating their capacity to make circularity a part of the routine decisions (Kintscher et al., 2020). Utilizing data-driven insights in circular design can enhance sustainability of goods by optimized use of resources (Garcia-Muina et al., 2018). Insights from machine learning through its prescriptive and predictive features can be useful in CE context for optimization of not only the raw materials and products but the related processes also (Bressanelli et al., 2018). Similarly, better inventory management is possible by accurate demand estimations through historical and real-time data, thereby reducing waste and improving sustainability in operations. DTs can also reduce waste by evaluating optimal techniques for remanufacturing and recycling. AI-driven picture identification can enhance e-waste recycling (Wilts et al., 2021). Consequently, researchers propose that DTs, including AI, are linked to systems that execute tasks that could be performed by only humans earlier and can furnish organizations with essential assistance to adopt CE principles.

A system that employs DTs to deliver intelligent functionalities for executing the circular initiatives has been referred as "digital circular economy (DCE)" of "smart circular economy" (Ghisellini et al., 2016). The utilization of various DTs, including AI and big data, can disrupt linear business paradigms by facilitating mass personalization, enabling enterprises to choose sustainable inputs that align with client needs (Kristoffersen et al., 2020). These technologies facilitate the prolongation of product lifespan by establishing predictive maintenance protocols, consequently improving customer experience and minimizing consumer waste (Rosa et al., 2020). Companies advancing towards digitalization can various prospects in production, distribution, by product management, and recovery of waste by enhancing visibility throughout all phases of the supply chain (Wang et al., 2020). The extraordinary integration of technology in this shift may enhance the sustainability at different level of economies along with productivity (Ingemarsdotter et al., 2019). Nonetheless, there are certain barriers in the adoption of DCE which is the merging of circular economy with digitalization that necessitate scholarly research. Integration of technology in CE context is prone to difficulties, and the advancement may not be consistent across industries (Ada et al., 2021). Managers can leverage specific enablers to achieve the expected improvements in circular performance through the utilization of technologies.

The integration of DTs into CE models, facilitated by smart and interconnected devices utilizing sensors for generating, processing, and analyzing data, will contribute to the formulation of strategies and the advancement of social processes (Rosa et al., 2020; Wang et al., 2020). Clearly, the implementation of the CE necessitates that companies transform their business strategies in light of I4.0 (Kristoffersen et al., 2020). The intricacies of transition have resulted in an increase in the research conducted on the convergence of CE with digitalization. Nonetheless, their results remain disjointed throughout multiple scientific fields. Consequently, this presents an opportunity and the appropriate timing to perform a thorough literature review and acquire a complete understanding of the principal barriers in the process of DCE implementation. Furthermore, the potential afforded by digitalization within the framework of the CE are many, necessitating the focus of diverse stakeholders, including the government, academicians, legislators, practitioners, managers, and the whole society. The domain has recently experienced an increase in scientific study. A literature review can facilitate the integration of findings from various studies and offer a thorough understanding of the significant barriers and offer implications for smooth transition from liner to DCE-based system. The chapter is organized in the following manner. First of all, general review of the concept of CE and DCE is provided. In the same section barriers related to the adoption of DCE are discussed thoroughly. Then the research methodology employed for the chapter is given. Next, study's implications are presented as well as the conclusion is given, which delineates the limitations and prospective avenues for further investigation.

14.2 Circular Economy

The three basic principles of circular economy (CE) on which it is founded include: (a) preservation and enhancement of the nature's resources by balancing the flow of renewable resource; (b) optimization of the value derived from the use of such resources through circulating the material as well as products or components to derive maximum utility during their life cycles; (c) enhancing the effectiveness of a system by identification and elimination of negative elements (Ghisellini et al., 2016). The CE focuses on restoration, renewal, and revitalization through the amalgamation of natural processes by societal and human endeavors. CE serves as the connective factor between an economy and environment, as noted by Cautitanu et al. (2018), ultimately fostering sustainability. CE is viewed to make minimal pollution, great efficiency, and low energy consumption. Various frameworks are established and developed for elucidating circularity and its associated measures. Centobelli et al. (2020) emphasized the ability of corporations to establish economic models for sustainable development that minimize the use of natural resource and protect the environment. Bianchini et al. (2019) examined various circular models based on different parameters including: (a) clarity of comprehension, (b) alignment with actual scenarios, (c) effective depiction of circular measures, (d) measurement of the circularity of an initiative and its adaptability a product, and (e) industrial sectors.

It has been asserted that the interpretation and application of the CE may vary across individuals, and this discourse frequently engenders uncertainty regarding the exact concept of the CE (Kirchherr et al., 2017). The CE has evolved over time and is grounded in various domains; countless researchers from diverse disciplines have contributed literature on the CE, shaping our comprehension of the concept. Garcés-Ayerbe et al. (2019) emphasized that the CE is innovative and still emerging as a concept. It remained initially centered on waste management along with considerations related to the product usage after its useful life is over, which is presently expanded to encompass a broad spectrum encompassing extension of useful life of a product and their circular designing. In contrast to the linear economy following the "cradle-to-grave approach," the CE prioritizes a "cradle-to-cradle approach" (Nascimento et al., 2019). CE advocates for cyclical thinking; rather than adopting a normal view of the value-added chain, i.e. open-ended system, through the reduced use of material. Ghisellini et al. (2016) believe that the growing worldwide focus on the CE, through the promotion of a closed-loop system, can enhance the efficiency of resources. Centobelli et al. (2020) stated that the CE transcends ecological emphasis and explicitly emphasizes the economic potential that may be realized from it while Singh et al. (2022) emphasized how social performance of an organization can be enhanced through increased circular activities with the support of human resources.

14.3 Digital Circular Economy

Digitalization enhances the interaction and coordination among different participants in the CE model. Through the exchange of data and knowledge, various stakeholders can collaboratively establish a highly effective and structured circular system (Neligan et al., 2023). Another paradigm for comprehending the function of DT in facilitating CE practices has been proposed by Liu et al. (2022), which is structured around two principal aspects. First is digital functions and second is digital mechanisms. The four digital functions for facilitating CE practices are tracking, marketing, designing, and control. These functions are subdivided into distinct mechanisms, including IoT, AI, BD, BT, and augmented reality. Neri et al. (2023) assert that within digital technologies, the IoT and BD stand out as the most promising, as they significantly facilitating the adoption of diverse CE methods.

CE models utilizing IoT data acquisition can be developed for assessment during the life cycle of specific devices, including electronic gadgets such as smartphones (Alcayaga et al., 2019). Smart and connected goods provide the monitoring of extensive functionalities and the product's external environment via sensors or other means (Han et al., 2020). Through the aggregation of data throughout the use phase, organizations can perpetually enhance their product designs; for instance, by augmenting durability, items incorporating digital components can readily upgrade their functionalities, hence extending their lifespan (Bressanelli et al., 2018). Data regarding the usage of product can enhance strategies related to recovery, recycling, and use. The capacity for data collection has predominantly facilitated the IoT application in endorsing the CE model while developing and implementing the systems (Lin et al., 2019). Atif (2023) examined the function of I4.0-based platforms for sharing data in facilitating service systems for various products within the context of the CE. The data generated through such utilized product can enhance recovery options, including recycling and remanufacturing (Khan et al., 2022). The inspection and testing of items entering the remanufacturing process can be enhanced by access to original design as well history of repair or refurbishment.

Big data (BD) is also gaining prominence due to its capacity to provide the substantial information regarding the quantities of waste material in the final stages of the supply chain especially in reference to food industry (Beretta et al., 2017). This analysis and subsequent optimization in decision-making processes related to food wastage in future can enable supply chains to minimize losses, resulting in substantial economic and social effects (Iqbal et al., 2020). The manufacturing industry is enhanced through BD and the IoT, as information regarding production, material resource consumption, product delivery, purchases, and optimal products for customers in particular regions is readily accessible. This results in the formulation of a strategy for recyclable packing material, limiting waste, determining which materials to procure or avoid, identifying materials availability after product's life ends, tackling challenges about logistics, and establishing targets for waste reduction; it is anticipated that such practices can be restructured to enhance performance with CE models (Awan et al., 2021).

Yildizbasi (2021) asserts that blockchain technology (BT) offers numerous benefits for energy sector management when assessed from a CE perspective, facilitating sustainability through its inherent mechanisms that enable shifting to a circular model from a linear one, characterized by transparency, immutability, and decentralization. Kouhizadeh et al. (2020) asserted that BT facilitate the security, visibility, and transparency of information along with its automation which can further enhance CE initiatives, as it allows for the disclosure of data regarding material and product sources, stakeholders, processes, energy consumption, and lifecycle endpoints within distributed ledgers. BT is recognized for their efficacy in trustless institutional settings, where technology facilitates efficient administration and enhances communication between producers and consumers (Pilkington, 2016). A business model utilizing DCE-based model incorporating IoT offers considerable benefits through BT, as a distributed ledger facilitates asset exchange among owners within the CE loop (Hatzivasilis et al., 2021). This approach enables monitoring of the product to prolong the lifespan of electronic assets, effective management of electronic waste, and augments the efficacy of organizations in economic terms.

14.4 Challenges

After a comprehensive analysis of existing research, following twelve barriers to I4.0 adoption in the CE are identified.

14.4.1 Lack of Resources and a Roadmap for the Effective DCE Adoption

A well-defined plan is essential for the efficient integration of I4.0 with CE principles. A clearly articulated plan is essential for comprehending the concepts and fundamentals of I4.0 deployment. A comprehensive evaluation of both financial as well as technological aspects is necessary when selecting the feasible approach. Inadequate planning is often turns out to be the primary cause of failing to implement appropriate processes and methodologies for the successful execution of DCE measures (Swan, 2018). Reliant on third-party technology suppliers is the main reason that disrupts the plans of an organization aiming to implement DCE. High-quality real-time data is essential for the effective implementation of DTs. The inadequacy of IT resources and infrastructure in adopting the latest technologies has heightened reliance on third-party suppliers (Yadav et al., 2020), thus raising numerous questions and concerns about system safety and security. I4.0 necessitates a continuous procurement and flow of data as and when required within businesses (Schröder, 2016). In this setting, a significant obstacle is encountered due to resource limitations along with the concern because of the top management of companies, as they exhibit greater reluctance towards the adopting the DTs. Kandasamy et al. (2023) argued that the adoption of I4.0 constitutes a strategic decision, and that hesitations among senior management may hinder the formulation of a digital strategy for executing I4.0-based activities in the organization.

14.4.2 Insufficient Infrastructure

With the rise of advanced technical standards, an upgraded infrastructure system is necessary for not only managing but also integrating various devices (Jaeger & Upadhyay, 2020). The deficiency of such advanced infrastructure significantly impacts the incorporation of I4.0 in business level CE measures (Rajput & Singh, 2019). The insufficient infrastructure complicates the integration of many components, consequently hindering interaction with the physical world and creating impediments, resulting in delays within the supply chain (Biswas & Gupta, 2019), which then increases the risk of internal software assaults, altering the information received by the user. Comprehensive broadband infrastructure is essential for the realization of I4.0. A significant drawback is the limited access to a high-speed broadband connection, particularly for SMEs. Only major entities having extensive funds presently possess advanced type of technology (Schröder, 2016). I4.0 necessitates the interaction of all members; hence, an advanced and updated digital infrastructure is an indispensable aspect. Buntz (2016) indicated that around one-third of organizations reported inadequate infrastructure as a barrier to IoT adoption. It is also revealed that numerous enterprises are now cooperating rather than competing in the essential infrastructure development required for Industry 4.0.

14.4.3 The Negative Perception Regarding the Introduction of New Technology

The adverse view primarily centers on public image, which significantly influences the adoption of new technologies for CE development. Recently, there has been a negative view regarding incorporating latest DTs; this is basically due to security deficiencies, which subsequently fosters illicit activities and results in organizational reluctance to embrace new technology (Trevisan et al., 2023). Moreover, a substantial correlation exists among the obstacles, security, and the adoption of new technologies, necessitating considerable focus (Lohmer & Lasch, 2020). Although security serves as a mediating barrier, it is widely recognized that inadequate IT resources may pose a significant difficulty in mitigating the adverse effects of adopting new technologies and disparity is considered one among such effects. It is pointed that I4.0 will cause instability in market as a societal unrest. The impact may be either beneficial or detrimental; nonetheless, these DTs will divide the job market into two segments first with poor-skill and hence low-pay while second with advanced-skill and high-pay segment, leading to societal discord (Schwab, 2017). It is additionally contended that I4.0 will advantage the proprietors with cognitive assets, while exacerbating the disparity between labor-dependent individuals and capital-dependent individuals, hence amplifying inequality. Disruption to current employment is another aspect of negative perception. The displacement of human-employees is again a potential concern within organizations. Schwab (2017) contended that the progress in I4.0 may exacerbate inequality, potentially destabilizing the labor market. As technology advances, employees are apprehensive about acquiring new skills because of the fear of losing the job, resulting in a reluctance to adapt to evolving technological standards, which consequently hinders the DCE adoption (Luthra et al., 2019). A deficiency in awareness regarding the benefits related to sustainability through the adaptation of digital technologies may contribute to employment insecurity and disruption.

14.4.4 Insufficient Expertise and Awareness Regarding the Adoption of DTs

It is emphasized that developing technologies significantly impact all the operations from production to logistics and necessitate advanced skillset, intellectuals, and workforce management among many stakeholders (Klerkx & Rose, 2020). Given its nascence, only a limited number of individuals has profound knowledge and expertise regarding the efficient adoption of technology in the circular economy (Gopalakrishnan et al., 2021). Therefore, companies must enhance their knowledge base and technological support to guarantee the effective deployment of DCE. Proactive strategies are necessary to execute digitalization across all organizational levels, and the absence of standardized tools is a hurdle for implementing I4.0 in any context (Manavalan & Jayakrishna, 2019). Numerous organizations in the international market fail to leverage their opportunities due to their deficiency in the foundational principles of I4.0 (Jabbour et al., 2018), consequently missing the benefits of its implementation, which are reduced negative impacts the environment and could potentially result in significant cost savings (Jaeger & Upadhya, 2020; Kumar et al., 2021). Breunig et al. (2016) contended that numerous organizations acknowledge their lack of requisite experience or capabilities for exploiting the promise of I4.0 applications to the full capacity. The authors asserted that insufficient digital abilities and experience hinder organizations' capacity to interact with the suppliers of software solution because the employees lack the proficiency to utilize the purchased products. It is noted that inadequately qualified individuals constitute the second significant obstacle to achieving I4.0, as organizations transition to an upgraded data-driven and advanced system, necessitating a highly efficient workforce with advanced skills (Geissbauer et al., 2014).

14.4.5 Security Dilemma

Enhanced connectedness, characterized by intricate relationships across value-chain partners, raises widespread apprehensions over the risks associated with the security of information shared among the partners of a network (Geissbauer et al., 2014). Organizations' cyber security apprehensions as well as the anxiety around potential data loss to either service providers or third-party software in I4.0 (Breunig et al., 2016). It has been identified that this issue as a significant challenge in IoT implementation, as hackers present substantial hazards (Lee & Lee, 2015). Hacking and misinformation present significant security concerns due to their incorporation of key information. Technical vulnerabilities also constitute significant technological impediments while ensuring the safety of information. Researchers indicated that numerous transactions within the whole network of supply chain possess details related to the authentication that are, consequently, unsafe. This may potentially result in considerable privacy concerns for users. Progress in commercial settings is constrained by insufficient security therefore, enhanced security is among the primary challenges to ensure a secure flow of information as well as data (Mathivathanan et al., 2021).

14.4.6 Insufficient Maturity of Technology

Lee and Lee (2015) highlighted additional barriers regarding the possibility for turmoil if nascent and inadequately validated technology is implemented in the organization. These technologies might pose significant concerns related to the privacy and security of the data, while the proliferation of inadequately tested devices could lead to disorder. The research contends that although this does not provide a substantial issue in an isolated context, it could severely impact an interconnected technological system. The inadequacy of technological advancement results in software vulnerabilities, precipitating cyber-attacks and consequent loss of information, money, and other resources. Given the global nature of the business, it is unrealistic to anticipate that all participants will possess the most advanced DTs (Luthra & Mangla, 2018). This raises the question of how to implement these DTs on a wide scale. Large, complicated transactions necessitate the employment of high-powered systems, while low-powered systems may result in significant delays (Karmaker et al., 2021). Numerous industries have not yet adopted the latest technologies, while others remain oblivious to contemporary tools, resulting in indirect repercussions on security difficulties and a poor view of technology (Kouhizadeh et al., 2021).

14.4.7 Inadequate Policy Support from Government

Support from the government is crucial for the CE adoption in relation to sustainability. Generally, there is a need to understand the different expectations held several stakeholders along with explicit regulatory standards for the implementation of CE practices (Ada et al., 2021). It has been seen that governments sometimes fails to promote the use of advanced technologies for achieving the goals related to sustainability (Luthra et al., 2019). Governments neglect to promote I4.0, despite the abundance of resources, due to numerous laws and restrictions. Moreover, in underdeveloped nations, the absence of governmental incentives and assistance hinder the advancement of I4.0 across industries to a great extent (Mittal et al., 2018). Schröder (2016) discusses the apprehensions of SMEs concerning the implementation of I4.0, because of unstandardized norms and lack of proper laws. The author emphasized that absence of norms poses difficulties for SMEs in participating in value-creation activities and networks. Proper policy framework and legislation must harmonize economic and environmental objectives to promote the CE models that safeguard the nature and promoting sustainable practices while optimizing the returns on investment through DTs (Steenmans et al., 2021). In case of a BT, smart contracts are based on algorithms established by programmers who dictate all the aspects related to the software such as design, application, and architecture. Consequently, it leads to various challenges in complying with the industry standards and regulations. The failure to adjust to evolving preferences of CE stakeholders regarding smart contracts that link to the physical realm and authenticate information on the blockchain presents legal issues for these platforms (Esmaeilian et al., 2020). The absence of policies and laws is anticipated to hinder the proliferation of DTs for CE.

14.4.8 Restricted Scalability

Restricted scalability is a significant obstacle that hinders the technology's extensive application in circular practices. The limited scalability results in lack of security and enhanced vulnerability of the present DTs for future utilization. Numerous company models see scalability as the primary difficulty. A primary factor contributing to restricted scalability may be the insufficient advancement in technologies pertaining to DCE implementation (Tönnissen & Teuteberg, 2020). Here, scalability denotes the ability of a system to sustain its efficiency of performing functions with an increase in store capacity. Within the realm of circular economy, for example blockchain scalability represents a significant obstacle to the technology's adoption (Erol et al., 2021). This problem arises when the increased volume of transactions leads to deficiencies of mechanism equilibrium (Esmaeilian et al., 2020). The study stated that communication failures in the network among records, participants, and data storage exemplify scalability challenges. Erol et al. (2021) observed that BT might lack sufficient robustness in terms of latency and data flow, presenting a significant challenge for an integrated system such as the solar energy sector. The implementation of CE adoption by enterprises may generate a substantial amount of transactions, hence potentially diminishing performance of BT marginally. Consequently, inadequate scalability may cast doubt on the viability of this technology in the CE (Kouhizadeh et al., 2019). Therefore, design and architectural factors are essential for optimizing the ledger's lightweight nature and maintaining adequate performance.

14.4.9 Challenges in Transforming Organizational Culture

Organizations are encountering significant challenges in altering their present culture or rules for the deployment of DCE. Certain companies face issues and delays while transforming the corporate culture to facilitate the CE implementation, as these models necessitate a culture oriented towards collaboration throughout the supply chain (Ada et al., 2021). An organization's capacity to digitize the activities related to customer engagement is contingent upon its preparedness to foster a suitable culture for such transformation (Nandi et al., 2021). Organizations increasingly prioritize short-term goals that demand fast outcomes, so constraining policy modifications and inhibiting changes in their workflows (Tönnissen & Teuteberg, 2020). Culture of any organization encompasses the conduct of its employees as well as a framework specifying the norms and procedures to be adhered to within their work environment. Organizations must formulate policy changes or establish new rules and cultivate a deeper internal understanding to facilitate efficient implementation of technology to support circular activities. The swift progression of technologies may necessitate alterations in organizational systems (Swan, 2018). To use the advantages of I4.0, it is essential to possess robust internal competencies of a team and cultivate a culture that encourages creativity and embraces innovation (Breunig et al., 2016). There would be a need to foster a digital culture and facilitate the collaboration among experts from several fields. Hence, the effective implementation of DCE is contingent upon a conducive culture encouraging the employees as well as management to embrace risks and expedite the technology's adoption.

14.4.10 High Cost

The cost of investment poses a big hurdle in the adoption of I4.0 for CE. Significant investment is the basic need for the adoption of cutting-edge technology associated with Industry 4.0 (Raj et al., 2020). When investment costs increase, investors frequently express concerns regarding transaction security, resulting in disinvestment. Many sectors are constrained by financial investments and assets, thereby hindering their access to the latest resources. The adoption of new technology necessitates substantial assistance in both hardware and software sectors. The expenses related to maintenance escalate with further expenditures, resulting in significant financial burdens for companies, as well as for the associated personnel and investors (Luthra & Mangla, 2018). Geissbauer et al. (2014) asserted that companies aiming to adopt I4.0 activities must augment their projected annual capital spending by 50% indicating that companies must overhaul their current strategy and invest significantly to attain DCE objectives. Further, substantial investment in personnel, procedures, and technology is necessary at different levels including corporate as well as supply chain levels (Kache & Seuring, 2017). In fact, the majority of enterprises are hesitant to invest in research related to I4.0 mainly due to the very high expenditure demands.

14.4.11 Ambiguous Advantages

The productivity enhancements and economic advantages of investing in digital technology have consistently been scrutinized. The productivity paradox associated with technology implementation creates ambiguity in accurately evaluating the economic advantages of technological investment (Stolarick, 1999). Top management exhibits hesitance in providing advantages and supporting I4.0 initiatives due to insufficient financial resources and inadequate planning (Biswas & Gupta, 2019). The absence of commitment on the part of higher level of management towards environment protection initiatives poses challenges for individuals and disrupts the cycle of supply chain. One factor for the lack of enough advantages is the utilization of disruptive technology, mostly affecting organizations who are inclined to take risks in embracing DTs. The absence of a recognition mechanism may greatly contribute to ambiguous advantages. The backing and dedication of upper management are crucial for fostering a robust connection among personnel, which is vital for I4.0 implementation (Mittal et al., 2018).

14.4.12 Resistance to Change and Ineffectual Change Management

A significant obstacle to the introduction of DCE in enterprises is the reluctance of employees to alter their work practices. These employees oppose the new technology and their accompanying behaviors (Haddud et al., 2017). Lee and Lee (2015) introduce a further aspect of resistance, contending that many IoT devices produce vast quantities of personal data, including information linked to home, health, and financial behaviors, which several corporations may exploit to enhance their operations. This discourages both companies and individuals from embracing DTs due to concerns over privacy violations. The data quality encompasses four dimensions: accuracy, wholeness, uniformity, and repetition (Chen et al., 2014). In the era of fully developed BD, when companies must be networked, a substantial volume of data is generated, and the intricate nature and heterogeneity of this data complicate the assessment of accuracy and wholeness. This elevates the likelihood of erroneous discovery (Lohr, 2012). Moreover, when data undergoes frequent alterations and is disseminated across numerous contributors, maintaining data integrity and consistency presents a significant difficulty. Also, the chance of unforeseen operational difficulties lowers the technology acceptance in CE practices. In the past decade, technology has significantly transformed the process of change. Previously, change was perceived as transactional and therefore manageable; it has now evolved into a revolutionary, continuous, and complex phenomenon. Consequently, it turned out to be transformative and exceedingly challenging to execute effectively (Anderson and Anderson, 2010). In the context of I4.0, systems will be more complicated than past decades, making management of change a critical task for enterprises.

14.5 Research Methodology

Notwithstanding the significance of DTs of I4.0 for the circular economy, numerous barriers persist. We performed an exhaustive literature review for determining the current status of research on such barriers. The article establishes a comprehensive framework to offer an integrated perspective on the digitalization-driven circular economy shift. After an in-depth analysis of the previous studies, the barriers to the adoption of DCE are identified. The study ultimately offers several implications for dealing with the barriers.

14.6 Discussion and Implications

This study made an attempt to discover the barriers to the implementation of CE combined with DTs of I4.0 referred as DCE. This research identified twelve key obstacles hindering the proper implementation of DCE. On the basis of literature, recommendations are offered for practitioners, professionals, stakeholders, policymakers, and government agencies to develop successful policies for organizations seeking to use DTs for CE during the transition process. A significant challenge faced by enterprises in adopting DCE is the engagement of stakeholders in innovative frameworks that may need the utilization of the technology. Integrating DTs into the CE may necessitate enterprises altering the business strategies and reassess their methods of delivering products and services. So, the companies must update their way of operations and establish protocols for authenticating information. The transition to a new D CE system is a complex matter that requires prudent management to foster employee engagement, increase knowledge, and avert the failure of the adoption process. Lack of infrastructure to support digitalization poses a great challenge in the adoption of DCE practices. The organization should invest in building an infrastructure that ease the process of digitalization in CE context. Software-related aspects must be given full consideration along with hardware aspects. Inspiration from senior management can assist in incorporating these requirements in the current system.

Negative perception towards technology also hinders the DCE adoption. The potential harm in terms of employment opportunities due to high skills requirements should be mitigated with appropriate training programs offered by governmental or private entities to empower the employees in executing DCE initiatives within firms. Among the non-technical difficulties confronting DCE is the less availability of personnel possessing the requisite knowledge and skills. Therefore, companies must enhance their knowledge base and technological support to guarantee the effective deployment of DCE. Individuals possessing these competencies would experience heightened demand and significantly reduced supply in such a context. Consequently, companies must be ready for relocating to those regions where such employees are easily available. Security risk is another concern in the way of DCE regarding sensitive company information. It is recommended that there should be systems for validating data uploaded to the digital platforms for ensuring that digital records and physical entities are connected in correct way, as well as to guarantee the accuracy of the information uploaded. There is a need to encourage CE stakeholders to implement secure methods for enhancing security and system resilience to avoid cyber threats. Security must be optimized while integrating heterogeneous systems associated with various CE parties.

Technological underdevelopment is another barrier as the technology maturity is crucial for its adoption and diffusion. Companies seeking to use technology-based circular economy solutions should identify the concrete benefits to the business and demonstrating the value of DTs in their circular economy initiatives through sufficient awareness of its applicability. The government should frame policies for promoting the DCE adoption and offer regulatory support for smooth transitioning to circular system. The government may introduce various incentives for organizations implementing the DCE. By providing financial support and incentives to establish DCE the government can facilitate the achievement of sustainability goals. Government authorities and company stakeholders shall work together to develop policies that incentivize the application of DCE. Collaborative activities can build a cultural revolution inside the organization to foster awareness. Within the CE context, scalability of DTs represents a significant obstacle in the form of communication failures. There is need of sufficient robustness in terms of latency and data flow for increased scalability of different types of DTs. Additionally, design and architectural factors are essential for optimizing the performance of these technologies.

Organizational culture can profoundly affect DCE implementation, hence, the companies should develop a system to include operations such as recycle, reuse, and trash management. Further, fostering a culture of collaboration to expedite the transition from processes and policies as components of a company's comprehensive strategy. The technology adoption process consists of multiple phases like designing, development, and implementation, as well as later on it also includes maintenance, so the associated investment costs may deter companies from embracing DCE. Consequently, managers must recognize the development expenses incurred both internally and with their business associates. The expenses associated with blockchain and IoT systems must be evaluated, encompassing the original expenditure for designing and acquiring sensors as well as the operational costs, including energy and maintenance. Thus, investments in DCE activities must be substantiated by justifiable returns on investment. Also, there is a necessity to modify current organizational structures and policies to facilitate the CE transition. It is observed that certain level of opposition is faced before the integration of new technologies but the change should align with current company practices and rules that demands creativity, transparency, risk-taking, and a well-defined organizational plan. Substantial executive commitment can enhance DCE adoption and ensure the allocation of suitable resources for better change management towards adapting the DCE directed by organizational structures and policies to enhance process efficiency and resource savings. By adhering to the suggestions outlined in this research, the organization can reap the advantages of DCE adoption.

14.7 Conclusion

This study thoroughly examined the barriers to implementing DT-enabled CE or DCE by a literature review. A thorough knowledge of the obstacles to DCE adoption encompass implementation costs, resistance to transformation, challenges in management, absence of policy regulations, skill shortages and lack of a trained workforce, data security challenges which necessitate increased focus. Organizational, infrastructural, regulatory, standardization, investment, technological, stakeholder, and data-related obstacles to the implementation of DCE necessitate effective solutions. The research findings provide crucial insights that assist in recognizing barriers, allowing experts, firms, and policymakers to devise more effective strategies for the successful integration of DCE. By recognizing and prioritizing the essential barriers impeding DCE adoption, companies can make decisions while keeping important points in mind regarding resource allotment, investment, and strategic planning. The research offers a thorough comprehension of the aspects affecting DCE implementation. This enables managers and policymakers to formulate targeted actions and allocate resources more efficiently, therefore expediting the shift towards sustainable practices in the industry.

The study's focus is confined to data obtained from literature. A further study should include expert viewpoints and conversations to elucidate the diverse obstacles discovered in the advancement of DCE. Moreover, subsequent research might focus on determining effective techniques to surmount the mentioned barriers and successfully execute DCE. This research underscores the imperative for additional investigation into digital infrastructure and its influence on the implementation of CE.

References

Ada, N., Kazancoglu, Y., Sezer, M. D., Ede-Senturk, C., Ozer, I., & Ram, M. (2021). Analyzing barriers of circular food supply chains and proposing Industry 4.0 solutions. *Sustainability, 13*(6812). https://doi.org/10.3390/su13126812

Ajwani-Ramchandani, R., Figueira, S., Torres de Oliveira, R., Jha, S., Ramchandani, A., & Schuricht, L. (2021). Towards a circular economy for packaging waste by using new technologies: The case of large multinationals in emerging economies. *Journal of Cleaner Production*. https://doi.org/10.1016/j.jclepro.2020.125139

Alcayaga, A., Wiener, M., & Hansen, E. (2019). Towards a framework of innovative-circular systems: An integrative literature review. *Journal of Cleaner Production, 221*, 622–634. https://doi.org/10.1016/j.jclepro.2019.02.085

Anderson, D., & Anderson, L. A. (2010). *Beyond change management: How to achieve breakthrough results through conscious change leadership* (Vol. 36). John Wiley & Sons.

Antikainen, M., Uusitalo, T., & Kivikytö-Reponen, P. (2018). Digitalisation as an enabler of circular economy. *Procedia CIRP*. https://doi.org/10.1016/j.procir.2018.04.027

Atif, S. (2023). The role of Industry 4.0-enabled data-driven shared platform as an enabler of the product-service system in the circular economy context: A systematic literature review and future research directions. *Business Strategy & Development, 6*(275–295). https://doi.org/10.1002/bsd2.238

Awan, U., Sroufe, R., & Shahbaz, M. (2021). Industry 4.0 and the circular economy: A literature review and recommendations for future research. *Business Strategy and the Environment, 30*, 2038–2060. https://doi.org/10.1002/bse.2731

Beretta, C., Stucki, M., & Hellweg, S. (2017). Environmental impacts and hotspots of food losses: Value chain analysis of Swiss food consumption. *Environmental Science & Technology, 51*, 11165–11173. https://doi.org/10.1021/acs.est.6b06179

Bianchini, A., Rossi, J., & Pellegrini, M. (2019). Overcoming the main barriers of circular economy implementation through a new visualization tool for circular business models. *Sustainability, 11*(6614). https://doi.org/10.3390/su11236614

Biswas, B., & Gupta, R. (2019). Analysis of barriers to implement blockchain in industry and service sectors. *Computers & Industrial Engineering, 136*, 225–241. https://doi.org/10.1016/j.cie.2019.07.006

Bressanelli, G., Adrodegari, F., Perona, M., & Saccani, N. (2018). Exploring how usage-focused business models enable circular economy through digital technologies. *Sustainability, 10*(3), 639.

Breunig, M., Kelly, R., Mathis, R., & Wee, D. (2016). Getting the most out of Industry 4.0. *McKinsey & Company*. Retrieved from https://www.mckinsey.com/business-functions/operations/our-insights/industry-40-looking-beyond-the-initial-hype

Brous, P., Janssen, M., & Herder, P. (2020). The dual effects of the Internet of Things (IoT): A systematic review of the benefits and risks of IoT adoption by organizations. *International Journal of Information Management, 51*, 101952. https://doi.org/10.1016/j.ijinfomgt.2019.05.008

Buntz, B. (2016). Top 10 barriers for adoption of the Internet of Things. *IoT Institute*. Retrieved from http://www.ioti.com/security/top-10-reasons-people-arent-embracing-iot

Cautitanu, C., Asandului, L., Borza, M., & Turturean, C. (2018). Quantitative approach to circular economy in the OECD countries. *Amfiteatru Economic, 20*, 262–277.

Centobelli, P., Cerchione, R., Chiaroni, D., Del Vecchio, P., & Urbinati, A. (2020). Designing business models in circular economy: A systematic literature review and research agenda. *Business Strategy and the Environment, 29*, 1734–1749. https://doi.org/10.1002/bse.2466

Chauhan, C., Sharma, A., & Singh, A. (2019). A SAP-LAP linkages framework for integrating Industry 4.0 and circular economy. *Benchmarking: An International Journal*. https://doi.org/10.1108/BIJ-10-2018-0310

Chen, Z. K., Yang, S. Q., Tan, S., Zhao, H., He, L., Zhang, G., & Yang, H. Y. (2014). The data allocation strategy based on load in NoSQL database. In *Applied mechanics and materials* (Vol. 513, pp. 1464–1469). Trans Tech Publications.

Erol, I., Peker, I., Ar, I. M., Turan, I., & Searcy, C. (2021). Towards a circular economy: Investigating the critical success factors for a blockchain-based solar photovoltaic energy ecosystem in Turkey. *Energy for Sustainable Development, 65*, 130–143. https://doi.org/10.1016/j.esd.2021.05.011

Esmaeilian, B., Sarkis, J., Lewis, K., & Behdad, S. (2020). Blockchain for the future of sustainable supply chain management in Industry 4.0. *Resources, Conservation and Recycling, 163*. https://doi.org/10.1016/j.resconrec.2020.105064

Garcés-Ayerbe, C., Rivera-Torres, P., Suárez-Perales, I., & Leyva-De La Hiz, D. I. (2019). Is it possible to change from a linear to a circular economy? An overview of opportunities and barriers for European small and medium-sized enterprise companies. *International Journal of Environmental Research and Public Health, 16*(851). https://doi.org/10.3390/ijerph16050851

Garcia-Muina, F. E., González-Sánchez, R., Ferrari, A. M., & Settembre-Blundo, D. (2018). The paradigms of Industry 4.0 and circular economy as enabling drivers for the competitiveness of businesses and territories: The case of an Italian ceramic tiles manufacturing company. *Social Sciences, 7*(12), 255. https://doi.org/10.3390/socsci7120255

Geissbauer, R., Schrauf, S., & Koch, V. (2014). Industry 4.0: Opportunities and challenges of industrial PricewaterhouseCoopers. https://www.pwc.nl/en/assets/document

Ghisellini, P., Cialani, C., & Ulgiati, S. (2016). A review on circular economy: The expected transition to a balanced interplay of environmental and economic systems. *Journal of Cleaner Production, 114*, 11–32. https://doi.org/10.1016/j.jclepro.2015.09.007

Gopalakrishnan, P. K., Hall, J., & Behdad, S. (2021). Cost analysis and optimization of blockchain-based solid waste management traceability system. *Waste Management, 120*, 594–607. https://doi.org/10.1016/j.wasman.2020.11.026

Haddud, A., DeSouza, A., Khare, A., & Lee, H. (2017). Examining potential benefits and challenges associated with the internet of things integration in supply chains. *Journal of Manufacturing Technology Management, 28*(8), 1055–1085. https://doi.org/10.1108/JMTM-05-2017-0094

Han, J., Heshmati, A., & Rashidghalam, M. (2020). Circular economy business models with focus on servitization. *Sustainability, 12*(21), 8799. https://doi.org/10.3390/su12218799

Hatzivasilis, G., Sotiris, I., Konstantinos, F., & George, S. (2021). The green blockchains of circular economy. *Electronics, 10*(16), 2021. https://doi.org/10.3390/electronics10162008

Ingemarsdotter, E., Jamsin, E., Kortuem, G., & Balkenende, R. (2019). Circular strategies enabled by the internet of things—a framework and analysis of current practice. *Sustainability, 11*(20), 5689. https://doi.org/10.3390/su11205689

Iqbal, R., Doctor, F., More, B., Mahmud, S., & Yousuf, U. (2020). Big data analytics: Computational intelligence techniques and application areas. *Technological Forecasting and Social Change, 153*, 119253. https://doi.org/10.1016/j.techfore.2018.03.024

Jabbour, L. S., Jabbour, A. B., Filho, C. J. G., & Roubaud, D. (2018). Industry 4.0 and the circular economy. A proposed research agenda and original roadmap for sustainable operations. *Annals of Operations Research, 270*(1), 273–286. https://doi.org/10.1007/s10479-018-2772-8

Jaeger, B., & Upadhyay, A. (2020). Understanding barriers to the circular economy: Cases from the manufacturing industry. *Journal of Enterprise Information Management*. https://doi.org/10.1108/JEIM-12-2019-0396

Kache, F., & Seuring, S. (2017). Challenges and opportunities of digital information at the intersection of big data analytics and supply chain management. *International Journal of Operations & Production Management, 37*(1), 10–36. https://doi.org/10.1108/IJOPM-02-2015-0078

Kandasamy, J., Venkat, V., & Mani, R. S. (2023). Barriers to the adoption of digital technologies in a functional circular economy network. *Operations Management Research, 16*(3), 1541–1561.

Karmaker, C. L., Ahmed, T., Ahmed, S., Ali, S. M., Moktadir, M. A., & Kabir, G. (2021). Improving supply chain sustainability in the context of the covid-19 pandemic in an emerging economy: Exploring drivers using an integrated model. *Sustainable Production and Consumption, 26*, 411–427. https://doi.org/10.1016/j.spc.2020.10.007

Khan, S., Ali, S. S., & Singh, R. (2022). Determinants of remanufacturing adoption for circular economy: A causal relationship evaluation framework. *Applied System Innovation, 5*(4), 62. https://doi.org/10.3390/asi5040062
Kintscher, L., Lawrenz, S., Poschmann, H., & Sharma, P. (2020). Recycling 4.0: Digitalization as a key for the advanced circular economy. *Journal of Communications, 15*(9), 652–660. https://doi.org/10.12720/jcm.15.9.652-660
Kirchherr, J., Reike, D., & Hekkert, M. (2017). Conceptualizing the circular economy: An analysis of 114 definitions. *Resources, Conservation and Recycling, 127*, 221–232. https://doi.org/10.1016/j.resconrec.2017.09.005
Klerkx, L., & Rose, D. (2020). Dealing with the game-changing technologies of Agriculture 4.0: How do we manage diversity and responsibility in food system transition pathways? *Global Food Security, 24*, 100347. https://doi.org/10.1016/j.gfs.2020.100347
Kortuem, G., Kawsar, F., Sundramoorthy, V., & Fitton, D. (2010). Smart objects as building blocks for the Internet of Things. *IEEE Internet Computing, 14*, 44–51. https://doi.org/10.1109/MIC.2009.143
Kouhizadeh, M., Saberi, S., & Sarkis, J. (2021). Blockchain technology and the sustainable supply chain: Theoretically exploring adoption barriers. *International Journal of Production Economics, 231*, 107831. https://doi.org/10.1016/j.ijpe.2020.107831
Kouhizadeh, M., Sarkis, J., & Zhu, Q. (2019). At the nexus of blockchain technology, the circular economy, and product deletion. *Applied Sciences, 9*(1712). https://doi.org/10.3390/app9091712
Kouhizadeh, M., Zhu, Q., & Sarkis, J. (2020). Blockchain and circular economy: Potential tensions and critical reflections from practice. *Production Planning & Control, 31*(11–12), 950–966. https://doi.org/10.1080/09537287.2019.1695925
Kristoffersen, E., Li, Z., Li, J., Jensen, T. H., Pigosso, D. C., & McAloone, T. C. (2020). *Smart circular economy: CIRCit workbook 4*. DTU, Kongens Lyngby, Denmark.
Kumar, P., Singh, R. K., & Kumar, V. (2021). Managing supply chains for sustainable operations in the era of Industry 4.0 and circular economy: Analysis of barriers. *Resources, Conservation and Recycling, 164*, 105215. https://doi.org/10.1016/j.resconrec.2020.105215
Lee, I., & Lee, K. (2015). The Internet of Things (IoT): Applications, investments, and challenges for enterprises. *Business Horizons, 58*(4), 431–440. https://doi.org/10.1016/j.bushor.2015.03.008
Lin, K.-P., Yu, C.-M., & Chen, K.-S. (2019). Production data analysis using novel process capability indices-based circular economy. *Industrial Management & Data Systems, 119*(8), 1655–1668. https://doi.org/10.1108/IMDS-07-2018-0266
Liu, Q., Trevisan, A., Yang, M., & Mascarenhas, J. (2022). A framework of digital technologies for the circular economy: Digital functions and mechanisms. *Business Strategy and the Environment, 31*(5), 2171–2192. https://doi.org/10.1002/bse.3028
Lohmer, J., & Lasch, R. (2020). Blockchain in operations management and manufacturing: Potential and barriers. *Computers & Industrial Engineering, 149*, 106789. https://doi.org/10.1016/j.cie.2020.106789
Lohr, S. (2012). The age of big data. *New York Times, 11*. Retrieved from https://www.nytimes.com/2012/02/12/sunday-review/big-datas-impact-in-the-world.html
Luthra, S., & Mangla, S. K. (2018). Evaluating challenges to Industry 4.0 initiatives for supply chain sustainability in emerging economies. *Process Safety and Environmental Protection, 117*, 168–179. https://doi.org/10.1016/j.psep.2018.04.021
Luthra, S., Mangla, S. K., & Yadav, G. (2019). An analysis of causal relationships among challenges impeding redistributed manufacturing in emerging economies. *Journal of Cleaner Production, 225*, 949–962. https://doi.org/10.1016/j.jclepro.2019.03.356
Manavalan, E., & Jayakrishna, K. (2019). An analysis on sustainable supply chain for circular economy. *Procedia Manufacturing, 33*, 477–484. https://doi.org/10.1016/j.promfg.2019.04.059
Mathivathanan, D., Mathiyazhagan, K., Rana, N. P., Khorana, S., & Dwivedi, Y. K. (2021). Barriers to the adoption of blockchain technology in business supply chains: A total interpretive structural modelling (TISM) approach. *International Journal of Production Research, 59*(11), 3338–3359. https://doi.org/10.1080/00207543.2020.1868597

Mittal, S., Khan, M. A., Romero, D., & Wuest, T. (2018). A critical review of smart manufacturing & Industry 4.0 maturity models: Implications for small and medium-sized enterprises (SMEs). *Journal of Manufacturing Systems, 49*, 194–214. https://doi.org/10.1016/j.jmsy.2018.10.005

Nandi, S., Sarkis, J., Hervani, A., & Helms, M. (2021). Do blockchain and circular economy practices improve post covid-19 supply chains? A resource-based and resource dependence perspective. *Industrial Management & Data Systems, 121*(2), 333–363. https://doi.org/10.1108/IMDS-09-2020-0500

Nascimento, D. L. M., et al. (2019). Exploring Industry 4.0 technologies to enable circular economy practices in a manufacturing context: A business model proposal. *Journal of Manufacturing Technology Management, 30*(3), 607–627. https://doi.org/10.1108/JMTM-03-2018-0071

Neligan, A., Baumgartner, R. J., Geissdoerfer, M., & Schöggl, J. P. (2023). Circular disruption: Digitalisation as a driver of circular economy business models. *Business Strategy and the Environment, 32*(3), 1175–1188. https://doi.org/10.1002/bse.3086

Neri, A., Negri, M., Cagno, E., Franzò, S., Kumar, V., Lampertico, T., & Bassani, C. A. (2023). The role of digital technologies in supporting the implementation of circular economy practices by industrial small and medium enterprises. *Business Strategy and the Environment, 32*(5), 4693–4718. https://doi.org/10.1002/bse.3164

Pilkington, M. (2016). Blockchain technology: Principles and applications. In F. X. Olleros, & M. Zhegu (Eds.), *Research handbook on digital transformations* (pp. 225–241). Edward Elgar. https://doi.org/10.4337/9781784717766.00019

Raj, A., Dwivedi, G., Sharma, A., de Sousa Jabbour, A. B. L., & Rajak, S. (2020). Barriers to the adoption of Industry 4.0 technologies in the manufacturing sector: An inter-country comparative perspective. *International Journal of Production Economics, 224*, 107546. https://doi.org/10.1016/j.ijpe.2019.107546

Rajput, S., & Singh, S. P. (2019). Connecting circular economy and Industry 4.0. *International Journal of Information Management, 49*, 98–113. https://doi.org/10.1016/j.ijinfomgt.2019.03.002

Rosa, P., Sassanelli, C., Urbinati, A., Chiaroni, D., & Terzi, S. (2020). Assessing relations between circular economy and Industry 4.0: A systematic literature review. *International Journal of Production Research, 58*(6), 1662–1687. https://doi.org/10.1080/00207543.2019.1680896

Schröder, C. (2016). The challenges of Industry 4.0 for small and medium-sized enterprises. *Friedrich-Ebert-Stiftung, Bonn.*

Schwab, K. (2017). *The fourth industrial revolution.* Crown Business.

Singh, R., Khan, S., & Centobelli, P. (2022). Investigating the interplay between social performance and organisational factors supporting circular economy practices.

Singh, R., Khan, S., & Dsilva, J. (2023). A framework for assessment of critical factor for circular economy practice implementation. *Journal of Modelling in Management, 18*(5), 1476–1497.

Steenmans, K., Taylor, P., & Steenmans, I. (2021). Blockchain technology for governance of plastic waste management: Where are we? *Social Sciences, 10*(4), 434. https://doi.org/10.3390/socsci100100131

Stolarick, K. (1999). *IT spending and firm productivity: Additional evidence from the manufacturing sector*. Center for Economic Studies, US Department of Commerce, Bureau of the Census.

Swan, M. (2018). Blockchain for business: Next-generation enterprise artificial intelligence systems. *Advances in Computers, 111*, 121–162. https://doi.org/10.1016/bs.adcom.2018.03.006

Tönnissen, S., & Teuteberg, F. (2020). Analysing the impact of blockchain technology for operations and supply chain management: An explanatory model drawn from multiple case studies. *International Journal of Information Management, 52*, 101953. https://doi.org/10.1016/j.ijinfomgt.2019.05.011

Trevisan, A. H., Lobo, A., Guzzo, D., de Vasconcelos Gomes, L. A., & Mascarenhas, J. (2023). Barriers to employing digital technologies for a circular economy: A multi-level perspective. *Journal of Environmental Management, 332*, 117437.

Wang, N., Ren, S., Liu, Y., Yang, M., & Wang, J. (2020). An active preventive maintenance approach of complex equipment based on novel product-service system operation mode. *Journal of Cleaner Production, 277*, 123365. https://doi.org/10.1016/j.jclepro.2020.123365

Wilts, H., Garcia, B. R., Garlito, R. G., & Gómez, L. S. (2021). Artificial intelligence in the sorting of municipal waste as an enabler of the circular economy. *Resources, 10*(4), 28. https://doi.org/10.3390/resources10040028

Yadav, G., Luthra, S., Jakhar, S. K., Mangla, S. K., & Rai, D. P. (2020). A framework to overcome sustainable supply chain challenges through solution measures of industry 4.0 and circular economy: An automotive case. *Journal of Cleaner Production, 254*, 120112. https://doi.org/10.1016/j.jclepro.2020.120112

Yildizbasi, A. (2021). Blockchain and renewable energy: Integration challenges in the circular economy era. *Renewable Energy, 176*, 183–197. https://doi.org/10.1016/j.renene.2021.05.053

Chapter 15
Digital Technologies for Circular Economy: Environmental and Ethical Considerations for Sustainable Business Practices

Gowri B. Chanal, Rakesh Suryadevara, A. Arun Kumar, and C. Yaduvamsi Sai Krishna

Abstract This study investigates how digital technologies, including big data, blockchain, IoT, and artificial intelligence (AI), can help move from a linear to a circular economy. These technologies are essential to attaining environmental sustainability because they let businesses optimize resource use, improve supply chain transparency, and cut waste. However, their adoption has several ethical and environmental issues, such as labour displacement, energy consumption, e-waste generation, and data privacy concerns. This essay looks at these issues and how companies can balance the advantages of digital innovations with their moral obligations. The potential for blockchain to enhance supply chain transparency, AI's capacity to maximize resource management, IoT's function in real-time monitoring, and big data's influence on decision-making are some of the major conclusions. Subsequent investigations ought to concentrate on creating energy-efficient digital tools, ethical AI models, and legislative frameworks that promote responsible innovation. The study's conclusion highlights the significance of integrating digital technologies with ethical and sustainable practices, and it makes the case that stringent legislative and regulatory measures are necessary to guarantee the circular economy's long-term viability.

Keywords Digital · Technologies · Circular economy · Environmental · Ethical · Practices

G. B. Chanal · R. Suryadevara · A. A. Kumar (✉) · C. Y. S. Krishna
ICFAI Law School, ICFAI Foundation for Higher Education (Deemed to Be University), Hyderabad, Telangana, India
e-mail: arunkumar@ifheindia.org

G. B. Chanal
e-mail: drgowribchanal@ifheindia.org

R. Suryadevara
e-mail: rakeshsuryadevara@ifheindia.org

C. Y. S. Krishna
e-mail: saikrishna@ifheindia.org

R. Singh and V. Kumar (eds.), *Sustainable Innovations and Digital Circular Economy*,
https://doi.org/10.1007/978-981-96-1064-8_15

15.1 Introduction

The circular economy idea is coming into sharper focus as the need to address the world's environmental problems becomes more pressing. By encouraging the practices like reducing, reusing, and recycling, the circular economy prioritizes resource efficiency in contrast to the conventional linear economy, which uses a take, make, and dispose model. The aim is to establish a sustainable system that minimizes waste and environmental harm by continuously cycling materials and resources. However, adopting this new economic paradigm calls for more than just adjustments to corporate strategy and policy. Making use of digital technologies is necessary for the environmental sustainability (Syed Abdul Rehman Khan, 2022).

However, for this technological revolution to be truly sustainable, it raises important ethical issues that must be resolved. Corporate decision-making must consider the ethical challenges posed by issues like resource extraction for digital devices, data privacy in interconnected systems, and labour practices in implementing these technologies. To ensure digital tools support environmental sustainability and moral responsibility, businesses must address these issues. The potential applications of cutting-edge digital technologies like big data, blockchain, Internet of Things (IoT), and Artificial Intelligence (AI) speed up this shift. These technologies have enormous potential to change consumer interactions, supply chains, and business models. They will also make it simpler to track materials throughout their lifecycle and manage resources more effectively. Blockchain can guarantee supply chain transparency, and real-time monitoring and predictive analytics from AI and IoT can maximize resource efficiency and reduce waste.

This research will look at how businesses can successfully incorporate digital technologies to meet the objectives of the circular economy while juggling ethical and environmental concerns. The research aims to offer a framework for companies looking to align their operations with a sustainable, digital circular economy by examining the advantages and disadvantages of these innovations. The transition to a circular economy can be significantly aided by digital technologies like blockchain, AI, the IoT, and big data, but they also bring with them several challenges that need to be resolved to ensure their ethical and successful application. These problems have ethical, environmental, and technical aspects. Many digital solutions for a circular economy, particularly those based on blockchain and IoT, depend on supply chains sharing real-time data. Transparency raises serious concerns about data privacy and the potential misuse of sensitive corporate and customer information, even though it is essential for tracking resources and products (Van Buren et al., 2016).

Businesses are more susceptible to cyberattacks because of the IoT' increased connectivity between systems and devices. Security lapses may result in lost confidence in digital circular systems, data theft, and operational disruptions. AI and blockchain technologies require many resources. Blockchain uses much energy, especially proof-of-work systems, which can be detrimental to environmental goals by increasing carbon emissions (Han, 2023).

The production of electronic devices and sensors needed for the IoT network is increasing due to the rapid pace of digitalization. This undermines the goal of the circular economy, which is to reduce waste and adds to the global e-waste problem. Raw materials like rare earth metals, frequently extracted using unfavourable labour practices and environmentally damaging methods, are needed to produce digital technologies. Dependence on these resources may make problems with resource depletion and human rights abuses in the mining sector worse (Mugge et al., 2017).

Automation and AI may lead to job losses or displacement in industries where machines replace labour, even though they increase efficiency in circular business models. This presents moral questions about the necessity of worker reskilling and fair employment practices. Digital platforms and standards may be adopted differently by different industries and geographical areas, resulting in a fragmented ecosystem that hinders interoperability between systems. As a result, businesses find it challenging to maintain consistent material tracking across international supply chains. Many companies still use antiquated technology incompatible with recent digital innovations. The technical complexity and cost of integrating blockchain, AI, or IoT into these antiquated infrastructures can be high (Blomsma & Tennant, 2020).

When implementing advanced digital technologies, it is frequently necessary to make large upfront investments in infrastructure, skilled labour, and system integration. The financial burden may prevent Small and medium-sized Enterprises (SMEs) from implementing digital solutions for models of the circular economy. Even though digital technologies offer long-term sustainability benefits, many companies find it difficult to project their short- to medium-term Return on Investment (ROI), especially considering the unpredictability of regulatory frameworks and consumer demand for circular practices. The regulatory environments about sustainability and digital technologies are continually changing. Businesses may find it challenging to match their digital operations with ethical and environmental standards if there are unclear global standards or inconsistent regional regulations concerns about data ownership include who owns, accesses, and makes money off data produced by IoT and AI systems (Okorie, 2018).

Algorithms utilizing AI and machine learning for supply chain optimization and resource management may be biased by nature, producing unjust or ineffective results. Biased data inputs have the potential to unfairly prioritize specific suppliers or practices, thereby undermining the inclusivity and fairness of circular economy practices. Concerns about the ethical ramifications of AI-driven decision-making are raised when it increasingly takes over circular systems. Without human supervision, AI systems might put efficiency ahead of justice or environmental sustainability (Trevisan, 2023).

The rapid production, use, and disposal of electronic devices has resulted in an increasing global problem known as e-waste or electronic waste. In a circular economy, reducing e-waste is essential to build a more environmentally conscious and sustainable economy. Manufacturers can design more robust and repairable electronics, extending their lifespan and lowering the frequency of replacements. This involves using premium components and modular designs that facilitate simple component replacement and repair. Fair Phone is a company that makes smartphones with easily swappable parts so that users can fix their gadgets instead of throwing them away. By encouraging laws like the Right to Repair movement, consumers can choose to fix their gadgets rather than throw them away, which will cut down on the quantity of electronics that are thrown away. Businesses and governments can fund more easily accessible and effective e-waste recycling initiatives. These initiatives guarantee the proper disassembly of electronic devices and the option to reuse some important materials.

Apple's recycling program allows consumers to trade in their old devices for recycling, ensuring valuable components are extracted and reused. Encouraging customers to dispose of their e-waste properly can lead to a rise in recycling program participation. Encouraging authorized e-waste recycling facilities guarantees that old electronics are disposed of in an eco-friendly manner (Ellen MacArthur Foundation, 2013).

Software as a Product (SaaS) is a one-time item, businesses can start providing them as a service. Under this model, the product's manufacturer maintains ownership and is responsible for upkeep, upgrades, and recycling at the end of the product's life. HP provides printing services where clients rent printers, and HP takes care of the printers' maintenance, recycling, and repair after the lease. Take-back program manufacturers can establish take-back programs where customers return old devices when purchasing new ones. These initiatives ensure that goods are reused or repaired instead of being thrown away. Take-back programs are provided by companies such as Dell and Samsung to encourage the responsible recycling of used electronics. By integrating recycled components into their manufacturing processes, companies can lower the demand for new raw materials (Geissdoerfer et al., 2017).

Recycling glass, plastic, and metals lessens the need to extract resources and lessens the environmental impact of making new electronics. Recycled plastic laptop models made from e-waste are part of Dell's Latitude laptop lineup. Lead, mercury, and cadmium are hazardous materials that manufacturers should try to eliminate or use less of in electronic devices. These substances can impede recycling efforts and worsen the effects of improperly disposed electronics on the environment. With the help of eco-labels and certifications, consumers can make more ethical purchases by having electronics with clear labelling that complies with sustainability and recyclability standards. By encouraging people to use their devices longer before replacing them, you can reduce the amount of e-waste produced by increasing consumer awareness of the environmental effects of regularly upgrading to the newest gadgets (Pagoropoulos et al., 2017).

Businesses can create refurbishment programs that fix returned or defective goods before reselling them as approved pre-owned electronics. This lessens the need to create new gadgets by giving electronics a second chance at life. By selling refurbished electronics, Amazon's Renewed Program helps reduce the amount of e-waste produced by still functional electronics. E-waste can be decreased by encouraging customers to sell or donate outdated electronics instead of throwing them away. Numerous organizations recycle outdated electronics for use in communities or schools that do not have access to more modern technology (Trevisan, 2023).

Tracking their lifecycle with blockchain technology can ensure more accountability and transparency in the recycling and disposal of electronic devices. Enhancing e-waste's traceability and preventing its unlawful dumping can help guarantee that it is disposed of in approved recycling facilities. Circulariser's blockchain platform helps with more responsible waste management by tracking products and materials in a circular economy. Governments should impose more stringent laws on producer responsibility and the disposal of e-waste. The Waste Electrical and Electronic Equipment (WEEE) Directive from the European Union is one law that ensures manufacturers take liability for the complete cycle of material usage (Hofmann, 2019).

Extended Producer Responsibility (EPR) also falls under a similar category where a complete life cycle of material sourcing to disposal is considered. This encourages businesses to create products that are easier to repair and last longer and invest in effective recycling systems. A multi-pronged strategy involving companies, consumers, and governments is needed to reduce e-waste. The world community can drastically reduce electronic waste's ethical and environmental effects by adopting circular economy models, designing more robust and repairable products, enhancing recycling infrastructure, and increasing consumer awareness. Blockchain technology is one digital solution that can improve transparency and guarantee responsible e-waste management. In a digital world, cutting down on e-waste is essential to sustainability (Bocken et al., 2016).

15.1.1 The Role of Digital Technologies in Advancing Circular Economy Goals

The objectives of a Circular Economy (CE), which maximize resource utilization, eliminate waste, and extend product lifecycles, are greatly advanced by digital technologies. Advancements in AI and innovative IoT technologies can greatly help this cause. Blockchain technology offers a decentralized, unchangeable ledger that can track materials from their sourcing to reuse. This guarantees that materials are sourced and recycled sustainably and ethically and improves logistics transparency. AI algorithms can analyze large datasets to maximize production resource utilization and reduce waste and energy usage. Additionally, it can predict demand more precisely, guaranteeing the effective use of resources and products. Predictive maintenance, made possible by AI, allows companies to replace or repair products and equipment before they break down. This promotes the circular concept of longevity by extending the life of equipment and goods (Geissdoerfer et al., 2017).

IoT devices, like sensors and networked systems, offer real-time data on resource utilization, assisting companies in streamlining operations like waste management, water use, and energy use. This encourages waste reduction and resource efficiency. Product-as-a-service and take-back programs, which allow for product returns, refurbishments, and reuse, are examples of circular business models that businesses can implement thanks to IoT's ability to track product performance and condition (Chiara Colombi, 2023).

Businesses can use big data to analyze patterns, trends, and insights about consumer behavior, resource usage, and environmental impact. This ensures that sustainability is incorporated into business strategies by providing information for decisions on waste reduction, supply chain management, and product design. Comprehensive Lifecycle Assessments (LCAs), which enable companies to assess a product's environmental impact from inception to disposal and verify that it conforms to circular economy principles, can be facilitated by big data.

Digital technologies present several ethical challenges that businesses must carefully manage, even though they offer significant environmental benefits in advancing circular economy goals. Digital technologies like IoT and blockchain require rare earth metals and other materials, contributing to resource extraction and e-waste generation. Businesses can use recyclable materials, create robust and repairable devices, and apply circular design principles to their digital products. Mitigating environmental harm can also be achieved by encouraging the recycling of e-waste and lowering the frequency of technology upgrades.

AI and blockchain, especially proof-of-work systems (Pow), consume much energy and have a large carbon impact. Companies can incorporate renewable energy sources into their operations and use energy-efficient technologies. For instance, shifting to proof-of-stake (PoS) blockchain models and optimizing AI algorithms for energy efficiency can reduce their environmental impact. Concerns regarding data privacy and potential cybersecurity risks are raised by the necessity of gathering and sharing enormous amounts of data when using IoT and big data in circular economy models. Businesses need to establish robust data governance frameworks that guarantee consent, security, and transparency when handling data to address these concerns. Compliance with data protection laws like General Data Protection Regulation (GDPR) is essential when using digital technologies to protect consumer privacy.

AI and IoT-driven automation and digitization may result in the loss of jobs and raise moral questions about labour practices, especially in sectors of the economy that primarily depend on human labour. Companies ought to spend money retraining and upgrading their employees so they can use the latest digital technologies. A culture of human–machine collaboration and the implementation of inclusive employment policies can help allay ethical worries about labour. AI algorithms that are not well-designed may display bias and result in unfair treatment or decisions, especially when it comes to supply chain management and resource allocation. Companies should put an emphasis on ethical AI development by making sure algorithmic design is transparent, equitable, and accountable. These risks can be reduced by creating guidelines for the moral application of AI and conducting routine audits of AI systems for biases. The objectives of the circular economy are significantly advanced by digital technologies, which provide solutions for resource optimization, waste reduction, and sustainability. However, businesses must carefully handle the moral dilemmas raised by new technologies, including labour issues, resource extraction, and data privacy. Businesses can fully realize the potential of digital innovation and incorporate sustainability and ethical responsibility (Trevisan, 2023).

15.2 Review of Literature

The main aim of CE is to bifurcate the economic growth from the material consumption metrics by reusing, repairing, remanufacturing, and recycling products to close the loop on their lifecycles. This is done by optimal usage of material sourced and minimizing wastage by adopting reusability in contrast to the conventional linear economy, which takes a take-make-dispose approach (Ellen MacArthur Foundation, 2013).

The CE has drawn attention because of its potential to address issues related to global sustainability, such as resource depletion, environmental degradation, and mounting pressure on waste management systems. Geissdoerfer et al. (2017) claim that the move towards circularity offers economic and environmental advantages by lowering carbon emissions and reliance on virgin materials. It also opens new business opportunities by utilizing creative product lifecycle models. Minimizing waste production and reducing material consumption through better product design and effective resource management (Ghisellini et al., 2016). By encouraging businesses to reconsider how resources are obtained, used, and disposed of, this principle helps to ensure long-term sustainability. Extending the life of products by encouraging their reusability and secondary markets by refurbishing or remanufacturing them. Material premature entry into the waste stream is avoided through reuse (Kirchherr et al., 2017); once they reach, it is the end of their lifecycle. Recycling lessens the need for virgin resources and lessens the impact of manufacturing on the environment (Ghisellini et al., 2016).

Creating long-lasting, repairable, and upgradable products through design results in products last longer and require fewer replacements over time (Stahel, 2016), and minimizing waste using circularity-focused, efficient design, production, and consumption techniques. At each phase of a product's lifecycle, businesses seek to reduce resource loss (Kirchherr et al., 2017).

A decentralized, unchangeable ledger that can track goods and materials throughout the whole supply chain is made possible by blockchain technology. This feature guarantees transparency and traceability, facilitating businesses to confirm the source of materials, track product lifecycles, and guarantee that products are recycled and sourced ethically (Saberi et al., 2019). To help cut waste and enhance food safety, IBM's Food Trust blockchain platform, for instance, allows stakeholders to track food products from farm to table (IBM, 2020).

AI is essential for prolonging product lifecycles and optimizing resource usage. AI can monitor machinery and determine when maintenance is needed through predictive analytics, which lowers the chance of malfunctions and increases asset lifespan (Antoniou et al., 2021). AI-powered decision-making aids businesses in streamlining production procedures to cut down on waste. Schneider Electric has incorporated AI into its energy management systems to optimize energy usage and cut waste, enabling real-time data analysis (Schneider Electric, 2020).

The IoT's ability to connect systems and devices makes real-time resource and product monitoring possible. IoT sensors help businesses manage their assets more effectively and cut waste by tracking the usage and condition of products. In waste management, IoT-enabled smart bins help cities like Barcelona optimize waste collection routes, reducing fuel consumption and operational costs. Businesses can use big data analytics for better decision-making. Businesses can more accurately predict demand and cut waste and overproduction by analyzing consumer behavior. To support its larger sustainability goals, Nike uses big data to improve the recyclability of its products and streamline its supply chain (Nidumolu et al., 2009).

Although digital technologies present numerous ethical and environmental challenges, they also hold great promise for advancing the goals of the circular economy. Energy consumption is one of the leading environmental issues related to digital technologies. According to Sedlmeir et al. (2020), blockchain, proof-of-work systems like those used in Bitcoin mining, require significant computational power, raising energy consumption and carbon emissions. Similarly, a large amount of data processing is needed for AI and IoT systems, which can raise energy requirements, especially in data centres (Xu et al., 2020).

Another urgent problem is e-waste, especially considering the growing IoT device market. Shorter product lifecycles and a rise in discarded electronics, many of which contain hazardous materials, result from the rapid proliferation of connected devices. Less than 20% of the 53.6 million metric tons of global e-waste produced in 2019 were formally recycled, according to Forti et al. (2020). Mining rare materials used to produce modern technological devices presents ethical questions about environmental damage and human rights abuses in mining communities. Child labour and hazardous working conditions have been connected to the extraction of cobalt, which is used in smartphone batteries, in nations like the Democratic Republic of the Congo (Amnesty International, 2016).

The increased automation made possible by AI and IoT also raises labour-related issues. Even though these technologies can increase productivity, workers may be replaced by them, especially in the manufacturing and logistics industries. Businesses must fund reskilling initiatives and make sure that technology innovations do not worsen inequality to address these issues (Autor, 2015). Data privacy is a concern with the growing use of IoT and big data in circular business models. IoT devices continuously gather data, and without sound governance, this can result in privacy violations, unauthorized access to data, and surveillance (Ziegeldorf et al., 2014). Furthermore, ethical governance of algorithms becomes essential as businesses embrace AI-driven decision-making to prevent biases and guarantee transparency in decision-making processes (Cath et al., 2018).

To guarantee a sustainable and moral transition, a complete analysis of issues such as data privacy, resource extraction, energy consumption, and the creation of e-waste is required. Future studies and the creation of new policies are necessary to address these issues and fully utilize digital tools to further the objectives of the circular economy.

15.3 Data Analysis and Interpretation

Blockchain technology makes end-to-end supply chain traceability possible, which lets companies monitor raw materials and finished goods through to their end. This is especially helpful in circular economy models where recycling and reuse are encouraged, and precise materials tracking is essential. According to an IBM study on its Food Trust blockchain platform, supply chain transparency reduced food waste for participating companies by 30% by streamlining logistics and shortening spoilage times.

A PwC report claims that because blockchain enhances transparency and immutability of transaction records, it can cut supply chain fraud by 50–60%. Blockchain allows companies to verify raw materials' origin and authenticity, improving accountability and trust throughout supply chains. This stops waste from leaking and guarantees that products are recycled and sourced ethically. For instance, in the fashion industry, blockchain-enabled platforms like provenance allow consumers to trace the lifecycle of garments, ensuring ethical sourcing and reduced waste. The energy usage of blockchain, however, is a serious issue. Strategies like integration with renewable energy sources and PoS protocols are being investigated to overcome these obstacles.

AI-powered systems maximize resource utilization by anticipating malfunctions, planning maintenance, and increasing production efficiency. AI has helped reduce material waste by 20% and increase operational efficiency by 15–25% in sectors like automotive manufacturing, where accuracy is crucial to cutting back on excess material use. AI extends product life cycles and minimizes downtime by analyzing massive datasets and forecasting equipment failures. This emphasizes keeping products in circulation for as long as possible, which is critical for circular economy models. As an illustration, businesses such as Siemens have employed AI to optimize their manufacturing processes, resulting in a notable decrease in waste and resource consumption. However, careful thought must be given to the ethical issues raised by AI. The automation of AI is creating a growing problem of job displacement. Upskilling employees and ensuring AI deployment benefits businesses and employees are necessary to address this issue (Kristoffersen et al., 2020).

IoT devices collect real-time data on resource flows, enabling businesses to monitor and optimize their environmental impact. According to Accenture's report on IoT-driven sustainability, IoT-enabled utilities and manufacturing sectors can cut their energy consumption by 10–25%. According to Cisco, the increased monitoring and analytics provided by IoT could result in a 15% decrease in waste and inefficiencies across a range of industries by 2025. IoT sensors enable real-time tracking of resources, including raw materials, energy, and water, allowing businesses to act before excessive waste or inefficiency happens. For instance, by helping utilities better balance energy supply and demand, IoT-powered smart grid systems have decreased energy waste and consumption. However, the extensive use of IoT devices prompts worries regarding e-waste. One of the biggest challenges facing the growing number of IoT devices is figuring out how to properly dispose of and recycle them so that their contribution to the circular economy goals does not cause more environmental damage (Garcia-Muina, 2018).

Big data analytics give companies resource management insights, enabling them to make wise decisions that cut waste and increase sustainability. According to Deloitte, big data analytics companies have improved resource planning and optimized production processes to reduce material consumption by 15–20% (Ranta et al., 2018). According to a Navigant Research report, big data in the energy sector has allowed utilities to optimize resource allocation, resulting in a 10% decrease in energy waste. Businesses can use big data analytics to find inefficiencies, streamline supply chains, and use fewer resources by utilizing a wealth of information (Kristoffersen et al., 2020).

However, there are ethical issues with data monopolization and data governance. Businesses with access to enormous data resources may control markets as big data becomes a crucial factor in decision-making, stifling competition and innovation. Data privacy problems also occur when businesses utilize customer data for resource management without obtaining sufficient consent or transparency.

15.3.1 Environmental Considerations of Digital Technologies

The environmental impact of AI, blockchain, and IoT technologies cannot be disregarded in the circular economy's transition towards digitalization. The energy consumption of these technologies and the growing problem of e-waste resulting from the proliferation of electronic devices are two important factors. A data-driven analysis of these environmental issues and potential fixes can be found below.

Blockchain is infamously energy-intensive, especially when proof-of-work (PoW) mechanisms are used (like in Bitcoin). If fossil fuels are used to power this enormous energy consumption, there will be significant carbon emissions. The mining of Bitcoin alone consumes about 127 TWh yearly (as of 2023), with a 22–23 megaton carbon footprint. This energy consumption level is like that of Malaysia or Sweden. According to the CBECI, specific blockchain networks, such as Bitcoin, release 118.9 Mt of CO_2 yearly (Rosa et al., 2019).

Energy-related issues have also arisen with the rise of deep learning and Machine Learning (ML). A University of Massachusetts study found that training one AI model could release up to 284 tons of CO_2, or the same number of emissions throughout five typical American cars' lifetimes. Training AI models can require up to 300,000 kWh of electricity, depending on the complexity and amount of data involved. The need for AI systems is expanding quickly; according to Open AI, training models are expected to require 300 times more processing power than models created in 2012. Energy consumption is rising in several industries due to the usage of IoT applications. These gadgets, including industrial machinery, smart home systems, and sensors, require energy. Even though many IoT devices are energy-efficient, billions of devices being online always have a significant cumulative impact.

Many AI and blockchain companies are pledging to use renewable energy. After switching from PoW to PoS, Ethereum's annual electricity consumption dropped from 112 TWh to less than 0.01 TWh. AI and IoT systems use energy-efficient algorithms more often to reduce their environmental impact. Federated learning, for instance, lessens the requirement for energy-intensive centralized data centers by enabling AI training across decentralized systems. Energy impacts can also be reduced by increasing data center efficiency through low-power servers, advanced cooling methods, and renewable energy sources. Google and Amazon have pledged to achieve carbon neutrality in their data centers (Kamble & Gunasekaran, 2019).

E-waste issues are greatly exacerbated by developing connected devices or other IoT applications. IoT devices have contributed to this growth because the e-waste problem is made worse by their short product lifecycles and rapid technological obsolescence. 53.6 Mt, of which only 17.4% are effectively recycled. Up to 30% of the growth in e-waste is expected to come from industrial sensors, smart home appliances, and IoT devices by 2030.

Reducing the need to throw away entire devices is achieved by using modular design in IoT and AI devices, where components can be easily upgraded or replaced. Fair Phone creates modular smartphones that let customers upgrade or swap out parts, significantly prolonging the device's useful life. Recovering valuable materials like gold, silver, and rare earth elements can help reduce the need for raw material extraction by improving access to and the effectiveness of e-waste recycling programs. The current recycling rate of e-waste worldwide is just 17.4%, indicating a significant unrealized potential for recovering and repurposing electronic components. This encourages producers to develop longer-lasting, more sustainable products and invest in infrastructure for recycling. For businesses aiming to implement circular economy models, the environmental impact of digital technologies poses significant challenges, particularly energy consumption and the generation of e-waste. Blockchain's energy consumption, AI's carbon footprint from training, and IoT's contribution to energy use and e-waste underscore the urgent need for sustainable innovations in these fields. These tactics need to be scaled as digital adoption increases to guarantee that technology's environmental impact is consistent with the objectives of the circular economy (Trevisan, 2023).

15.3.2 *Ethical Considerations in Digital Technologies for Circular Economy*

The main topics of this section are the ethical issues surrounding data privacy governance, labour practices in the digital economy, and the extraction of materials for digital technologies. Statistics and secondary data are analyzed to evaluate these areas' effects on sustainability and moral business conduct. There is a growing need for rare earth metals like lithium, cobalt, and nickel due to the growing demand for digital technologies, particularly AI, IoT, and blockchain applications. The manufacturing of batteries, semiconductors, and electronic components require these materials. According to a report by the World Bank, the demand for key metals (cobalt, lithium, and rare earth elements) is expected to grow by 500% by 2050 due to the increased use of digital devices and renewable energy technologies. Although supply chain transparency can be increased by technologies like blockchain, a 2023 PwC survey revealed that only 23% of worldwide businesses had fully incorporated blockchain into their supply chain management, indicating a lag in adoption despite the potential advantages. Concerns about labour exploitation, environmental degradation, and unsustainable mining practices are ethical issues brought up by the high demand for rare earth materials. These problems are made worse by global supply chains' lack of transparency. The blockchain has the potential to bring much-needed transparency, but its adoption is slow, which suggests that businesses aren't taking full advantage of it to guarantee responsible sourcing.

Global labour markets are changing due to the rise of automation and AI, which will increase productivity while also eliminating jobs. A 2020 World Economic Forum report states that by the end of 2025, AI could compensate for the loss of existing jobs with more digital technology jobs. According to ILO estimates, automation may result in the loss of 14% of jobs worldwide by 2030, with manufacturing, logistics, and retail workers being the most affected. Automation fuelled by AI tends to disproportionately affect lower-skilled jobs, increasing unemployment in some industries and creating income inequality. Businesses, however, have taken their time implementing reskilling initiatives. Merely 30% of multinational corporations have invested substantially in reskilling initiatives, according to McKinsey & Company (McKinsey Report, 2024).

Data collection has reached previously unheard-of levels due to adopting IoT and AI technologies in corporate operations. However, this raises moral questions about ownership, governance, and data privacy. Gartner predicts that 75 billion IoT devices will be connected globally by 2025, producing 79.4 zettabytes of data. According to a Statista report, vulnerabilities in IoT and AI systems were the leading cause of data breaches in 2022 for 47% of global organizations. According to a 2023 Pew Research Centre survey, 79% of participants expressed concern about how businesses gather and utilize their personal data, pointing to a growing lack of confidence in corporate data practices (Kamble & Gunasekaran, 2019).

Businesses face significant challenges in ensuring that their use of digital technologies supports both environmental sustainability and ethical business practices. These challenges are revealed by analyzing secondary data in the three key ethical areas of resource extraction, labour practices, and data privacy. The unsustainable demand for rare earth materials emphasizes the need for more transparent supply chains and responsible sourcing; although blockchain technology is currently used sparingly, it may offer a solution. AI-driven automation is reshaping the labour market, but the potential for job displacement calls for urgent investment in reskilling and inclusive employment strategies. Robust data governance policies are imperative in safeguarding consumer privacy and preserving credibility as IoT and AI systems become increasingly interwoven into corporate operations (García-Muiña, 2018).

15.3.3 Case Studies

This section explains the effect of digital technologies on circular economy models. We will evaluate how these technologies optimize resource usage, enhance transparency, and improve waste management practices using real-world case studies from businesses like Cisco, IBM Food Trust, Schneider Electric, and Walmart.

1. **Block chain in Circular Supply Chains: Walmart and IBM Food Trust**

Businesses are increasingly using blockchain technology to improve supply chains' transparency and traceability, which guarantees more efficient resource management and lowers waste. Walmart and IBM Food Trust have used blockchain technology to track the flow of food goods from suppliers to customers, increasing accountability and decreasing inefficiencies. A blockchain ensures that goods and raw materials can be tracked all the way through the supply chain, which helps companies find inefficiencies and enhance their reuse and recycling procedures. Walmart, keeps track of the provenance, freshness, and safety of food items, minimizing waste and enabling effective recalls when needed. Blockchain reduces environmental effects by guaranteeing that materials are recycled and used again.

The time it took to track down the food source was cut from seven days to 2.2 s thanks to Walmart's implementation of block chain in its food supply chain. This degree of openness is essential to guaranteeing effective recall procedures and cutting down on food waste (Walmart, 2024). As demonstrated by IBM's Food Trust blockchain platform, improved supply chain tracking and transparency can prevent 80% of food waste. The global blockchain market projections show that supply chain applications are expected to grow from $253 million in 2020 to $3.27 billion by 2026, indicative of the growing dependence on this technology to enhance supply chain efficiency and transparency (IBM, 2024).

2. **AI for Predictive Resource Optimization: Schneider Electric and Siemens**

AI-powered solutions analyze massive datasets and make real-time adjustments, allowing businesses to optimize operations and predict resource usage. Schneider Electric and Siemens have incorporated AI into their energy management and manufacturing systems to increase productivity and cut waste in manufacturing processes.

AI helps companies minimize resource consumption by using predictive analytics to schedule maintenance, monitor machinery, and identify potential inefficiencies before they lead to waste. Additionally, it supports smart production systems, which minimize waste and overproduction by allocating materials according to current demand.

Energy savings of upto 30% in industrial operations have been achieved by clients. Thanks to Schneider Electric's AI-driven energy efficiency solutions, which also lower waste and carbon emissions. AI-based predictive maintenance has reduced waste by 15% and increased operational efficiency by 20% in Siemens' manufacturing facilities. The machinery can be serviced before major failures occur, saving waste and downtime. By 2030, AI is expected to help reduce global CO_2 emissions by 1.5–2.5 gigatons annually through more sustainable practices and optimized resource use (Siemens, 2024).

3. **IoT for Smart Waste Management: Cisco and Smart Cities**

Businesses and cities can more effectively manage waste disposal. Thanks to IoT's real-time monitoring of resource flows and waste. Cisco and several municipal governments have implemented IoT solutions through their Smart City initiatives to track waste production, manage recycling, and optimize resource use. Real-time resource tracking and data on waste generation's location and timing are possible with IoT devices. This makes it possible to implement dynamic waste management systems that cut down on inefficiencies, maximize recycling routes, and guarantee the sustainable use of resources.

With real-time data monitoring, Cisco's IoT-enabled smart waste management systems have been implemented in cities such as Barcelona, resulting in a 20% decrease in waste collection expenses. IoT sensors installed in trash cans in New York City have increased recycling rates by 30% and decreased the amount of waste dumped in landfills overall by 15%. As more cities and businesses adopt these technologies, the global implementation of IoT in waste management, an annual growth of 24.7% with a market size of $4.10 billion by 2025, is expected.

15.4 Policy and Regulatory Implications

Integrating digital technologies into circular economy models presents opportunities and challenges for businesses, governments, and society. The main legislative and regulatory ramifications for encouraging the ethical application of digital technologies in the circular economy are examined in this section. The circular economy is by its very nature international, with supply chains frequently spanning several nations. Nevertheless, the regulatory environment for digital technologies like IoT, blockchain, and AI is still disjointed. Global standards that harmonize laws controlling the application of these technologies in circular business models are desperately needed.

The Sustainable Development Goals (SDGs) and Global Reporting Initiative (GRI) provide frameworks for sustainability reporting, but they should be expanded to explicitly address digital tools' role in fostering circularity. To encourage companies to implement circular practices, policymakers must incorporate the circular economy concepts into local and national regulatory frameworks. The Circular Economy Action Plan from the European Union offers companies a road map for incorporating circular practices that could be applied in other areas. The adoption of eco-labels and digital certifications, which show whether a company's digital supply chains and systems adhere to circular economy principles, could be encouraged by governments and international organizations. Due to increased transparency, customers would be encouraged to select more sustainably produced goods. Extended Producer Responsibility (EPR) would encourage businesses to create digital goods that are more resilient, reusable, and repairable, hence lessening the amount of e-waste produced.

Governments should set energy-efficiency guidelines for digital technologies, especially for tools that consume much energy, like AI and blockchain. Incentives for using renewable energy sources in blockchain networks or adopting more energy-efficient algorithms can help mitigate the environmental impact of these technologies. Governments should enact stricter laws to address the environmental risks posed by the growing use of IoT and other connected devices. Policies promoting digital device recycling and appropriate disposal are required to handle the e-waste produced by circular economy technologies. By offering funds for innovation, governments can encourage Research and Development (R&D) into digital technologies that advance circularity. This can entail investing in blockchain technology, circularity-focused IoT devices, or more energy-efficient AI.

15.5 Future Scope and Limitations of the Study

Subsequent investigations should concentrate on surmounting the ecological and moral dilemmas linked to digital technologies within the circular economy. Research on proof-of-stake systems and energy-efficient algorithms, two more sustainable blockchain models, may help lower the carbon footprint of blockchain applications in the circular economy. AI decision-making's ethical ramifications and biases, particularly in resource allocation, supply chain management, and labour displacement, require more research. Ensuring transparency and fairness in AI algorithms will be critical to ethically integrating AI into circular economy practices.

Conscientious corporate behavior can be encouraged by looking into policy frameworks that address ethical and environmental issues and provide incentives for using digital technologies in the circular economy. Future studies should examine how carbon taxes, green certifications, and EPR influence adopting sustainable business practices. For the circular economy to succeed in the long run, digital innovation must align with ethical and sustainable frameworks.

15.6 Conclusion

The revolutionary potential of digital technologies is important to achieving the objectives of the circular economy. Companies can use these technologies to ensure supply chain transparency, minimize waste, increase product lifecycles, and maximize resource utilization. Block chains can improve supply chain transparency through material tracking and environmental and ethical standards compliance. While IoT allows for real-time resource flow monitoring and waste reduction, AI provides predictive capabilities for resource optimization. Big data analytics gives companies even more ability to make sustainable decisions. The incorporation of these digital tools is not without difficulties, though. The research underscores several environmental concerns, including the energy consumption of blockchain and AI systems and the growing e-waste problem from digital devices, particularly IoT sensors. Ethical issues, such as data privacy, resource extraction for the manufacture of digital technology, and possible labour displacement from automation, also present significant obstacles. To ensure sustainable and ethical practices, businesses must weigh the environmental benefits of digital technologies against their ethical obligations.

References

Amnesty International. (2016). *This is what we die for human rights abuses in the democratic republic of the Congo power the global trade in cobalt.* Amnesty International.

Antoniou, G., Gupta, S., & van Hillegersberg, J. (2021). Blockchain technology and circular economy: Current research and future directions. *Sustainability, 13*(5), 2567.

Autor, D. H. (2015). Why are there still so many jobs? The history and future of workplace automation. *Journal of Economic Perspectives, 29*(3), 3–30.

Blomsma, F., & Tennant, M. (2020). Circular economy: Preserving materials or products? Introducing the resource states framework. *Resources, Conservation, and Recycling, 156*, 104698.

Bocken, N. M., De Pauw, I., Bakker, C., & Van Der Grinten, B. (2016). Product design and business model strategies for a circular economy. *Journal of Industrial and Production Engineering, 33*(5), 308–320. https://doi.org/10.1080/21681015.2016.1172124

Cath, C., Wachter, S., Mittelstadt, B., Taddeo, M., & Floridi, L. (2018). Artificial Intelligence and the good society: The US, EU, and UK approach. *Science and Engineering Ethics, 24*(2), 505–528.

Colombi, C., & D'Itria, E. (2023). Fashion digital transformation: Innovating business models toward circular economy and sustainability. *Sustainability, 15*(6), 4942. https://doi.org/10.3390/su15064942

Ellen MacArthur Foundation. (2013). Towards the circular economy: Economic and business rationale for an accelerated transition.

Forti, V., Baldé, C. P., Kuehr, R., & Bel, G. (2020). *The global E-waste monitor 2020: Quantities, flows and the circular economy potential.* United Nations University.

García-Muiña, F. E. (2018). Identifying sustainability patterns in manufacturing supply chains: An empirical approach based on circular economy practices. *Journal of Cleaner Production, 195*, 1085–1096.

Ghisellini, P., Cialani, C., & Ulgiati, S. (2016). A review on circular economy: The expected transition to a balanced interplay of environmental and economic systems. *Journal of Cleaner Production, 114*, 11–32.

Geissdoerfer, M., Savaget, P., Bocken, N. M. P., & Hultink, E. J. (2017). The circular economy—A new sustainability paradigm? *Journal of Cleaner Production, 143*, 757–768.

Han. (2023). *Explore how digital technologies such as AI enable circular economies, particularly in product lifecycle management, helping reduce waste and improve sustainability practices*. Cambridge University Press.

Hofmann, F. (2019). Circular business models: Business approach as driver or obstructer of sustainability transitions? *Journal of Cleaner Production, 224*, 361–374. https://doi.org/10.1016/j.jclepro.2019.03.115

IBM. (2020). *What is a circular economy?* Retrieved from https://www.ibm.com/think/topics/circular-economy

Kamble, S. S., & Gunasekaran, A. (2019). *Analyzing the role of industry 4.0 technologies and circular economy practices in improving sustainable performance in Indian manufacturing organizations*. Preprint, https://www.researchgate.net/publication/337001164_Analyzing_the_role_of_industry_40_technologies_and_circular_economy_practices_in_improving_sustainable_performance_in_Indian_manufacturing

Khan, S. A. R. (2022). Circular economy and digital technologies: An evolving trend in environmental research. *Integrated Environmental Assessment and Management, 18*, 853–854. https://doi.org/10.1002/ieam.4629

Kirchherr, J., Reike, D., & Hekkert, M. (2017). Conceptualizing the circular economy: An analysis of 114 definitions. *Resources, Conservation and Recycling, 127*, 221–232.

Kristoffersen, E., Blomsma, F., Mikalef, P., & Li, J. (2020). The smart circular economy: A digital-enabled circular strategies framework for manufacturing companies. *Journal of Business Research, 120*, 241–261.

McKinsey Report. (2024). *Technology trends outlook*. Retrieved from https://www.mckinsey.com/~/media/mckinsey/business%20functions/mckinsey%20digital/our%20insights/the%20top%20trends%20in%20tech%202024/mckinsey-technology-trends-outlook-2024.pdf

Mugge, R., et al. (2017). Product design and consumer behavior in a circular economy. *Sustainability, 9*(12), 2239.

Nidumolu, R., Prahalad, C. K., & Rangaswami, M. R. (2009). Why sustainability is now the key driver of innovation. *Harvard Business Review, 87*(9), 56–64.

Okorie, O. (2018). Digital manufacturing technologies for a circular economy: A review. *Sustainability, 10*(9), 3000.

Pagoropoulos, A., Pigosso, D. C. A., & McAloone, T. C. (2017). The emergent role of digital technologies in the circular economy: A review. *Procedia CIRP, 64*, 19–24.

Rosa, P., et al. (2019). Digital transformation of circular economy through Industry 4.0 technologies: Insights from the Italian automotive industry. *Sustainability, 11*(3), 709.

Saberi, S., Kouhizadeh, M., Sarkis, J., & Shen, L. (2019). Blockchain technology and its relationships to sustainable supply chain management. *International Journal of Production Research, 57*(7), 2117–2135.

Schneider Electric. (2020). *Circular economy principles for building services*. Retrieved from https://www.se.com/ww/en/download/document/Circular_Economy_for_Buildings/

Sedlmeir, J., Buhl, H. U., Fridgen, G., et al. (2020). The energy consumption of blockchain technology: Beyond myth. *Business & Information Systems Engineering, 62*, 599–608. https://doi.org/10.1007/s12599-020-00656-x

Stahel, W. R. (2016). The circular economy. *Nature, 531*(7595), 435–438.

Suchek, N., Fernandes, C. I., Kraus, S., Filser, M., & Sjögrén, H. (2021). Innovation and the circular economy: A systematic literature review. *Business Strategy and the Environment, 30*(8), 3686–3702. https://doi.org/10.1002/bse.2834

Trevisan. (2023). *Analyze the barriers faced by industries when employing digital technologies for circular economy initiatives, presenting a multi-level perspective on organizational challenges.* Cambridge University Press.

Van Buren, N., et al. (2016). Corporate social responsibility in circular economies: The role of digital technologies at IBM. *Journal of Environmental Management, 162*, 162–169.

Xu, M., David, J. M., & Kim, S. H. (2020). The fourth industrial revolution: Opportunities and challenges. *International Journal of Financial Research, 11*(3), 90–94.

Ziegeldorf, J. H., Morchon, O. G., & Wehrle, K. (2014). Privacy in the internet of things: Threats and challenges. *Security and Communication Networks, 7*(12), 2728–2742.

Chapter 16
SDGs for Organizational Sustainability: Integrating Digital Technologies for a Digital Circular Economy

Rajeev Sijariya, Vikas Garg, and Pooja Tiwari

Abstract In the present scenario, majority of the organizations wants to attain sustainable development goals in their various business practices. The main focus of the current chapter is to explore the digital technologies can be integrated with the various practices of the organization for the attainment of SDGs and sustainability advancement. This is initiated with an overview of the SDGs, highlighting their importance in addressing challenges at global level which includes change in climate, depletion of resources, and social inequality. The chapter defines the concept of the digital circular economy, highlighting how it encourages transition of business towards circular model of consumption and production form linear model. Key technologies in the digital space which includes blockchain, AI, and IoT are examined to test the capability for enhancing the resource efficiency, supply chain optimization, and facilitate sustainable practices. The discussion also addresses the challenges organizations face in adopting these technologies, such as organizational inertia, technological limitations, and regulatory hurdles. However, it emphasizes the opportunities presented by these advancements, including the potential for innovation, the creation of sustainable value, and developing the business models which are novel in nature. This chapter further outlines the future implications of digital technologies in fostering a circular economy, emphasizing the need for collaboration among governments, industries, and organizations to establish supportive regulatory frameworks. Ultimately, the current work as a call to action for both organizations and individual involved in policy making to embrace digital technologies as critical tools for fostering sustainability. By leveraging these technologies and collaborating

R. Sijariya
Atal Bihari Vajpayee School of Management and Entrepreneurship, Jawaharlal Nehru University, New Delhi, India

V. Garg (✉)
Symbiosis Institute of Business Management, Noida, Symbiosis International (Deemed University), Pune, India
e-mail: vikasgargsir@gmail.com

P. Tiwari
Sharda School of Business Studies, Sharda University, Greater Noida, India

R. Singh and V. Kumar (eds.), *Sustainable Innovations and Digital Circular Economy*,
https://doi.org/10.1007/978-981-96-1064-8_16

across sectors, organizations can not only contribute to the SDGs but also drive significant positive influences on the setting and society.

Keywords Sustainable Development Goals (SDGs) · Digital circular economy · Digital technologies · IoT · Blockchain · AI

16.1 Introduction

16.1.1 Overview of Sustainable Development Goals (SDGs) and Their Relevance to Organizational Sustainability

United Nation in 2015 has established the Sustainable Development Goals, facilitates and act as blueprint for all the organizations globally for the achievement of better and sustainable future. A total of 17 SDGs are there which starts from poverty elimination to addressing change in climate, which gives holistic framework for nations, people, and different organizations to handle the unrelenting challenges at the global level by 2030 (United Nations, 2015). For businesses, the SDGs present is not only confines to imperatives and environmental and social domain but takes into consideration the economic opportunities as well. By aligning their strategies with these global goals, contribution of organization towards sustainable development while enhancing their longstanding competitiveness, building trust with stakeholders, and opening up new markets. The SDGs have become increasingly relevant to organizational sustainability, as they push companies to integrate responsible practices into their operations, from reducing waste and emissions to ensuring fair labour practices.

Organizational sustainability can be defined as "company's ability to operate in an environmentally, socially, and economically sustainable manner over the long term." It goes beyond traditional financial metrics to contemplate the broader influence of the working of the companies on the environment and society. Integrating the SDGs into business strategies helps organizations ensure that they are not only profitable but also contribute in the challenges at the global level positively. This alignment between business objectives and sustainable development is seen as a critical driver for long-term organizational success, as businesses that operate sustainably are better prepared to adapt to governing changes, consumer preferences, and resource scarcity (Whelan & Fink, 2016).

16.1.2 Definition of the Digital Circular Economy

One of the emerging concepts is digital circular economy that integrates the various principles of digital technologies with circular economy. Circular economy is designed to benefit the various businesses, environment, and the society through an approach to economic development. Its objective is to diverge from the model of conservative linear economy "take, make, dispose" towards a better model of sustainability in which resources can be recycled, repaired and reused, which facilitates in waste minimization (Ellen MacArthur Foundation, 2013). The incorporation of digital technologies into this model has given rise to what is termed the "digital circular economy". Digital technologies for example blockchain, big data analytics, AI, and IoT help the organizations for increasing the efficacy of resources, material tracking by the help of value chain, product designing for longer lifespans. These technologies not only facilitate better resource management, but also allowing for better maintenance and eventual recycling or reuse (Geissdoerfer et al., 2017). The digital circular economy holds noteworthy capability for advancing sustainable development. By leveraging digital technologies, companies can transition from resource-intensive business models to better sustainable actions that makes the contribution towards the achievement of SDGs. For example, digital platforms can enable businesses to track carbon footprints more accurately, optimize logistics to reduce emissions, or use predictive analytics to extend product life cycles (Umasankar et al., 2022).

16.1.3 Purpose and Objectives of the Chapter

The major objective of the chapter is to discover the intersection of SDGs, digital technologies, and the circular economy within the context of organizational sustainability. This chapter's objective is to examine how digital innovations can help organizations align their operations with SDG targets, contributing to more sustainable business practices. By discussing key digital technologies and their applications, this chapter highlights the significant role digital transformation plays in facilitating a circular economy and enabling businesses to meet sustainability goals.

In particular, this chapter focuses on the following objectives:

1. **To provide a detailed information of the SDGs and its significance to businesses**: The chapter will explore how the SDGs serve as a roadmap for corporate sustainability and the potential for organizations to drive positive change by aligning their strategies with these goals.
2. **To define and analyse the notion of the digital circular economy**: By integrating digital technologies into the circular economy, organizations can create better models of business sustainability. The chapter will explain how this integration works and its implications for businesses and sustainability.
3. **To evaluate the role of digital technologies in meeting SDGs**: The chapter will investigate how emerging digital technologies contribute to achieving SDGs by facilitating sustainable operations, enhancing resource efficiency, and enabling new business models.

4. **To outline practical applications of digital technologies for organizational sustainability**: Finally, the chapter will provide insights into the practical integration of digital tools to drive sustainability initiatives, emphasizing the need for businesses to embrace digital transformation as part of their sustainability strategies.

16.1.4 Importance of Integrating Digital Technologies to Meet SDGs in Businesses

The amalgamation of digital technologies into different activities of business is becoming increasingly essential for companies striving to meet SDGs. In the modern economy, digital tools offer numerous advantages for tracking, managing, and improving sustainability efforts. For example, IoT sensors allow companies to evaluate resource usage in real-time, which leads to better efficient energy consumption and reduced waste. Similarly, blockchain can enhance supply chain transparency (Bai et al., 2020).

Incorporating digital technologies also offers a competitive edge. While making the decisions pertaining to purchase majority of the customers prioritize the sustainability and organization can showcase their dedication to the SDGs through digital innovation are more prone to build better relationships with their stakeholders (Stahel, 2016). Moreover, as governments worldwide implement stricter environmental regulations, businesses that adopt digital technologies to meet sustainability targets will be better positioned to comply with these policies and avoid penalties (Ghobakhloo, 2020).

16.2 Historical Background and the Emergence of the SDGs

In the year 2015, Sustainable Development Goals (SDGs) were established as a part of the United Nations' *2030 Agenda for Sustainable Development*. In continuation with Millennium Development Goals (MDGs), SDGs emerged and which focused on addressing poverty, hunger, and inequality between 2000 and 2015. While the MDGs laid a strong foundation for international development, their focus was narrower, primarily targeting developing countries and basic human needs. They encompass 17 interconnected goals and 169 targets, offering a comprehensive framework for handling various challenges at the global level which includes poverty eradication, inequality, change of climate, and degradation of environment (United Nations, 2015).

The development of the SDGs was the output of extensive discussions. The process reflected a global consensus on the requirement for a more inclusive, sustainable, and resilient development path that leaves no one behind. The SDGs emphasize shared responsibility, encouraging both developed and developing countries to contribute to achieving these targets. Importantly, the SDGs also recognize the crucial role that businesses play in driving sustainable development (Sachs et al., 2019).

16.2.1 Key SDGs Related to Sustainability and Economic Growth

While all 17 SDGs are interlinked and critical to global development, several goals stand out for their direct relevance to growth of the economy and sustainability. One of the key SDGs in this regard is **SDG 8**, which emphasizes the need for comprehensive and sustainable economic growth that creates jobs, fosters innovation, and reduces inequality. By promoting fair wages, productive employment, and equal opportunities, SDG 8 supports long-term economic stability while ensuring social equity.

Another crucial goal is **SDG 12**, which is focused on waste reduction, resource optimization, and promoting sustainable production and consumption patterns. This goal is particularly relevant to businesses, as it motivates the adoption of principles of circular economy, in which resources can be reused and recycled, minimizing environmental impact (Schroeder et al., 2019).

SDG 9 focuses on resilient building infrastructure, fostering innovation, and promoting sustainable industrialization. Digital transformation plays a critical role here, enabling businesses to adopt smarter, more efficient processes that reduce resource consumption and carbon footprints.

SDG 13 also holds significant importance for businesses, as it addresses the pressing need to battle change of climate and its impacts. Organizations are increasingly expected to adopt renewable energy and build climate resilience into their operations (United Nations, 2020).

Organizational Sustainability

Organizational sustainability is referred to as a company's capacity to work in a way that balances financial performance with environmental and social responsibility. Sustainable organizations are those that seek not only to maximize profits but also to contribute positively to the environment and society (Elkington, 1997).

At its core, organizational sustainability involves three interconnected dimensions: social, economic and environmental. The economic dimension emphasizes productivity and long-term financial performance, ensuring the business remains competitive and viable. The environmental dimension focuses on reducing the company's ecological footprint, managing resources efficiently, minimizing waste, and addressing issues like change in climate and biodiversity loss. The social dimension covers the company's role in society, including how it treats its people, how it operates in the various communities, and its influence on stakeholders such as customers, suppliers, and the wider public (Dyllick & Muff, 2016).

16.2.2 The Importance of Sustainability in Modern Business Practices

In the twenty-first century, sustainability has become a central consideration for businesses due to evolving regulatory landscapes, consumer expectations, and environmental pressures. Modern businesses increasingly recognize that operating sustainably is not just an ethical and moral perspective but also provides a strategic edge to the company. Companies that adopt sustainable practices are often seen as more responsible and trustworthy, which can strengthen their brand, attract investment, and foster customer loyalty (Whelan & Fink, 2016).

Furthermore, sustainability practices can lead to saving the cost. For example, optimizing energy use, reducing waste, and adopting circular economy principles can significantly cut operational costs. Additionally, businesses that adopt sustainable practices are better positioned to navigate regulatory requirements and avoid potential penalties. Governments across the world are enacting stricter environmental laws, and businesses that fail to conform risk not only financial penalties but also reputational damage (Ghobakhloo, 2020).

16.2.3 Role of SDGs in Guiding Corporate Social Responsibility (CSR)

The SDGs is important framework for guiding corporate social responsibility (CSR) initiatives, helping businesses align their social and environmental and social activities with global development priorities. CSR can be defined as "a company's commitment to operating in an ethical and sustainable manner, ensuring that its operations positively impact society and the environment." Traditionally, CSR efforts have focused on philanthropy and compliance. However, with the emergence of the SDGs, companies are increasingly adopting a more strategic approach to CSR, aligning their initiatives with specific SDG targets to maximize impact (Scheyvens et al., 2016).

It provides a global benchmark for companies to measure their progress on sustainability issues. By aligning CSR initiatives with SDGs, businesses can enhance their credibility, demonstrating to stakeholders that their sustainability efforts are not only local or sector-specific but also contribute to a broader, global agenda. For instance, companies can align their CSR strategies with **SDG 3 by** promoting employee wellness programs or supporting health initiatives in communities where they operate. Moreover, the SDGs offer a common language that helps businesses, governments, and civil society collaborate more effectively on sustainability issues. By using the SDGs as a guiding framework, companies can structure their CSR activities around shared global goals, enabling more meaningful partnerships and impactful outcomes (Rosati & Faria, 2019).

16.2.4 Digital Technologies and Their Role in a Circular Economy

Digital technologies have transformed the way organizations operate, empowering them to improve efficiency, improve decision-making, and create innovative business models. Among the most impactful digital technologies are the blockchain, AI, big data analytics, and IoT.

1. **Internet of Things (IoT)**: IoT is the different devices are interconnected to each other through network that facilitates the communication and exchange of data through internet. In business scenario, IoT empowers establishments to monitor and manage assets in real-time, providing valued acumens into operational performance and resource usage (Gonzalez et al., 2019). For example, smart sensors can track energy consumption in manufacturing facilities, allowing for immediate adjustments to reduce waste and improve efficiency.
2. **Artificial Intelligence (AI)**: AI incorporates a variety of technologies that simulate human intelligence, which includes vision of the computer, machine learning and NLP. It can analyse gigantic quantities of data for pattern identification and prediction making enabling organizations to optimize processes and improve decision-making (Zhao et al., 2020). In the context of sustainability, AI can help businesses forecast demand more accurately, thereby reducing overproduction and minimizing waste.
3. **Blockchain**: Blockchain is a technology which is decentralized in nature and possess secured digital ledger that helps in tamper proof and transparent record-keeping. Blockchain can enhance traceability by recording every transaction and movement of goods in supply chains, ensuring that products are sourced sustainably and ethically (Kamble et al., 2020).
4. **Big Data Analytics**: Big data analytics is the process where large datasets can be examined and it facilitates in uncovering the hidden patterns, insights and correlation among each other. Businesses can influence big data analytics to assess their sustainability performance, identify inefficiencies, and track progress towards environmental goals (Duan et al., 2021).

16.2.5 Digital Transformation and Innovation in Organizations

Digital transformation is a process in which in all the different aspects of business digital technology is integrated which fundamentally changes how organizations operate. This transformation often encompasses a cultural shift, requiring organizations to embrace novel ways of thinking and working (Bharadwaj et al., 2013).

Innovation is an important component of digital transformation. Organizations that leverage digital technologies can lead to new products and services development, and models of business that meet evolving consumer demands and sustainability goals (Bessant & Tidd, 2015).

Furthermore, digital transformation fosters collaboration and knowledge sharing within organizations and across industries. Digital platforms enable stakeholders to connect, share insights, and collaborate on sustainability initiatives. For example, businesses can collaborate on joint sustainability projects, leveraging shared data and expertise to address common challenges (Kane et al., 2015).

16.2.6 Definition and Principles of the Circular Economy

The circular economy is an economic model that focuses on waste minimization and by creating the closed loop system it utilizes maximum resources. In contrast to the traditional linear economy, which follows a "take-make-dispose" pattern, the circular economy highlights sustainability and resource efficiency (Ellen MacArthur Foundation, 2013).

Main principles of the circular economy include:

1. **Longevity**: Products should be intended to have lengthier life cycles, allowing for reuse and refurbishment. This involves selecting durable materials and designing for repairability.
2. **Resource Recovery**: When the product's life ends, materials should be recycled and recovered to create new products. This principle emphasizes the importance of recycling and resource recovery infrastructure.
3. **Product as a Service**: As an alternative of selling products outright, businesses can offer them as services. This model inspires companies to take responsibility for their lifecycle, promoting better maintenance and reuse (Lacy & Rutqvist, 2015).
4. **Systems Thinking**: The circular economy requires a complete approach that contemplates the entire lifecycle of products and services, recognizing interdependencies between different systems (Bocken et al., 2016a, 2016b).

16.3 How Digital Technologies Support Circular Business Models

Digital technologies play a critical role in enabling and ascending circular business models. By providing real-time data and insights, these technologies help organizations optimize resource use, enhance transparency, and facilitate innovative practices.

1. **IoT and Resource Monitoring**: IoT devices can be used in real-time, providing organizations with data on energy use, water consumption, and material flows. This data allows businesses to identify inefficiencies and implement targeted

interventions to reduce waste (Gonzalez et al., 2019). For example, smart meters can track energy usage in buildings, allowing establishments to enhance consumption patterns and decrease costs.

2. **Blockchain for Supply Chain Transparency**: Blockchain technology improves supply chain transparency by providing a secure and tamper-proof transaction records. This capability is essential for circular economy practices, as it allows businesses to trace the origin and lifecycle of materials, confirming that they are tracked sustainably and ethically (Kamble et al., 2020). By using blockchain, organizations can build trust with consumers and comply with regulatory requirements related to sustainability.
3. **AI for Predictive Analytics**: Artificial intelligence can analyse massive quantities of data to recognize trends and make predictions about future demand, resource needs, and waste generation. By leveraging AI-driven analytics, organizations can improve production processes, reducing overproduction and minimizing waste (Zhao et al., 2020). This predictive ability is predominantly valuable for corporations transitioning to circular business models, as it allows them to align production with actual demand.
4. **Big Data for Circularity Assessment**: Big data analytics enables organizations to assess their circularity performance by analysing data linked to material flows, waste generation, and resource recovery. This assessment helps businesses identify areas for upgrading and track progress toward circular economy goals (Duan et al., 2021).
5. **Collaboration through Digital Platforms**: Digital technologies facilitate collaboration between businesses, consumers, and other stakeholders in the circular economy. Online platforms can connect companies with suppliers, recyclers, and consumers, enabling efficient resource sharing and circular practices (Kane et al., 2015).

16.3.1 Integration of Digital Technologies for Sustainable Development

The IoT is a technology that is transformative in nature and connects physical devices to the internet, analysis, and real-time monitoring of systems. This capability is predominantly relevant for enhancing resource efficiency, which is a critical component. IoT devices can significantly improve resource efficiency in various sectors (Wolfert et al., 2017). This precision agriculture approach not only enhances yield but also minimizes environmental impacts by reducing over-fertilization and water wastage. In the industrial sector, consumption of energy can be observed in real-time by using IoT technology, permitting companies to recognize inadequacies and implement energy-saving measures. For instance, smart meters can track energy usage patterns, leading to adjustments in production schedules to minimize peak demand, which can help reduce costs and carbon footprints (Zhao et al., 2020). Moreover,

IoT-enabled waste management systems can enhance collection routes and schedules based on real-time data from smart bins that monitor fill levels. This can reduce fuel consumption and recover the efficiency of waste collection, contributing to SDG 11: Sustainable Cities and Communities (Jiang et al., 2018).

16.3.2 Blockchain for Sustainable and Transparent Supply Chains

Blockchain technology is gradually identified for its capability to improve transparency and sustainability in supply chains. It operates as a distributed digital ledger that securely records transactions across multiple parties, ensuring that all participants have access to the same information.

One of the primary benefits of blockchain in supply chain management is its ability to enhance traceability. Each transaction on a blockchain is immutable, meaning that after the data is recorded, it is not possible to alter the data without mutual consent all parties involved (Kamble et al., 2020). For instance, companies like Everledger use blockchain to track the lifecycle of diamonds, assuring consumers about the ethical sourcing of these precious stones.

Additionally, blockchain enables collaboration among stakeholders. Different parties in a supply chain can share data on a single platform, enhancing communication and coordination (Wang et al., 2019). This collaborative approach fosters trust and facilitates joint sustainability initiatives, such as collective waste reduction efforts or shared resource management (Hazen et al., 2014).

16.3.3 Artificial Intelligence (AI) for Process Optimization

Artificial Intelligence (AI) is transforming several industries by optimizing processes, cultivating decision-making, and driving effectiveness. The integration of AI into business operations aligns closely with the SDGs. AI algorithms can analyse massive quantities of data to identify inefficiencies and predict outcomes, permitting organizations to enhance their operations. For example, AI-powered predictive maintenance in manufacturing can analyse data from machinery sensors to forecast failures of equipment's before they occur. This proactive approach reduces downtime, extends the lifespan of equipment, and minimizes waste associated with unplanned repairs (Zhao et al., 2020).

AI can enhance energy consumption patterns in buildings. Smart energy management systems powered by AI can analyse historical usage data and adjust heating, cooling, and lighting based on occupancy patterns, leading to significant energy savings and reduced carbon emissions (Mohammad et al., 2020). By aligning energy consumption with actual needs, organizations can contribute to **SDG 7: Affordable and Clean Energy**.

Additionally, AI can enhance resource efficiency in supply chains. By analysing demand patterns and inventory levels, AI can help businesses enhance their stock levels, reducing excess inventory and curtailing waste (Wang et al., 2019).

16.3.4 Big Data Analytics for Sustainability Performance Tracking

Big data analytics comprises the systematic examination of huge datasets to extract valuable insights, permitting establishments to arrive at the decision which is based on data. One of the key applications of big data analytics in sustainability is performance tracking. Organizations can collect and analyse data related to their consumption of resources, generation of wastes and emission of carbon to evaluate their sustainable performance. For example, companies can use big data analytics to track their progress toward sustainability goals and identify areas for improvement (Duan et al., 2021).

Additionally, big data analytics can enhance transparency by providing stakeholders with real-time access to sustainability metrics. This transparency is critical for building trust with consumers, investors, and regulatory bodies. By sharing sustainability performance data, organizations can demonstrate their commitment to responsible practices and compliance with sustainability standards (Wang et al., 2019). Moreover, big data analytics can facilitate benchmarking against industry peers, enabling organizations to identify best practices and adopt strategies that enhance their sustainability performance. By comparing their data with industry benchmarks, organizations can set realistic targets and monitor their progress toward achieving sustainability objectives (Zhao et al., 2020).

16.3.5 Enabling Sustainable Business Models with Digital Technologies

Digital tools play a crucial role in facilitating circular product design and recycling processes, aligning closely with **SDG 12 l**everaging digital technologies, organizations can design products with sustainability in mind, confirming that materials can be reused and refurbished. One of the key digital tools for circular product design is computer-aided design (CAD) software. CAD permits designers to create detailed digital models of products, enabling them to simulate performance, evaluate material usage, and identify potential design flaws before physical prototypes are produced (García-Muiña et al., 2021). This iterative design process minimizes waste by reducing the need for excessive prototyping and material waste during the development phase. Additionally, digital platforms that facilitate collaborative design can enhance circularity in product development. Collaborative tools allow designers, engineers, and stakeholders to work together in real-time, sharing insights and feedback throughout the design process. This collaborative approach can lead to more innovative designs that prioritize sustainability (Ness & Rausch, 2021).

Moreover, digital technologies enable organizations to implement recycling and take-back programs effectively. For instance, digital tracking systems can monitor the return of end-of-life products, allowing companies to manage collection and recycling logistics efficiently. These systems can also analyse data on returned products to identify opportunities for refurbishment and resource recovery (Lacy & Rutqvist, 2015).

3D printing technology has emerged as a key enabler of circular product design. 3D printing allows companies to produce components and products on demand, reducing excess inventory and minimizing waste. Additionally, materials used in 3D printing can often be recycled or reused, aligning with circular economy principles (Gao et al., 2015).

16.3.6 Digital Technologies for Resource Management, Reuse, and Waste Reduction

Digital technologies play an important role in enhancing resource management, motivates reuse, and waste reduction in various sectors. By providing real-time data and insights, these technologies permit organizations to optimize their resource. One of the major applications of digital technologies in resource management is the use of smart sensors and IoT devices. These devices can observe consumption of resources, such as raw material, water and energy, in real-time. For example, in manufacturing, IoT sensors can track energy usage across production lines, allowing organizations to identify inefficiencies and implement targeted interventions to reduce consumption (Zhao et al., 2020). This data-driven approach enhances resource efficiency and supports sustainability goals.

Moreover, digital platforms can simplify the sharing and reuse of resources among organizations. Collaborative consumption platforms enable businesses to share assets, such as tools or equipment, reducing the need for individual ownership and reduction of waste (Lacy & Rutqvist, 2015). For instance, companies can leverage online platforms to connect with other businesses and share resources, promoting a circular economy approach.

Additionally, big data analytics can support waste reduction efforts by analysing data on waste generation patterns. Organizations can identify areas where waste is generated and implement strategies to minimize it. For example, data analytics can reveal inefficiencies in production processes that lead to excess waste, allowing organizations to optimize operations and reduce their environmental impact (Duan et al., 2021).

Advanced sorting technologies, such as AI-driven vision systems, can identify and separate recyclable materials more efficiently. (Zhao et al., 2020).

16.4 Challenges and Opportunities in the Digital Circular Economy

The change to a digital circular economy often encounters significant organizational barriers. One of the primary obstacles is resistance to change, which can stem from established organizational cultures that prioritize traditional business models over innovative practices. Employees may be accustomed to conventional workflows and may absence the necessary skills or knowledge to adapt to new digital technologies (Bocken et al., 2014). Additionally, insufficient leadership commitment can impede the adoption process. Without strong advocacy from top management, initiatives to implement digital technologies may lack the necessary resources and strategic focus (Wang et al., 2020). Organizations may find it difficult to prioritize initiatives or allocate resources effectively, which can slow down progress toward a fully integrated digital circular economy (Kumar et al., 2021).

16.4.1 Technological Limitations

Technological limitations pose significant challenges in the journey towards a digital circular economy. One of the primary concerns is data interoperability and integration (Rogers et al., 2019). For instance, integrating IoT devices with existing enterprise resource planning (ERP) systems may need substantial investment in upgrades or the implementation of new platforms, which can be a barrier to adoption. Ensuring the security of sensitive data is paramount, predominantly in industries such as healthcare and finance, where privacy of data regulations is stringent (Luthra et al., 2020). This necessitates robust cybersecurity measures, which can be resource-intensive and may divert attention from sustainability initiatives.

Additionally, the high cost of implementing advanced technologies can deter organizations from pursuing digital transformation. Many small and medium-sized enterprises (SMEs) lack the monetary resources to invest in cutting-edge solutions, limiting their ability to compete in the digital circular economy (Cohen & Winn, 2020). As a result, the technological divide between larger corporations and SMEs may widen, exacerbating existing inequalities in sustainability practices.

16.4.2 Regulatory and Compliance Challenges

Regulatory and compliance challenges can significantly impact the acceptance of digital technologies in the context of a circular economy. Different regions may have varying regulations governing data privacy, environmental standards, and product lifecycle management (González et al., 2021). Additionally, the lack of harmonization between regulations across different jurisdictions can create uncertainty, discouraging organizations from investing in digital solutions that align with circular economy principles (Mishra et al., 2021). regulatory compliance can present a formidable barrier to the acceptance of digital technologies, requiring organizations to invest significant resources in understanding and adhering to various regulations while attempting to innovate sustainably (Singh et al. 2021).

16.4.3 Innovation Potential and New Business Models

Despite the challenges, the digital circular economy presents significant opportunities for innovation and new model of businesses are developed. The integration of digital technologies fosters a culture of creativity, enabling organizations to rethink traditional practices and explore novel approaches to sustainability (Bocken et al., 2016a, 2016b). This shift from ownership to access not only diminishes resource consumption but also creates new revenue streams for companies.

Additionally, data analytics and AI can empower businesses to enhance their operations and improve product offerings. By leveraging real-time data insights, organizations can identify inefficiencies in their supply chains and develop strategies to minimize waste. For example, predictive analytics can assist businesses anticipate demand fluctuations, allowing for more accurate inventory management and reducing excess production (Wang et al., 2020). This proactive approach brings into line with the principles of the circular economy by promoting resource efficiency and minimizing environmental impact. In contrast of selling products, companies can offer them as services, encouraging customers to return items for refurbishment or recycling at the end of their lifecycle. This model not only improves customer loyalty but also allows businesses to retain ownership of materials, facilitating resource recovery and reducing waste (Lacy & Rutqvist, 2015).

16.4.4 Creation of Sustainable Value Through Technology

Digital technologies play a critical role in creating sustainable value for organizations and society as a whole. By integrating these technologies into their operations, businesses can enhance their sustainability performance while simultaneously driving profitability. For instance, IoT devices can monitor energy usage and optimize resource allocation, resulting in significant cost savings for organizations (Jiang et al., 2018). This dual benefit of economic and environmental value is a key driver for organizations seeking to align their operations with sustainability goals.

Moreover, digital technologies facilitate transparency and accountability, which are essential for building trust with stakeholders. By leveraging blockchain technology, organizations can provide verifiable information about their supply chains, demonstrating their commitment to ethical sourcing and sustainability practices (Kamble et al., 2020). This transparency not only enhances brand reputation but also catches the attention of consumers who prioritize sustainability in their purchasing decisions.

By tracking key performance indicators related to resource consumption, waste generation, and carbon emissions, companies can demonstrate their progress toward sustainability goals. It enables organizations to communicate their sustainability efforts to stakeholders, including investors, customers, and regulators, ultimately creating a competitive advantage (Zhao et al., 2020).

16.4.5 Best Practices for the Integration of Digital Technologies

To successfully navigate the challenges associated with the digital circular economy, organizations can adopt several best practices for integrating digital technologies.

1. **Leadership Commitment and Vision**: Strong leadership commitment is essential for driving digital transformation efforts. Leaders should establish a clear vision for how digital technologies can enhance sustainability and match with the organization's strategic objectives (Wang et al., 2020). By nurturing a culture of innovation and reassuring employees to embrace change, organizations can overcome resistance and build momentum for digital initiatives (Khan et al., 2022).
2. **Skills Development and Training**: Investing in skills development and training programs is crucial for equipping people with the knowledge and capabilities to leverage digital technologies effectively. Organizations should provide ongoing training opportunities to enhance digital literacy and adoptive a culture of continuous learning (Kumar et al., 2021).
3. **Collaboration and Partnerships**: Collaborating with external partners, including technology providers, industry associations, and research institutions, can enhance an organization's ability to implement digital solutions effectively (Bocken et al., 2016a, 2016b).
4. **Agile Implementation**: Organizations should adopt an agile approach to implementing digital technologies, allowing for iterative development and continuous

improvement. By piloting new technologies in controlled environments, organizations can gather feedback, identify challenges, and refine their strategies before scaling up implementations (Luthra et al., 2020).

5. **Monitoring and Evaluation**: To assess the digital initiatives it is essential to establish robust monitoring and evaluation mechanisms. Organizations should regularly track key performance indicators pertaining to sustainability and digital transformation, enabling data-driven decision-making and continuous optimization of processes (Zhao et al., 2020).

16.5 Future of Digital Technologies in a Circular Economy for SDGs

The future of digital technologies in the background of a circular economy is marked by rapid advancements that promise to enhance sustainability efforts and contribute to the achievement of Sustainable Development Goals (SDGs). These technologies can optimize processes, enabling organizations to predict resource needs accurately, reduce waste, and improve efficiency across the supply chain (Mohammad et al., 2020). Additionally, IoT is expected to play a pivotal role in enabling real-time monitoring of resource consumption and waste generation. By deploying IoT sensors across production processes and supply chains, organizations can gather detailed insights into their operations, facilitating a more circular approach to resource management (Duan et al., 2021). For instance, IoT-enabled devices can track the lifecycle of products, providing data that can inform decisions about repair, refurbishment, or recycling, thus extending the life of materials (Rubee Singh et al., 2022). Emerging technologies such as 3D printing and biotechnology offer innovative ways to minimize waste and create sustainable products. For instance, 3D printing allows for on-demand production, which reduces excess inventory and material waste (Gao et al., 2015). Meanwhile, biotechnology can contribute to the circular economy by developing biodegradable materials and processes that minimize environmental impact (Petersen et al., 2021).

16.5.1 The Role of AI and Automation in the Circular Economy

AI and automation are central to realizing the potential of a circular economy. These technologies can meaningfully improve operational efficacy and facilitate sustainable practices. By automating monotonous tasks and optimizing processes, organizations can decrease resource consumption and minimize waste. For instance, AI-driven analytics can help businesses identify inefficiencies in production, leading to reduced energy and material usage (Mohammad et al., 2020). Moreover, AI can support predictive maintenance, where machine learning algorithms analyse equipment performance data to forecast failures before they occur. This approach not only reduces downtime but also extends the lifespan of machinery, aligning with circular economy principles (Kamble et al., 2020).

Furthermore, automation technologies, such as robotics, can enhance recycling efforts by improving sorting processes. Automated systems equipped with AI can quickly and accurately identify and separate recyclable materials, increasing recycling rates and reducing contamination (Zhao et al., 2020). In addition, AI can enable the development of innovative business models. This model inspires manufacturers to design for durability and recyclability, fostering a more sustainable approach to consumption (Lacy & Rutqvist, 2015).

16.5.2 The Need for Collaboration Between Governments, Industries, and Organizations

The successful implementation of digital technologies in a circular economy relies heavily on collaboration among various stakeholders, including governments, industries, and organizations. Governments play a vital role in generating a conducive policy situation that encourages innovation and investment in sustainable practices. Policymakers must establish clear frameworks that encourage the adoption of digital technologies and incentivize organizations to changeover towards circular business models (Mishra et al., 2021).

Moreover, industry collaborations can facilitate knowledge sharing and best practices, driving collective efforts toward sustainability. By forming partnerships across sectors, organizations can pool resources and expertise to develop innovative solutions that address common challenges (Rogers et al., 2019). For instance, cross-industry collaborations can lead to the establishment of shared digital platforms that enable resource sharing and collaborative consumption, reducing overall resource use and waste generation (Rubee Singh et al., 2023).

Non-governmental organizations (NGOs) and civil society groups can play an essential role in advocating for sustainable practices and holding businesses accountable for their environmental impact. Engaging with these stakeholders can foster a sense of shared responsibility and encourage organizations to align their operations with broader sustainability goals (Kumar et al., 2021).

16.5.3 Future Regulatory Frameworks and Their Impact on Sustainability and Circular Economy Initiatives

As the digital landscape continues to evolve, future regulatory frameworks will need to adapt to support sustainability and circular economy initiatives effectively. One critical area of focus will be the establishment of regulations that promote data sharing and interoperability among digital platforms. Ensuring that data can be accessed and utilized across different organizations will be essential for enhancing collaboration and enabling a more circular economy (González et al., 2021). Additionally, regulations must address the need for transparency and traceability in supply chains. Future frameworks could require organizations to disclose information about the sourcing of materials, production processes, and end-of-life management of products. Such transparency will empower consumers to make informed choices and encourage businesses to adopt sustainable practices (Kamble et al., 2020). Moreover, regulatory frameworks should incentivize the adoption of digital technologies that contribute to sustainability. This could include tax breaks, grants, or subsidies for organizations that invest in technologies that enhance resource efficiency, waste reduction, and circularity (Mishra et al., 2021).

16.6 Conclusion, Limitations and Future Scope

The integration of digital technologies into organizational practices presents a pivotal opportunity to advance sustainability and achieve the Sustainable Development Goals (SDGs). Throughout this chapter, we have explored the significance of the SDGs in guiding organizational sustainability, defined the digital circular economy, and examined how various digital technologies can enhance sustainability initiatives. The chapter began with an overview of the SDGs, emphasizing their historical context and the pressing necessity for sustainable practices in light of global challenges such as climate change and resource reduction. The SDGs provide a complete framework for organizations to match their strategies with global sustainability objectives. We then delved into the concept of the digital circular economy, which represents a transformative approach to resource management and production processes. This paradigm shift encourages businesses to rethink traditional linear models and adopt practices that promote recycling, reusing, and reducing waste. Digital technologies, including the Internet of Things (IoT), artificial intelligence (AI), and blockchain, play critical roles in facilitating this transition by optimizing resource use, enhancing transparency, and enabling innovative business models.

Moreover, we explored the key digital technologies that contribute to the SDGs, highlighting their potential in improving resource efficiency, creating sustainable supply chains, and tracking sustainability performance. The chapter also addressed the challenges and opportunities associated with integrating these technologies, underscoring the need for organizations to adopt best practices and foster a culture of innovation.

Looking ahead, digital technologies will continue to play an increasingly vital role in fostering sustainability and driving progress toward the SDGs. As these technologies advance, they will provide organizations with powerful tools to enhance their operational efficiencies, reduce their environmental footprints, and innovate sustainably. For instance, AI and machine learning will enable businesses to make data-driven decisions that optimize resource utilization, while IoT will facilitate real-time monitoring of resource consumption and waste generation. the importance of transparency in supply chains will grow, with blockchain technology enabling traceability and accountability. Consumers and stakeholders are becoming more informed and conscious of sustainability issues, and organizations that leverage digital technologies to provide transparency will likely gain a competitive advantage. As we move forward, it is imperative for organizations and policymakers to embrace the potential of digital technologies in their sustainability efforts. Organizations should prioritize the integration of these technologies into their operations, investing in training and resources that empower employees to innovate sustainably. Collaborative initiatives between sectors can drive knowledge sharing and accelerate the adoption of best practices. Policymakers, on the other hand, must create an enabling environment that fosters innovation and investment in digital solutions for sustainability. This includes developing regulatory frameworks that incentivize the adoption of sustainable practices, promoting research and development in digital technologies, and facilitating public–private partnerships that advance circular economy initiatives.

References

Bai, C., Sarkis, J., Behnam, S., & Dou, Y. (2020). Corporate sustainability development in the context of the circular economy: A review. *Journal of Cleaner Production, 273*, 123134. https://doi.org/10.1016/j.jclepro.2020.123134

Bessant, J., & Tidd, J. (2015). *Innovation and entrepreneurship*. Wiley.

Bharadwaj, A., El Sawy, O. A., Pavlou, P. A., & Venkatraman, N. (2013). Digital business strategy: Toward a next generation of insights. *MIS Quarterly, 37*(2), 471–482. https://doi.org/10.25300/MISQ/2013/37.2.13

Bocken, N. M. P., de Pauw, I., Bakker, C., & van der Grinten, B. (2016a). Product design and business model strategies for a circular economy. *Journal of Industrial and Production Engineering, 33*(5), 308–320. https://doi.org/10.1080/21681015.2016.1201716

Bocken, N. M. P., Short, S. W., Rana, P., & Evans, S. (2014). A literature and practice review to develop sustainable business model archetypes. *Journal of Cleaner Production, 65*, 42–56. https://doi.org/10.1016/j.jclepro.2013.11.039

Bocken, N. M., Short, S. W., Rana, P., & Evans, S. (2016b). A literature and practice review to develop sustainable business model archetypes. *Journal of Cleaner Production, 65*, 42–56. https://doi.org/10.1016/j.jclepro.2013.11.039

Cohen, B., & Winn, M. I. (2020). Market imperfections, opportunity, and sustainable entrepreneurship. *Journal of Business Venturing, 21*(1), 103–125. https://doi.org/10.1016/j.jbusvent.2005.04.002

Duan, Y., Edwards, A., & Dwivedi, Y. K. (2021). Big data for sustainable development: A systematic review. *Sustainable Development, 29*(1), 62–75. https://doi.org/10.1002/sd.1971

Dyllick, T., & Muff, K. (2016). Clarifying the meaning of sustainable business: Introducing a typology from business-as-usual to true business sustainability. *Organization & Environment, 29*(2), 156–174. https://doi.org/10.1177/1086026615575176

Elkington, J. (1997). *Cannibals with forks: The triple bottom line of 21st-century business*. Capstone Publishing.

Ellen MacArthur Foundation. (2013). Towards the circular economy: Economic and business rationale for an accelerated transition. *Ellen MacArthur Foundation*. Retrieved from https://www.ellenmacarthurfoundation.org

Gao, W., Zhang, Y., & Ramakrishna, S. (2015). 3D printing of polymeric materials: A review of the current status and future challenges. *Journal of Materials Science, 50*(3), 1057–1070. https://doi.org/10.1007/s10853-014-8280-3

García-Muiña, F. E., González-Sánchez, R., & Padrón-García, J. A. (2021). Digital transformation in design and manufacturing: Current and future trends. *Procedia CIRP, 98*, 98–103. https://doi.org/10.1016/j.procir.2021.01.016

Geissdoerfer, M., Savaget, P., Bocken, N. M., & Hultink, E. J. (2017). The circular economy–A new sustainability paradigm? *Journal of Cleaner Production, 143*, 757–768. https://doi.org/10.1016/j.jclepro.2016.12.048

Ghobakhloo, M. (2020). Industry 4.0, digitization, and opportunities for sustainability. *Journal of Cleaner Production, 252*, 119869. https://doi.org/10.1016/j.jclepro.2019.119869

Gonzalez, E. S., Cabrera, J. A., & Morales, J. (2019). IoT technologies for smart cities: A survey of applications and challenges. *IEEE Access, 7*, 32830–32855. https://doi.org/10.1109/ACCESS.2019.2901068

González, M. R., da Silva, A. J., & Teixeira, A. (2021). Regulatory aspects and impacts of the circular economy in the European Union: A critical review. *Sustainability, 13*(1), 347. https://doi.org/10.3390/su130100347

Hazen, B. T., Boone, C. A., Ezell, J. D., & Jones-Farmer, L. A. (2014). Data quality for data science, predictive analytics, and big data in supply chain management: An introduction to the problem and suggestions for research and applications. *International Journal of Production Economics, 154*, 72–80. https://doi.org/10.1016/j.ijpe.2014.04.018

Jiang, P., Liu, C., & Li, Y. (2018). Smart waste management: A review of IoT-enabled solutions. *Journal of Cleaner Production, 215*, 669–682. https://doi.org/10.1016/j.jclepro.2018.12.103

Kamble, S. S., Gunasekaran, A., & Ghadge, A. (2020). Blockchain technology for sustainable supply chain management: A review and bibliometric analysis. *Sustainable Production and Consumption, 21*, 228–245. https://doi.org/10.1016/j.spc.2019.10.014

Kane, G. C., Palmer, D., Phillips, A. N., & Kiron, D. (2015). Strategy, not technology, drives digital transformation. *MIT Sloan Management Review, 57*(3), 1–25. Retrieved from https://sloanreview.mit.edu/article/strategy-not-technology-drives-digital-transformation/

Khan, S., Ali, S. S., & Singh, R. (2022). Determinants of Remanufacturing Adoption for Circular Economy: A Causal Relationship Evaluation Framework. *Appl. Syst. Innov., 5*, 62. https://doi.org/10.3390/asi5040062

Kumar, S., Singh, A. K., & Bhatia, S. (2021). Digital transformation and organizational performance: A review of the literature and future research directions. *Technological Forecasting and Social Change, 162*, 120380. https://doi.org/10.1016/j.techfore.2020.120380

Lacy, P., & Rutqvist, J. (2015). *Waste to wealth: The circular economy advantage*. Palgrave Macmillan.

Luthra, S., Mangla, S. K., & Kharat, M. (2020). An integrated approach for sustainable development in supply chains. *Journal of Cleaner Production, 259*, 120836. https://doi.org/10.1016/j.jclepro.2020.120836

Mishra, D., Sharma, S. K., & Dey, B. L. (2021). Sustainable business models in the context of circular economy: A systematic literature review. *Journal of Cleaner Production, 312*, 127735. https://doi.org/10.1016/j.jclepro.2021.127735

Mohammad, A., Ganaie, M. A., & Rahman, A. (2020). Artificial intelligence in energy systems: Current applications and future trends. *Journal of Cleaner Production, 259*, 120756. https://doi.org/10.1016/j.jclepro.2020.120756

Ness, D. A., & Rausch, M. (2021). A circular economy approach to product design: What is the role of digital tools? *Journal of Cleaner Production, 314*, 127990. https://doi.org/10.1016/j.jclepro.2021.127990

Petersen, M. A., Liu, H., & Wang, L. (2021). Biodegradable polymers and bioplastics in circular economy: Status and future trends. *Waste Management, 122*, 108–117. https://doi.org/10.1016/j.wasman.2021.03.031

Rogers, K., Garza-Reyes, J. A., & Alarcon, L. F. (2019). Industry 4.0 technologies and their influence on operational performance in the circular economy. *International Journal of Production Research, 57*(20), 6180–6197. https://doi.org/10.1080/00207543.2018.1488466

Rosati, F., & Faria, L. G. D. (2019). Business contribution to the Sustainable Development Agenda: Organizational factors related to early adoption of SDG reporting. *Corporate Social Responsibility and Environmental Management, 26*(3), 588–597. https://doi.org/10.1002/csr.1705

Sachs, J. D., Schmidt-Traub, G., Kroll, C., Lafortune, G., & Fuller, G. (2019). *Sustainable development report 2019.* Bertelsmann Stiftung and Sustainable Development Solutions Network (SDSN).

Scheyvens, R., Banks, G., & Hughes, E. (2016). The private sector and the SDGs: The need to move beyond "business as usual." *Sustainable Development, 24*(6), 371–382. https://doi.org/10.1002/sd.1623

Schroeder, P., Anggraeni, K., & Weber, U. (2019). The relevance of circular economy practices to the sustainable development goals. *Journal of Industrial Ecology, 23*(1), 77–95. https://doi.org/10.1111/jiec.12732

Singh, R., Khan, S., & Centobelli, P. (2022). Investigating the Interplay between social performance and organisational factors supporting circular economy practices. *Sustainability, 14*(24), 16781. https://doi.org/10.3390/su142416781

Singh, R., Dsilva, J., Centobelli, P., & Tripathi, V. (2021). A review on sustainable HRM: A study evaluating sustainability for organisational development. *The Empirical Economics Letters, 20*(2), 1–10.

Singh, R., Khan, S., & Dsilva, J. (2023). A framework for assessment of critical factor for circular economy practice implementation. *Journal of Modelling in Management, 18*(5), 1476–1497. https://doi.org/10.1108/JM2-06-2021-0145

Stahel, W. R. (2016). The circular economy. *Nature, 531*(7595), 435–438. https://doi.org/10.1038/531435a

Umasankar, M., Boopathy, S., Padmavathy, S., Fukey, L. N., & Singh, R. (2022). Disconnect to reconnect: Employee wellbeing through digital detoxing. *Journal of Positive School Psychology, 6*(2), 4663–4673.

United Nations. (2015). Transforming our world: The 2030 agenda for sustainable development. *United Nations*. Retrieved from https://sdgs.un.org/2030agenda

United Nations. (2020). Take action for the sustainable development goals. *United Nations*. Retrieved from https://www.un.org/sustainabledevelopment/sustainable-development-goals

Wang, H., Li, Z., & Zhang, Y. (2019). Big data analytics for sustainable manufacturing: A review. *Journal of Cleaner Production, 211*, 487–503. https://doi.org/10.1016/j.jclepro.2018.11.208

Wang, Y., Gunasekaran, A., & Ngai, E. W. T. (2020). Big data in logistics and supply chain management: An overview of the current state of play and future directions. *International Journal of Production Economics, 176*, 98–110. https://doi.org/10.1016/j.ijpe.2016.06.002

Whelan, T., & Fink, C. (2016). The comprehensive business case for sustainability. *Harvard Business Review*. Retrieved from https://hbr.org/2016/10/the-comprehensive-business-case-for-sustainability

Wolfert, S., Ge, L., Verdouw, C., & Bogaardt, M. J. (2017). Big data in smart farming—a review. *Agricultural Systems, 153*, 69–80. https://doi.org/10.1016/j.agsy.2017.01.023

Zhao, X., Yang, Y., & Chen, W. (2020). Artificial intelligence in sustainable supply chain management: A review and future directions. *Sustainable Production and Consumption, 22*, 352–364. https://doi.org/10.1016/j.spc.2020.01.014

The manufacturer's authorised representative in the EU is Springer Nature Customer Service Centre GmbH, Europaplatz 3, 69115 Heidelberg, Germany. If you have any concerns regarding our products, please contact ProductSafety@springernature.com

Printed and bound by CPI Group (UK) Ltd, Croydon, CR0 4YY
01/10/2026
02225942-0002